Advances in Low-Temperature Mineralogy and Geochemistry

Advances in Low-Temperature Mineralogy and Geochemistry

Guest Editors

Iuliu Bobos
Franck Bourdelle

MDPI

Basel • Beijing • Wuhan • Barcelona • Belgrade • Novi Sad • Cluj • Manchester

Guest Editors
Iuliu Bobos
Instituto de Ciências da Terra
Universidade do Porto
Porto
Portugal

Franck Bourdelle
Institut des Sciences de la
Terre de Paris (ISTeP)
CY Cergy Paris Université
Cergy
France

Editorial Office
MDPI AG
Grosspeteranlage 5
4052 Basel, Switzerland

This is a reprint of the Special Issue, published open access by the journal *Minerals* (ISSN 2075-163X), freely accessible at: https://www.mdpi.com/journal/minerals/special_issues/ALTMG.

For citation purposes, cite each article independently as indicated on the article page online and as indicated below:

Lastname, A.A.; Lastname, B.B. Article Title. *Journal Name* **Year**, *Volume Number*, Page Range.

ISBN 978-3-7258-4915-4 (Hbk)
ISBN 978-3-7258-4916-1 (PDF)
https://doi.org/10.3390/books978-3-7258-4916-1

Cover image courtesy of Franck Bourdelle

Contents

About the Editors

Iuliu Bobos

Iuliu Bobos is a Professor Chair of Crystallography and Mineralogy, Economic Geology, and Biogeochemistry, and is the former Director of the Master (Geomaterials and Geological Resources) and PhD (Science and Environmental Technology) programs at the Faculty of Sciences, University of Porto. He conducted PhD thesis post-PhD in Low-Temperature Geochemistry, Low-grade Metamorphism, Economic Geology (W, Sn, Li, U and Pb, Zn ores). He received honors, awards, and fellowships as Visiting Professor, Arizona State University, U.S.A (2010), Visiting Exchange Scientist, U.S.G.S.—Boulder, U.S.A. (2007), Invited Professor, University of Aveiro, Portugal (2001), Invited Professor, Université "Louis Pasteur"—Strasbourg, France (2000), NATO Research Fellowship (1998), Invited Scientist, Portuguese Foundation for Science (1995–1997). He is a member of the Mineralogical Society, Geochemical Society, Society for Applied Geology, and Clay Minerals Society.

Franck Bourdelle

Franck Bourdelle is a Professor at the Institut des Sciences de la Terre de Paris (ISTeP)—CY Cergy Paris Université, specializing in the study of clay minerals and fluid/rock interactions. For over 15 years, he has been developing thermometric, thermodynamic, and analytical tools to study diagenetic and hydrothermal systems, such as a new semi-empirical chlorite geothermometer specially dedicated to low-temperature systems (¡350 °C). He relies on three complementary approaches: compositional analysis at the submicrometric scale, thermodynamics, and experimentation (laboratory study of fluid/claystone interactions). A former co-director of the Master's program GeoResources, Geohazards, Geotechnics at the University of Lille, former deputy Director of the Laboratory of Geosciences and Environment Cergy (GEC), he currently heads the PRISME team at ISTeP, where fluid-rock interaction is a key theme. Through his research activities, he has led or co-led numerous projects of various scales, both nationally and internationally, on the chloritization processes in hydrothermal contexts, on the alteration of mining shale in northern France, on the mineralogical evolution of argillite in the context of deep radioactive waste storage, and on geothermal systems and the associated chlorite markers.

minerals

MDPI

Editorial

Editorial for the Special Issue of *Minerals*, "Advances in Low-Temperature Mineralogy and Geochemistry"

Iuliu Bobos [1,*] and Franck Bourdelle [2,*]

1 Institute of Earth Sciences, University of Porto, Rua do Campo Alegre 687, 4168-007 Porto, Portugal
2 CY Cergy Paris Université, Sorbonne Université, CNRS-INSU, Institut des Sciences de la Terre de Paris, ISTeP, F-95000 Cergy, France
* Correspondence: ibobos@fc.up.pt (I.B.); franck.bourdelle@cyu.fr (F.B.)

Low-temperature geochemical processes are ubiquitous in geological areas related to weathering, diagenesis, very low-grade metamorphism, and post-magmatic processes, and provide an explanation of the P–T–X history of the Earth's crust. These processes may provide challenging topics of the broad qualitative questions related to near-surface Earth processes involving chemical transfer, mineralogical reactions, recycling of chemical elements, mineralization, etc.

In this way, light (δ^7Li; δ^{11}B), stable (δ^{18}O, δD), and radiogenic isotope (^{144}Nd/^{143}Nd, ^{207}Pb/^{206}Pb, ^{176}Hf/^{177}Hf, etc.) ratios are potentially powerful tracers of fluid–mineral interactions, while thermo-barometer indicators (e.g., chlorite thermometer, phengite barometer, phengite–chlorite pairs, muscovite-paragonite solvus, etc.) based on fine mineralogical compositions are appropriate probes to reconstruct pro- and retrograde mineralogical reactions. Furthermore, remarkable advances have recently been made in understanding chemical element recycling within the diagenesis and low-grade metamorphism environment, hydrothermal systems, and fluid circulation. Over the past decade, analytical techniques (e.g., X-ray diffraction, secondary-ion mass spectroscopy, laser-ablation mass spectroscopy, high-resolution transmission electron microscopy, etc.) have also advanced and offered a new vision for scientists in mineralogy and geochemistry, making data available with an unparalleled analytical and spatial resolution.

Contributions on low-temperature mineralogy and geochemistry from a broad range of fields were selected and published. The Special Issue begins with a review of chlorite geothermometry and recent related analytical advances [1]. Chlorite, a ubiquitous clay mineral in metamorphic, diagenetic, and hydrothermal systems, exhibits a chemical composition partly dependent on temperature (T) and pressure (P) conditions. Recent analytical developments that allow for circumventing or responding to certain criticisms regarding the low-temperature application of thermometers are shown. Furthermore, new perspectives in terms of analysis (e.g., Mn redox in Mn-chlorite) and geothermometers (molecular solid-solution model, oxychlorite end-member) are addressed.

The ore-forming processes and subsequent processes of alteration–mineralization explain the role of fluids during mineralogical reactions and element recycling. A very interesting mineralogical reaction from hübnerite to scheelite and Fe,Mn-chlorite was studied in the magmatic–hydrothermal system of Borralha (Portugal) [2]. Replacing wolframite with scheelite is common among tungsten deposits where these two tungstates may coexist. Additionally, a genetic linkage between W (Cu, Mo) and chlorite minerals was recognized in this ore-system deposit [3,4]. The replacement of wolframite with scheelite in CO_2-rich fluids was identified in the Borralha magmatic–hydrothermal system, where fluid pressure decreases (or simple cooling) and the ratios of total Ca^{2+} (from feldspar albitization) to total Mn^{2+} (from hübnerite dissolution) contents increase in fluids. A decrease in fluid pressure from lithostatic to hydrostatic levels at a depth of 3–8 km causes CO_2 escape and pH rise and precipitates over 70% of tungsten [5]. Scheelite was formed from a low salinity CO_2-H_2O fluid dominated by CO_2, where the homogenization temperature decreased from 380 °C to 200 °C (average of 284 °C). The $\delta^{18}O_{Fluid}$ values calculated for quartz–water and wolframite–water fractionation fall within the calculated magmatic water range. The $\Delta_{quartz\text{-}scheelite}$ fractionation occurred at approximately 350–400 °C. The $\Delta_{chlorite\text{-}water}$ fractionation factor calculated was approximately +0.05‰ for 330 °C, dropping to −0.68‰ and −1.26‰ at 380 °C and 450 °C, respectively. In addition, estimated crystallizing temperatures based on semi-empirical chlorite geothermometers [6–8] show a narrower temperature range of 375 °C to 410 °C estimated for Fe,Mn-chlorite crystallization.

The next sequence of articles is dedicated to epithermal Au + Sb ores in Southwestern Guizhou (China) and the Baiyun gold deposit in the northeast Hubei Province (China). The Guizhou ore deposit is famous for hosting clusters of Carlin-type Au, Sb, and Hg-Tl deposits [9]. Stibnite, realgar, and orpiment coexist with calcite at the periphery of high-grade orebodies in Au deposits, while stibnite is generally intergrown with fluorite in Sb deposits. The rare earth element (REE) concentrations, Sm/Nd isotope ratios, and Sm–Nd isochron ages were analyzed on separated samples of calcite and fluorite from Au (Zimudang) and Sb (Dachang) deposits. The Sm–Nd isochron ages of 148.4 ± 4.8 and 141 ± 20 Ma obtained on fluorite and calcite are consistent with the age range constrained by the low-temperature thermochronology of zircon (132–160 Ma) and bring new evidence of low-temperature hydrothermal mineralization in Southwestern Guizhou.

In situ LA-ICP-MS was used for trace element and isotope analyses on selected sulfides to evaluate the genesis and evolution of ore-forming fluids from the Baiyun gold deposit in northeastern Hubei, China [10]. Based on in situ Fe and S isotope on pyrite and galena measurements, the authors inferred that the Baiyun gold deposit was formed via magmatic mineralization related to the subduction of the Pacific plate during the Yanshanian plate.

Emphasizing the diagenetic to low-grade metamorphism sequence, phyllosilicate chemistry was used for determining the geological evolution of low-grade metamorphic sequences in early Paleozoic metaclastic rocks in the Eastern Tauride Belt, Türkiye [11]. The authors show that the mineral chemistry of illite/mica and chlorites, together with the evaluation of textural data of low-temperature metaclastic rocks, plays an important role in determining their origin and metamorphic grade.

The kaolinite-to-chlorite conversion is one mineralogical reaction that occurs either in low-temperature diagenetic or hydrothermal systems, where its mechanism is still poorly understood. Direct transformation, conversion via berthierine as an intermediate phase, or direct formation of a berthierine/chlorite mix, either by dissolution–crystallization or by solid-state transformation (or a combination of both), are all hypotheses put forward by Bourdelle et al. [12]. The petrographical, mineralogical, and chemical data suggest that the Fe-chlorite results from the interaction between the shale, providing the Fe,Mg supply, and the Si,Al-rich veins, leading to the chloritization of the kaolinite at a small scale via at least one dissolution–recrystallization step. High-resolution observations highlight that neoformed Fe-rich chlorite contains some 7 Å isochemical layers, as relict of berthierine. The conversion mechanism, suggested by the authors, takes place either through the precipitation of berthierine, followed by a second step involving solid-state berthierine-chlorite conversion, or through the direct precipitation of a chlorite-rich/berthierine-poor mix driven by the Fe/(Fe + Mg) ratio, at low temperature and in reduced conditions. The authors put forward arguments in favor of the second hypothesis.

Active acid-sulfate environments associated with endogenous degassing (i.e., H_2O, CO_2, CH_4, H_2S, SO_2, HCl, HF, etc.) and hot water fluid circulation (hydrothermal/geothermal systems) are described around the world [13–17]. Sulphate efflorescent minerals precipitated on the argillic facies as the result of water–rock interaction and fumarole emission covering the surface of the Donnoe and Dachnoe fields of the Mutnovsky volcano are described by Zhitova et al. [18]. Ca, Ba, $(NH_4)^+$, Na-Fe^{3+}, $(NH_4)^+$-Al, $(NH_4)^+$-Fe^{3+}, Na-Al, K-Al, and K-Fe^{3+} sulphates were reported, where ammonia species were concentrated around fumaroles.

Finally, basalt-derived carnelian was identified in the Newark basin as medium- to coarse-size pebbles in fluvial gravel and alluvium and colluvium formed from erosion of Lower Jurassic Preakness Basalt [19]. Carnelian contains 97–98 wt.% SiO_2, ~1.0 wt.% Fe_2O_3, and 1.0–1.4 wt.% LOI, and trace element abundances are low except for Y, Nb, Ta, W, Th, and U. Rare earth element (REE) patterns display heavy REE enrichment and large negative Eu anomalies. Si was likely derived from the alteration of basaltic glass. $\delta^{18}O_{VSMOW}$ values measured on carnelian are high and range from +18.3 to +31.2‰, comparable to global occurrences of volcanic rock-derived chalcedony. According to authors, carnelian was precipitated from the mixing of hydrothermal fluid with meteoric water under conditions of low temperature (20–80 °C) and neutral to slightly alkaline pH.

Acknowledgments: We are grateful to all our dear collaborators, contributors, and authors who contributed to this Special Issue. Many thanks to Urs Klötzli (University of Wien, Austria) for reading the manuscript and making suggestions.

Conflicts of Interest: The authors declare no conflicts of interest.

References

1. Bourdelle, F. Low-Temperature chlorite geothermometry and related recent analytical advances: A review. *Minerals* **2021**, *11*, 130. [CrossRef]
2. Bobos, I.; Sá, C.; Noronha, F. Mineralogy, Fluid Inclusions, and Oxygen Isotope Geochemistry Signature of Wolframite to Scheelite and Fe,Mn-Chlorite Veins from the W, (Mo, Cu) Ore Deposit of Borralha, Portugal. *Minerals* **2022**, *12*, 24. [CrossRef]
3. Bobos, I.; Noronha, F.; Mateus, A. Fe-, Fe,Mn- and Fe,Mg-chlorite: A genetic linkage to W, (Cu,Mo)—Mineralization in the magmatic-hydrothermal system of Borralha, Northern Portugal. *Mineral. Mag.* **2018**, *82*, S259–S279. [CrossRef]
4. Bobos, I.; Stein, H.; Deng, X.D.; Sudo, M.; Noronha, F. U–Pb LA-ICP-MS and Re–Os dating of wolframite and molybdenite: Constraints on multiple mineralization and cooling history ($^{40}Ar/^{39}Ar$) for the magmatic–hydrothermal system at Borralha, northern Portugal. *Ore Geol. Rev.* **2024**, *168*, 106013. [CrossRef]
5. Liu, H.; Liu, X.C.; Zhang, D.H.; Zhou, Z.J.; Han, F.B. The precipitation mechanisms of scheelite from CO_2-rich hydrothermal fluids: Insight from thermodynamic modeling. *Appl. Geochem.* **2024**, *175*, 106187. [CrossRef]
6. Bourdelle, F.; Parra, T.; Chopin, C.; Beyssac, O. A new chlorite geothermometer for diagenetic to low-grade metamorphic conditions. *Contrib. Miner. Petrol.* **2013**, *165*, 723–735. [CrossRef]
7. Lanari, P.; Wagner, T.; Vidal, O. A thermodynamic model for di-trioctahedral chlorite from experimental and natural data in the system MgO–FeO–Al_2O_3–SiO_2–H_2O: Applications to P–T sections and geothermometry. *Contrib. Miner. Petrol.* **2014**, *167*, 968–984. [CrossRef]
8. Inoue, A.; Meunier, A.; Patrier-Mas, P.; Rigault, C.; Beaufort, D.; Vieillard, P. Application of chemical geothermometry to low-temperature trioctahedral chlorites. *Clays Clay Miner.* **2009**, *57*, 371–382. [CrossRef]
9. Wang, Z.; Tan, Q.; Xia, Y.; Liu, J.; Yang, C.; Li, S.; Li, T.; Chen, F.; Wang, X.; Pan, Q.; et al. Sm-Nd Isochron Age constraints of Au and Sb mineralization in the southeasterm Guizhou Province, China. *Minerals* **2021**, *11*, 100. [CrossRef]
10. Song, W.; Liu, J.; Zou, Y.; Liu, X.; Long, T.; Zhu, J.; Fu, S.; Chen, S.; Xiong, Y.; Zhou, R.; et al. Genesis of the Baiyun gold deposit in the northeastern Hubei Province, China: Insights from In-Site Trace elements and S-Fe isotopes of sulfides. *Minerals* **2024**, *14*, 517. [CrossRef]

11. Bozkaya, O.; Yalcin, H. Mineral Chemistry of Low-Temperature Phyllosilicates in Early Paleozoic Metaclastic Rocks, Eastern Tauride Belt, Türkiye. *Minerals* **2022**, *12*, 1088. [CrossRef]
12. Bourdelle, F.; Dubois, M.; Lloret, E.; Durand, C.; Addad, A.; Bounoua, S.; Ventalon, S.; Recourt, P. Kaolinite-to-chlorite conversion from Si, Al-rich fluid origin veins Fe-rich Carboniferous shale interaction. *Minerals* **2021**, *11*, 804. [CrossRef]
13. Giggenbach, W.F. Magma degassing and mineral deposition in hydrothermal systems along convergent plate boundaries. *Econ. Geol.* **1992**, *97*, 1927–1944.
14. Christenson, B.W.; Wood, C.P. Evolution of a vent-hosted hydrothermal system beneath Ruapehu crater lake, New Zeeland. *Bull. Volcanol.* **1993**, *55*, 547–565. [CrossRef]
15. Delmelle, P.; Bernard, A. Geochemistry, mineralogy, and chemical modelling of the acid crater lake of Kawah Ijen volcano, Indonesia. *Geochim. Cosmochim. Acta* **1994**, *58*, 2445–2460. [CrossRef]
16. Africano, F.; Bernard, A. Acid alteration in the fumarolic environment of Usu volcano, Hokkaido, Japan. *J. Volcanol. Geotherm. Res.* **2000**, *97*, 475–495. [CrossRef]
17. Bobos, I.; Gomes, C. Mineralogy and Geochemistry (HFSE and REE) of the present-day acid-sulfate types alteration from the active hydrothermal system of Furnas volcano, São Miguel Island, Azores Archipelago. *Minerals* **2021**, *11*, 335. [CrossRef]
18. Zhitova, E.S.; Khanin, D.A.; Nuzhdaev, A.A.; Nazarova, M.A.; Ismagilova, R.M.; Shilovskikh, V.V.; Kupchinenko, A.N.; Kuznetsov, R.A.; Zhegunov, P.S. Efflorescent sulfates with M^+ and M^{2+} cations from fumarole and active geothermal fields of Mutnovski volcano (Kamchatka, Russia). *Minerals* **2022**, *12*, 600. [CrossRef]
19. Volkert, R.A.; Gorring, M.L.; Peck, W.H.; Stanford, S.D. Characterization and Origin of Basalt-Derived Carnelian in the Mesozoic Newark Basin, New Jersey, USA. *Minerals* **2023**, *13*, 1249. [CrossRef]

minerals

MDPI

Review

Low-Temperature Chlorite Geothermometry and Related Recent Analytical Advances: A Review

Franck Bourdelle

Univ. Lille, Institut Mines-Télécom, Univ. Artois, Junia, ULR 4515-LGCgE, Laboratoire de Génie Civil et Géo-Environnement, F-59000 Lille, France; franck.bourdelle@univ-lille.fr

Abstract: Chlorite, a 2:1:1 phyllosilicate, has all the required attributes to form the basis of a geothermometer: this mineral is ubiquitous in metamorphic, diagenetic, and hydrothermal systems with a broad field of stability and a chemical composition partly dependent on temperature (T) and pressure (P) conditions. These properties led to the development of a multitude of chlorite thermometers, ranging from those based on empirical calibrations (linking T to Al^{IV} content) to thermodynamic or semi-empirical models (linking T to chlorite + quartz + water equilibrium constant). This present study provides an overview of these geothermometers proposed in the literature for low-temperature chlorite (T < 350 °C), specifying the advantages and limitations of each method. Recent analytical developments that allow for circumventing or responding to certain criticisms regarding the low-temperature application of thermometers are also presented. The emphasis is on micrometric and nanometric analysis, highlighting chemical intracrystalline zoning—which can be considered as evidence of a succession of local equilibria justifying a thermometric approach—and mapping ferric iron content. New perspectives in terms of analysis (e.g., Mn redox in Mn-chlorite) and geothermometer (molecular solid-solution model, oxychlorite end-member) are also addressed.

Keywords: low-temperature chlorite; thermometry; diagenesis; hydrothermal systems; low-grade metamorphism; nanoscale analytical techniques

1. Introduction

Determining the pressure, temperature, and chemical paleo-conditions (P-T-X) that have prevailed in geological contexts has long been a major issue, especially in systems of low temperature (up to epizone, i.e., T < 350 °C). In fundamental research, this type of study contributes to the understanding of the Earth system, its evolution, and the interactions between its compartments. In applied research, the quantification of paleo-conditions has been useful in the exploration of new resources, such as hydrocarbons ("oil and gas windows") for decades, or geothermal energy (deep geothermal energy) in the context of ecological transition. In this way, many thermobarometric tools have been developed that, in turn, use the properties of organic matter (reflectance of vitrinite, e.g., [1]), Raman spectroscopy of carbonaceous materials thermometry (RSCM, e.g., [2–4]), the microthermometric data of fluid inclusions trapped in neoformed crystals (e.g., [5–7]), isotopy (e.g., [8]), or even the crystallochemical characteristics of specific minerals, such as phyllosilicates (e.g., [9]).

Chlorite, a 2:1:1 phyllosilicate (also denoted as TOT O; [10]), is a mineral with all the required attributes to form the basis of a thermometer: it is present in all geological contexts, has a wide field of stability (from 80 °C to more than 700 °C), and its chemical composition is influenced by the chemistry of the system, the fugacity of oxygen, and temperature and pressure conditions (e.g., [11]). For decades (e.g., [12]), the chemical composition of chlorite has been studied and used as a marker of paleo-conditions and their evolution, in particular, in low-temperature–low-pressure contexts (LT and LP). However,

LT–LP chlorite thermometry is open to debate [13,14]. First, the method employed for LT chlorite thermometer calibration is questioned: Which composition to use? Is an empirical approach sufficient? What is the resulting applicability? Second, the nature of LT chlorites is debatable. Like all clay minerals, chlorite can exhibit crystallochemical defects or interstratification with other phases, and slow transformation kinetics when the temperature is below 350 °C can lead to the formation of metastable phases. The question then arises of the relevance of chlorite thermometers, especially since the formation of chlorite does not follow a single process but, on the contrary, can follow several paths, from precipitation from a fluid to the progressive chloritization of a precursor mineral [15]. LT chlorite thermometry is, therefore, confronted with two challenges: to offer a reliable geothermometer with a wide field of applicability (in terms of chlorite compositions), and to be able to apply it to compositions where it is certain they represent thermodynamic equilibrium, i.e., that they have recorded a stage in the evolution of P-T-X conditions.

The present study provides an overview of the existing chlorite geothermometers, from the first empirical calibrations to the recent semi-empirical tools, specifying the scope, advantages, and inconveniences of each method. In addition, the most innovative analytical techniques, which modified the approach of chlorite thermometry during the last decade, and made it possible to reach the chemical compositions containing the thermometric information, are exposed.

2. Compositional Variability of Chlorites

Chlorite is a phyllosilicate consisting of a 2:1 layer (talc-like layer) in which two tetrahedral sheets (denoted as IV) surround an octahedral sheet (denoted as VI), associated with an interlayer octahedral sheet (brucite-like sheet). This structure, 14 Å thick, can be defined as follows [10]:

$$(M3)_2M4(OH)_6\ M1(M2)_2[(T1)_2(T2)_2]O_{10}(OH)_2 \quad (1)$$

where M1 and M2 (three octahedral sites in the talc-like layer), and M3 and M4 (three octahedral sites in the brucite-like layer) sites are occupied by divalent (R^{2+}) cations such as Fe^{2+}, Mg^{2+}, Mn^{2+}, or trivalent cations (R^{3+}) such as Al^{3+} and Fe^{3+}, while T sites (four tetrahedral sites) are filled by Si^{4+}, Al^{3+}, and Ti^{4+}, giving the formula:

$$\left(R^{2+},R^{3+}\right)^{VI}_{2-3}\left(R^{2+},R^{3+}\right)^{VI}_{2-3}\left(Si_2\left(Si,\ R^{3+}\right)_2\right)^{IV}O_{10}(OH)_6 \quad (2)$$

These structural specificities favor cationic exchanges, which are carried out according to substitutions, making it possible to guarantee layer electroneutrality (i.e., 28+ charges corresponding to the standard $O_{10}(OH)_8$ anionic basis). Regarding this last point, it is often necessary to consider the presence of vacancies, i.e., of empty octahedral sites (usually denoted as □) in the structure. The most remarkable cationic substitutions occurring in chlorites are, therefore, as follows:

Tschermak substitution (TK) as $Si^{IV} + (Fe^{2+},Mg^{2+})^{VI} = Al^{IV} + Al^{VI}$
di/trioctahedral substitution (DT) as $2(Al^{3+},Fe^{3+})^{VI} + \square^{VI} = 3(Fe^{2+},Mg^{2+})^{VI}$
ferromagnesian substitution (FM) as $Mg^{2+} = Fe^{2+}$
trivalent cation substitution (AF) as $Al^{3+} = Fe^{3+}$

All these cationic substitutions can be combined to form a multitude of exchange vectors. Additional substitutions can also be examined, such as the "hydroxyl" substitution and the deprotonation process, written $Fe^{3+} + 2OH^- = Fe^{3+} + O^{2-} + OH^-$ and $Fe^{2+} + H^+ = Fe^{3+}$, respectively [11,16] (see Section 5). Consequently, octahedral sheets can theoretically exhibit three out of three occupied cationic sites (tri-tri octahedral structure; clinochlore, daphnite), two out of three occupied cationic sites (di-dioctahedral structure; dombassite), or one or other combination of these configurations, i.e., two out of three sites occupied for $M1(M2)_2$ and three out of three sites occupied for $(M3)_2M4$ (di-trioctahedral structure, sudoite), or the opposite (tri-dioctahedral structure) [17,18]. The first structure is the most common in nature, while the fourth structure has never been observed.

This great compositional variability, the diversity of elements that can be received in the structure, and the large number of cationic substitutions that can occur, sometimes simultaneously, lead to a wide compositional field for chlorite. The chlorite structure can thus be defined by a solid-solution model (e.g., [19–22]), knowing that most natural chlorite composition is delimited by Fe,Mg-sudoite, Fe,Mg-amesite, and Al-free chlorite end-members [23] (Figure 1), to which some Fe^{3+}-end-members can be added. Chlorites from LT systems, especially diagenesis contexts, display higher Si and lower R^{2+} contents than in metamorphic systems [24]. This supports the hypothesis that chlorite composition is, at least in part, linked to the temperature of formation.

Cathelineau and Nieva [12] and Cathelineau [25], studying chlorites from Los Azufres and Salton Sea, two geothermal systems, highlighted the increase in Al^{IV} and the decrease in octahedral vacancies with increasing temperature. They consequently defined the $Si_{-1}Al^{VI}{}_{-1}\square_{-1}Al^{IV}{}_{+1}R^{2+}{}_{+2}$ cationic exchange vector, resulting from a combination of the TK and DT substitutions, and used it as the base for thermometric empirical calibration (see Section 3). The same compositional variation was also observed in diagenetic and low-grade metamorphism contexts [24,26–29] which, at first, could provide reassurance for the application of empirical thermometers. To better constrain the links between cationic substitution, temperature, and the chemical environment, Bourdelle and Cathelineau [23] compiled hundreds of analyses of LT chlorites from hydrothermal, diagenetic, and low-grade metamorphism systems [25,26,28–40] for which the formation temperature is known (i.e., determined using independent geothermometers). It appears that LT chlorite composition (with the exception of sudoite (e.g., [41]) which often remains close to its ideal composition) is mainly restricted to the $1.5 < R^{3+} < 3$ range (apfu, atoms per formula unit) while showing Si contents up to 3.5 apfu. Regardless of the environment in which the chlorites were formed, each set of LT chlorite compositions—corresponding to all compositions of chlorites coming from the same geologic field—exhibits a deviation from the Al-free chlorite/amesite line to the pyrophyllite-gibbsite/Al-free chlorite line (representing the DT exchange line; Figure 1), like so many parallels with respect to the lines of constant R^{3+} values. This observation implies that (i) compositions of chlorites from the same geological field follow the $Si_{-1}\square_{-1}R^{2+}{}_2/Si_{-1.25}\square_{-0.75}AlR^{2+}$ exchange vectors according to T, as a 1/1 and 5/3 combination of TK and DT substitutions, respectively; (ii) chlorites with different compositions and from different geological fields can be linked to a single temperature, forming several isotherms subparallel to the TK exchange line (higher T isotherm closest to the Al-free chlorite/amesite line; Figure 1). These compositional parallels to the constant R^{3+} value lines (corresponding to each dataset; led by a decrease in Si content/increase in Al^{IV} content) are subperpendicular to defined isotherms and refer to the linear empirical thermometers established during the 1980s (see Section 3).

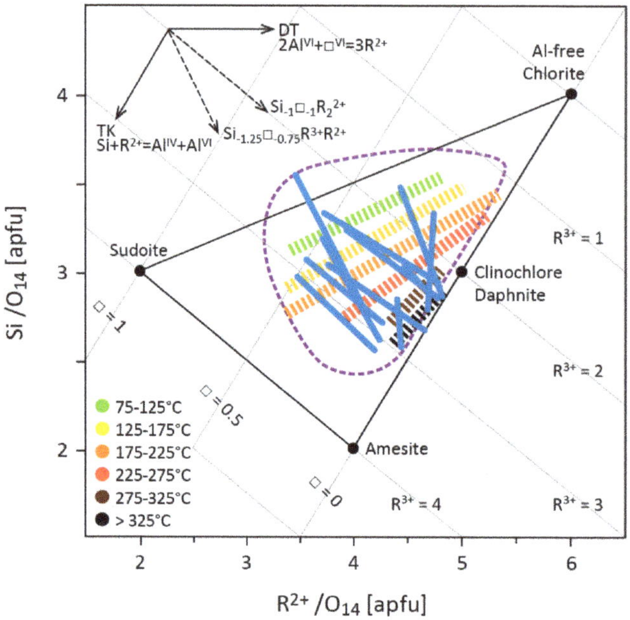

Figure 1. Representation of several compositional datasets of low-temperature chlorites from several geological settings ([25,26,29–36,42]; modified from [23]) in the R^{2+}–Si diagram of [17]. Compositional trends drawn by each dataset were identified and are shown as qualitative lines (in blue). The compositions were also assembled according to their temperature of formation (chlorite compositions from [37–40,43] were also considered), which was determined by each author using several independent methods (e.g., RSCM, vitrinite reflectance, microthermometric data of fluid inclusions, bottom hole temperature, isotopy), allowing identification of several isotherms (dashed colored lines; see [23]). The field covered by LT chlorite compositional datasets is approximated by the area circled with a purple dotted line. All Fe is regarded as Fe^{2+}. R^{2+} refers to divalent cations (Fe^{2+} and Mg^{2+}), R^{3+} to trivalent cations (Al^{3+}, Fe^{3+}), and □ to vacancies. apfu—atoms per formula unit.

3. Concepts of Chlorite Thermometers

Here, only the geothermometers based on the chlorite composition in siliceous contexts will be discussed, excluding chlorite polytypism and chlorite-carbonate [44] geothermometry.

In the 1980s, Cathelineau and Nieva [12] demonstrated a close relationship between the Al^{IV} content of chlorites in the Los Azufres geothermal system (Mexico) and the formation temperature determined by independent methods. Cathelineau [25] completed the study with a series of data acquired on chlorites from the Salton Sea hydrothermal system (United States) and confirmed this relationship, given in the following equation:

$$T(^\circ C) = -61.92 + 321.98\ Al^{IV} \quad (3)$$

This first chlorite geothermometer, which was purely empirical, gained ground, leading to a multiplication of proposals for empirical equations of the same type. Indeed, all the compositional lines (in Figure 1), parallel to the constant R^{3+} value lines in the compositional field of LT chlorites and perpendicular to the isotherms, are all possible empirical equations. This empiricism could not account for the varying influences of TK and DT substitutions on the variation of Al^{IV} content—some have tried to correct their proposed equation with a corrective factor based on the Fe/Mg ratio, i.e., on FM substitution. Thus, there are no less than five other empirical thermometric equations, in addition to that of Cathelineau (Table 1), proposed in the literature [24,38,45–48], in some cases being used as a thermometer when they had not been suggested by their authors for this [24].

In the absence of constrained thermodynamic models capable of taking into account all the substitutions occurring in chlorites—with the exception of the attempt by [11] which excluded Si-poor (Si < 3.2 apfu) and/or Al-rich (Al > 2.5 apfu) chlorites (Figure 2) in spite of their common occurrence—these empirical thermometers were widely used during the 1990s to estimate T conditions in LT environments. However, their multiplicity shows that an empirical thermometer has an extremely limited field of application, i.e., being restricted to the context in which it has been calibrated, and that a new equation must be established for each new system studied. Indeed, an empirical simplification associates a temperature with one—and only one—possible composition of chlorite, contrary to what is observed in nature. As a result, their systematic application, without taking into account these specific conditions, has failed or led to erroneous geological interpretations in many cases [13,14,23,43].

At the turn of the 2000s, the development of empirical thermometers was stopped, though they were still in use, and a thermodynamic approach was favored for LT contexts. This approach is based on the fact that the equilibrium condition of any balanced chemical reaction can be expressed by

$$\Delta G_r^0 + RT \ln K = 0, \tag{4}$$

with ΔG_r^0 as the Gibbs free energy of reaction, R the gas constant, and K the chlorite + quartz equilibrium constant, defined as

$$K = \prod_j (a_{ideal}\gamma_m)_j^{v_j}, \tag{5}$$

where v_j is the stoichiometric reaction coefficient of end-member j, a_{ideal} is the ideal part of the end-member activity, and γ_m is the activity coefficient accounting for non-ideal contributions (which can be obtained from Margules parameters)—i.e., the influence of interactions between cations exchanged at the same site on enthalpy, entropy, and volume. In this way, Vidal et al. [9] proposed, through an experimental approach coupled with the study of natural samples, a thermodynamic model for aluminous chlorites between 100 and 600 °C, based on an ordered distribution of cations in the structure (in T2, M1, M2, M3, and M4 sites) [19] and on a four-thermodynamic-component solid-solution model (Mg-sudoite, Mg-amesite, clinochlore, and daphnite), which makes it possible to account for all the substitutions influencing the chlorite composition (Figure 2). Symmetric Margules parameters and standard state properties of defined end-members were calibrated, allowing estimation of the pressure and temperature of chlorite formation for a wide range of P, T, and geological contexts. Vidal et al. [49,50], and then Lanari et al. [51], improved this model by adding the Fe-rich Si-poor end-member Fe-amesite, on the basis of the work of [52]. The authors also suggest that the Fe^{3+} content should be considered because it strongly influences the structural formula of chlorites and, consequently, the P-T estimates; until then, this parameter was always neglected, with Fe_{total} always assumed to be Fe^{2+}. To better take ferric iron into account, Vidal et al. [53] defined the thermodynamic properties of a Fe^{3+}-end-member as ferri-sudoite (Table 1) [53,54].

Table 1. Chlorite geothermometers.

Ref.	Year	Type	Equation (Oxygen Basis (14 or 28) in Brackets)/Equilibrium [1,2,3]	Elements/End-Members
Cathelineau and Nieva [12]	1985	empirical equation	$T = 212.3Al^{IV} + 17.5$ (14)	Al^{IV} content
Kranidiotis and McLean [47]	1987	empirical equation	$T = 106(Al^{IV} + 0.7XFe) + 18$ (28)	Al^{IV}, Fe, Mg contents
Cathelineau [25]	1988	empirical equation	$T = 321.98Al^{IV} - 61.92$ (14)	Al^{IV} content
Jowett [48]	1991	empirical equation	$T = 319(Al^{IV} + 0.1XFe) - 69$ (14)	Al^{IV}, Fe, Mg contents
Hillier and Velde [24]	1991	empirical equation	$T = 249.56Al^{IV} - 320.28$ (28)	Al^{IV} content
Zang and Fyfe [45]	1995	empirical equation	$T = 106.2(Al^{IV} - 0.88[XFe - 0.34]) + 17.5$ (28)	Al^{IV}, Fe, Mg contents
Xie et al. [38]	1997	empirical equation	$T = 321.98(Al^{IV} + 0.133[0.31 - XFe\ (<0.31)]) - 61.92$ (14)	Al^{IV}, Fe, Mg contents
Vidal et al. [9]	2001	thermodynamic model	2 clinochlore + 3 Mg-sudoite = 4 Mg-amesite + 7 quartz + 4 H_2O	clinochlore daphnite Mg-amesite Mg-sudoite
Vidal et al. [49,50]	2005 2006	thermodynamic model	+ 5 Fe-amesite + 4 clinochlore = 5 Mg-amesite + 4 daphnite [4] 16 daphnite + 15 Mg-sudoite = 20 Fe-amesite + 6 clinochlore + 35 quartz + 20 H_2O 4 daphnite + 6 Mg-sudoite = 5 Fe-amesite + 3 Mg-amesite + 14 quartz + 8 H_2O	+ Fe-amesite
Lanari et al. [51]	2014	thermodynamic model	+ 2 daphnite + 3 Fe-sudoite = 4 Fe-amesite + 7 quartz + 4 H_2O Mg-amseite + 2 Fe-sudoite = Fe-amesite + 2 Mg-sudoite 2 clinochlore + 5 Fe-sudoite = 2 daphnite + 5 Mg-sudoite	+ Fe-sudoite
Vidal et al. [53]	2016	thermodynamic model	+ 14 equilibrium implying quartz, O_2, and ferri-sudoite	+ Ferri-sudoite
Walshe [11]	1986	semi-empirical model	$T = 1626/(6.542 + 0.33(\log K) - 273$ 5 Al-free Mg-chlorite + 3 pyrophyllite-gibbsite = 6 clinochlore + 14 quartz + 8 H_2O	Al-free Mg-chlorite clinochlore chamosite (daphnite) pyrophyllite-gibbsite
Inoue et al. [28]	2009	semi-empirical model	$T = 1/(0.00293 - 0.000513(\log K) + 0.00003904(\log K)^2) - 273$ Al-free Mg-chlorite + 3 Mg-sudoite = 3 Mg-amesite + 7 quartz + $4H_2O$	Al-free Mg-chlorite Mg-corundophilite (amesite) Mg-sudoitechamosite (daphnite)
Bourdelle et al. [55]	2013	semi-empirical model	$T = -9400/(\log K - 23.40) - 273$ Al-free Mg-chlorite + 3 Mg-sudoite = 3 Mg-amesite + 7 quartz + $4H_2O$	Al-free Fe-, Mg-chlorite Fe-, Mg-amesite Fe-, Mg-sudoite

[1] XFe = Fe/(Fe + Mg). [2] T in °C. [3] K: chlorite + quartz equilibrium constant. [4] Only in particular P-T conditions.

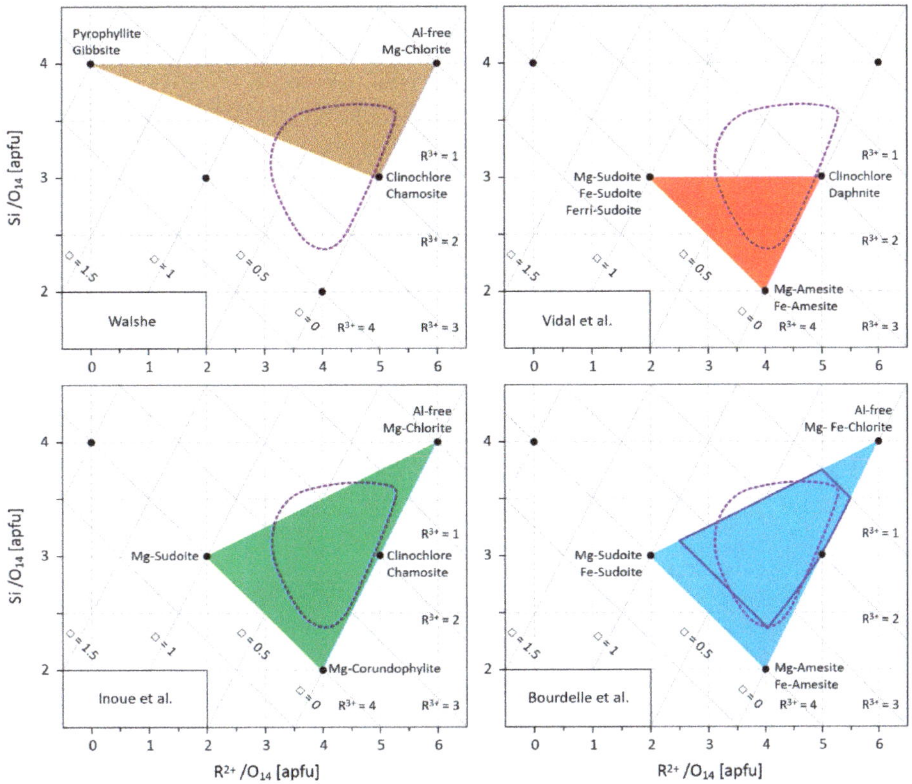

Figure 2. Projection field (R^{2+}–Si diagram) for chlorite compositions [17]. Values are given in atoms per formula unit (apfu) with a 14-oxygen basis. Several end-members are indicated and the compositional spaces defined by the models of Walshe [11], Vidal et al. [9,49,50,53], Inoue et al. [28], and Bourdelle et al. [55] are outlined. The field covered by LT chlorite compositional datasets (see Figure 1) is approximated by the area circled in purple dotted lines. R^{2+} refers to divalent cations (Fe^{2+} and Mg^{2+}), R^{3+} to trivalent cations (Al^{3+}, Fe^{3+}), and □ to vacancies. The area of optimal reliability for Bourdelle's thermometer is shown in dark blue [23].

Unlike empirical thermometers, the thermodynamic approach—based on the end-member properties determined experimentally and/or by the study of natural samples—is the most rigorous, but requires complex calculations and is dependent on the availability of thermodynamic data, which limits its field of application. As such, the lack of thermodynamic data on Si-rich end-members prevents taking into account chlorites whose Si content is greater than three atoms per formula unit (apfu), which is one of the characteristics of LT chlorites. Consequently, two studies [28,55] propose a trade-off between limited applicability/simplicity on one side, and rigor/complexity on the other by combining an empirical equation linking *K* equilibrium constant and T, and a multi-component solid-solution model. For this, Equation (5) is restricted to its ideal part as

$$K = \prod_j (a_{ideal})_j^{v_j}, \tag{6}$$

given the rearranged Equation (4) as

$$\ln \prod_j (a_{ideal})_j^{v_j} = \frac{-\Delta G_r^0}{RT} = \frac{A}{T} + B, \tag{7}$$

where A and B are constants. a_{ideal} is calculated for each end-member j as follows:

$$(a_{ideal})_j = \prod_s \prod_m \left(\frac{ns}{r_m} Xm\right)^{r_m}, \tag{8}$$

where $\boldsymbol{ns}$ is the multiplicity of site s, r_m and Xm are the number and the mole fraction of cation m on site s, respectively. This semi-empirical chlorite thermometry approach requires (i) the calibration of A and B constants of Equation (7) for chlorite + quartz equilibrium and (ii) to define the involved end-members and the cationic distribution in sites s (discussion on solid-solution models and cationic site occupancies can be found in [9,56]). Inoue et al. [28] chose the magnesian end-members of Al-free chlorite, of corundophilite (similar to amesite), and of sudoite, and the chamosite as the ferrous end-member, and assumed a random mixing of octahedral cations based on the fact that Fe-rich chlorites are commonly the Ib polytype for which details of the crystal structure are not clearly established. Bourdelle et al. [55] preferred the magnesian and ferrous end-members of Al-free chlorite, amesite, and sudoite, and an ordered cationic site repartition model, following the recommendations of [9,10,19]. In both cases, the choice of Al-free chlorite as an end-member (Table 1) allows accounting for Si-rich chlorites (Si > 3 apfu; Figure 2) which were not considered by [9,49–51,53]. Inoue et al. [28] and Bourdelle et al. [55] calibrated two Equation (7) for chlorite + quartz equilibrium (Table 1), i.e., defining A and B constants, from natural diagenetic and hydrothermal LT chlorite compositions, with a recommended upper limit of applicability around 250 and 350 °C, respectively. Therefore, and contrary to empirical equations, these two geothermometers predict that one chlorite composition corresponds to one temperature of formation, but that at any fixed temperature, a range of chlorite compositions exist and can be theoretically calculated, in agreement with observations of natural chlorites. These two geothermometers are easy to use, circumvent bulk rock composition effects, and perform well in diagenetic to low-grade metamorphic contexts (e.g., [55,57–62]), particularly in the T range where Si-rich chlorites are observed. They were later derived in a graphical way by [23] to estimate T only from Si and R^{2+} contents, making it as easy to use as empirical thermometers.

To achieve this, these geothermometers cannot dispense with simplifications. The effect of pressure is therefore neglected—the B constant being considered as a fixed ΔS—as well as the non-ideal part of the end-member activity. These assumptions are, however, acceptable within the restricted P-T range of diagenetic, hydrothermal, and low-grade metamorphism contexts [55]. Similarly, H_2O activity is assumed to be equal to 1 as it is often unknown. If this hypothesis seems reasonable in LT systems, it may remain locally questionable, especially in the presence of carbonates. The activity of quartz is also assumed to be equal to 1, which implies that the application of these geothermometers is restricted to quartz-bearing rocks. Generally, the conditions of simplification validation are reached when simplifications are identified and their impact on the T estimate is quantified. Indeed, the choice of the chlorite geothermometer will be driven by the P-T window studied, compositional range of the targeted chlorite, host rock (nature, chemistry, etc.), data (i.e., scale of the compositional analysis; see Section 4), desired precision for T estimates, and acceptable simplifications (according to the geological context).

4. Chemical Heterogeneity, Achievement of Equilibrium, and Related Analytical Advances

Hydrothermal chlorites result from precipitation or rock alteration due to hydrothermal fluid circulation: either propylitic alteration linked to heat transfer from igneous intrusions, or phyllic alteration in open systems involving low-pH fluid fracture-controlled infiltration. In this context, chlorite can crystallize as discrete phases, or as mixed-layer minerals in which chlorite content increases with T. Three conversion paths can therefore be defined: a Mg series from saponite to Mg-rich chlorite via corrensite as an intermediate phase, an Fe series from berthierine to Fe-rich chlorite, and an Al(Mg) series from kaolinite to sudoite via tosudite. Diagenetic chlorites may crystallize as neoformed rims on grains, as pore-filling material, or by conversion of a precursor including berthierine, smectite, kaolinite (defining three conversion series similarly to hydrothermal chlorites), Fe-, Mg-rich detrital grains, and volcanic rock fragments. Mg-rich chlorites are usually observed in terrestrial and marine sandstones, whereas Fe-rich chlorites occur preferentially in coastal environments [63]. Fe-rich chlorites are also reported as a result of the interaction between Fe-rich host rocks (shale) and veins of kaolinite from Si–Al-rich meteoric-origin fluids. Regardless of the context or the reaction pathway, the mechanism involved in chlorite formation can be direct precipitation (DP), solid-state transformation (SST), dissolution–recrystallization (DC), or a mix of all three [64]. Due to the plurality of processes leading to chlorite formation, chlorite often exhibits (i) layers of another phase and/or (ii) a chemical heterogeneity, at the sample scale, where crystallites with different compositions coexist, or at the crystal scale, where crystallites present internal chemical zoning. From these observations, many criticisms have been formulated of the application and applicability of chlorite geothermometers [14,65–67]. Each of these criticisms must be treated independently because each heterogeneity refers to different phenomena.

Compositional variations observed in LT chlorites may be attributed to contamination of the chlorite by kaolinite, smectite, or berthierine [14,66]. Chlorite analysis contaminated by smectite can be discarded by applying chemical criteria: smectite contains Ca, K, or Na interlayered cations while chlorite does not; only analysis presenting $Na_2O + CaO + K_2O < 0.5$ wt% must be considered. The interstratification with 7Å-phases as kaolinite and berthierine is, however, difficult to identify chemically from a true chlorite, but several criteria can be used to verify that the chlorite composition corresponds to a combination of defined end-members [9], discarding some of the contaminated analyses. The method for geothermometer calibration can also be used to address this issue. Indeed, a large database for calibration [55], covering a wide field of geological parameters, such as lithology, temperature gradient, kinetics, fluid regimes, oxidation state, chlorite composition, chlorite precursors, impurities, and analytical contaminations, can be considered the best way to limit the impact of each parameter making the residual contamination without effects on the T estimate. Finally, the presence of smectite, berthierine, or kaolinite layers can be detected thanks to a high-resolution spatial control using transmission electron microscopy (TEM).

The chemical heterogeneity beyond analysis contamination often has a true significance in terms of paleo-condition record. Generally, it is considered that sluggish kinetics in LT contexts preclude the achievement of global equilibrium, leading to the crystallization of heterogeneous–metastable phases. In other words, the formation of LT chlorites would be more likely due to kinetic effects than to a re-equilibration process with conditions. In fact, chlorite formation occurs during reactions which are transient disequilibrium states governed by local chemical potential gradients and kinetic processes. A re-equilibration process may proceed through diffusion or, more probably, dissolution–recrystallization (DC) [68,69], but at a very small scale due to the slow kinetics of reaction and diffusivity at low temperatures. In high-grade metamorphic contexts, thermodynamic equilibrium can sometimes be approached not globally but locally at the micrometric scale, leading to large chemical zoning [50,70,71], but in diagenetic and hydrothermal domains, the equilibrium is almost always reached very locally. In this way, compositions in equilibrium with the P-T-X conditions in the system at a given time and, therefore, the only ones able to justify application of the thermometric approach, are found at very small spatial scales, probably coexisting with nonequilibrated compositions, as described by Ostwald ripening. This view explains the apparent heterogeneity observed in LT chlorites, each composition recording the evolution of a very local fluid–solid system, which can sometimes be interpreted as a part of the rock history. Discussions on local equilibrium, solid-state transformation, and chlorite formation processes can be found in [15,53,64,72].

This local equilibrium, or the succession of local equilibria, requires high-resolution spatial analysis to be highlighted in LT domains and used in thermometric perspectives. In this respect, TEM equipped with energy dispersive X-ray spectroscopy (TEM–EDX) presents advantages. Jahren and Aagaard [27,32,73] have therefore observed lateral variation of the composition from crystal core and crystal rims for diagenetic illite–chlorite assemblages in arenites (Norway, 90–180 °C). For the first time, chemical zoning at the nanoscale was described for LT chlorites. However, the analysis of such a nanozoning is difficult as the texture of the crystallite must be preserved. Bourdelle et al. [74] proposed a protocol to couple TEM–EDX, providing the proper spatial resolution in analysis, with a focused ion beam (FIB), making the in situ preparation of TEM samples with high spatial control and preservation of mineral microtextures possible. FIB is a technique for ultrathin section milling [75] using a Ga^+ ion beam which had been developed in the field of geosciences throughout the 2000s. Bourdelle et al. [29,43] have applied this analytical protocol to the diagenetic chlorites of the Gulf Coast (Texas) and to the low-grade metamorphism chlorites of Glarus (Switzerland), distinguishing between analyses made at the rims of nano- to micro-crystals and those made at the core. In the first case, TEM and FIB techniques highlight consistent within-grain patterns of major-element variations in chlorites sampled at different depths and locations of the Gulf Coast area (100–230 °C). Here, the nanozoning is due to various crystallization stages during burial, the crystal core composition referring to detrital chlorite and the crystal rim compositions to neo-formed chlorites. The chlorite rim compositions show a T dependence with depth and location, in agreement with the diagenetic history described at a regional scale [29]. In the second case, quantitative TEM–EDX analysis carried out on FIB sections cut in chlorites from Glarus (200–350 °C) displays a compositional trend from crystal core to crystal rim (Figure 3), consistently observed along a north–south transect through the Glarus area in all samples, with calculated rim temperatures systematically lower than core temperatures. This chlorite nanozoning is testament to the retrograde path of the system, with the crystal rims recording a later step of this exhumation than the crystal cores. In both cases, the temperature dependence of the chlorite rim and/or core compositions supports the idea that the chemical zoning of LT chlorites, observable only at the nanoscale, is not controlled by a kinetic effect but rather by the local equilibria themselves influenced by regional-scale processes. The thermodynamic/thermometric approach therefore makes sense on the condition of correctly identifying—i.e., on an appropriate spatial scale—the composition(s) that are in equilibrium, or have been, with the P-T-X conditions. The adapted nanoscale analytical protocol allows the application of an approach similar to that used for metamorphic rocks at the micro- to millimetric scale [50,76]. However, the TEM–EDX/FIB protocol is complex to implement routinely. The analytical protocol should be chosen according to the targeted chlorites: electron microprobe (EMP) for homogeneous, large grain chlorites (often formed via direct precipitation or long-term dissolution–recrystallization), TEM–EDX/FIB for heterogeneous, small grain chlorites (often formed via successive local equilibria). This choice of the micrometric or the nanometric approach will often be based on petrological indices, i.e., the shape of the crystallites, the size, the orientation with respect to foliation or fractures, and the preferred zones of crystallization. Thus, a chlorite exhibiting all the characteristics of a detrital phase will require a nanoscale analysis, as chlorites of small sizes, linked to another phase, or fitting closely the foliation. On the other hand, a large chlorite having all the characteristics of a phase neoformed from a fluid or from a recrystallization may be the subject of a micrometric analysis. It will be the same if the petrological criteria show that several generations of chlorites coexist in the rock: in this case, the information on a regional scale will be easy to obtain, and one could be satisfied with a micrometric analysis. In case of doubt, it will be necessary to couple the two approaches on a few selected chlorites, and to adapt the analytical methods for the following analyses. It should be noted that (i) the TEM–EDX/FIB protocol also enables checking the absence of smectite, kaolinite, or berthierine layers in chlorite, and (ii) the TEM–EDX spot size is around 50 nm, i.e., pointed analyses target between 30 and 40 chlorite layers, limiting undetectable interstratification with layers of other phases.

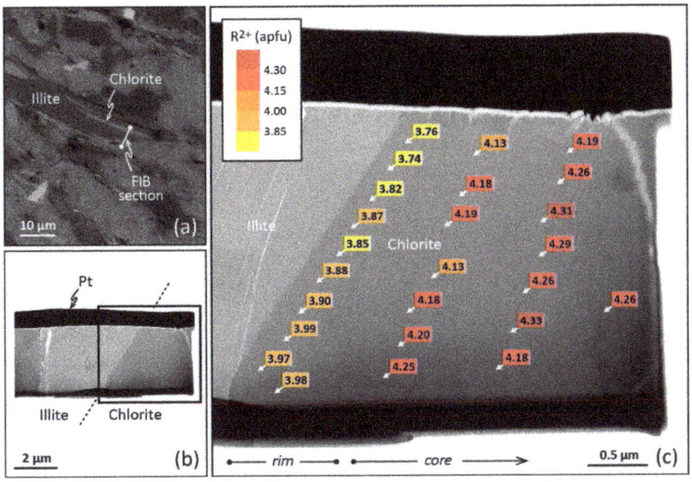

Figure 3. TEM–EDX analysis obtained on a FIB cross-section of a chlorite from Glarus. (**a**) Scanning electron microscopy image of the petrographic thin section showing the emplacement of the FIB foil to be cut across the chlorite. (**b**) Bright-field TEM image of the FIB foil extraction, with the platinum strap on top. (**c**) Enlargement of part b showing the R^{2+} (Fe^{2+} + Mg^{2+}) contents of chlorite (apfu) and their variation from crystal core to crystal rim (TEM–EDX analysis). Reproduced with permission from [43], published by Elsevier B.V.

The Fe^{3+} content in LT chlorites, and its consideration for thermometric estimates, are also subjects of debate. If a part of Fe_{total} is considered as Fe^{3+} instead of Fe^{2+}, the octahedral vacancy content in the structural formula will increase, influencing the calculation of end-member activities and, finally, leading to a decrease in the temperature estimates. In this respect, three approaches are conceivable: (i) to calibrate a thermometer intrinsically considering Fe^{3+} influence, making it unnecessary to quantify the Fe^{3+} content; (ii) to propose a model that predicts the Fe^{3+} content in parallel with the chlorite formation temperature; (iii) to develop analytical techniques able to determine the Fe^{3+} content for each chlorite analyses, accounting for chemical zoning.

The first approach was adopted by Bourdelle et al. [55], noting that Fe^{3+} content is rarely issued from direct analysis (e.g., by electron microprobe). This thermometer is calibrated on a large database to include the possible effects of Fe^{3+}, knowing that Fe^{3+} content has even less influence the lower the temperature [23,53]. The effect of the non-consideration of Fe^{3+} has a subordinate role because temperature estimates are, in the first place, controlled by TK and DT substitutions [23,55], in particular at the Fe_{total} content and XFe^{3+} ranges usually observed for LT chlorites. Studies that compare T estimates from the thermometer of Bourdelle et al. [55] with those of Inoue et al. [28], Vidal et al. [49,50], or Lanari et al. [51], which require prior quantification of Fe^{3+} content, conclude at the convergence of the two approaches [23,59,60,62,77], especially when T < 250 °C.

The second approach was favored by Inoue et al. [28], Vidal et al. [49,50,53], and Lanari et al. [51]. Despite being a model without a ferric end-member, Inoue et al. [28] advised considering Fe^{3+} content in applying the proposed geothermometer to avoid T overestimates. In this respect, Fe^{3+} content must be quantified independently. On their side, the thermometers of Vidal et al. [44,45] and Lanari et al. [51] require integrating the Fe^{3+} content in the chlorite structural formula calculation but, for this, propose estimating the Fe^{3+}/Fe_{total} ratio together with T by the convergence of Chl + Qz equilibria

(Table 1) [78,79]. These models were completed by the addition of a ferric end-member, i.e., ferri-sudoite, by Vidal et al. [53]. Recently, by combining Inoue et al. [28] and Walshe [11] models, Inoue et al. [80] presented a method to estimate Fe^{3+} content and T, and evaluate redox conditions (fO_2). Applied on chlorites from the hydrothermal system of Noboribetsu (Japan), this method gave promising results for qualitative discussion on redox condition variations.

The third approach therefore consists of the development of an analytical method to precisely quantify Fe^{3+} content and its variation through crystallites, either at the micrometric scale for metamorphic chlorite zoning, or at the nanoscale for LT chlorite zoning. Muñoz et al. [76] mapped the Fe^{3+}/Fe_{total} ratio (XFe^{3+}) in metamorphic chlorites for the first time using X-ray absorption near-edge structure spectroscopy data (XANES) acquired at the Fe K-edge (7100–7200 keV) in fluorescence mode (synchrotron light). A spot size of 10 μm allowed the authors to map an area 390 × 180 microns in size to highlight XFe^{3+} zoning in chlorite and to refine thermometric predictions [50]. This kind of map can be correlated to EMP maps given exhaustive characterization of the chlorite chemistry. This spectacular advance in the estimation of ferric iron in HT chlorites is, however, not suitable for LT chlorites due to the scale of XFe^{3+} variation. Several years later, Bourdelle et al. [81] developed an analytical procedure to quantify the Fe^{3+}/Fe_{total} ratio in silicates using Fe $L_{2,3}$-edge (700–730 eV) XANES spectra, i.e., at 10× less energy. This authorizes a spot size of 30 nm, an unequal spatial resolution. Data were acquired thanks to scanning transmission X-ray microscopy (STXM; in transmission mode), on FIB ultrathin sections already serving as a support for the TEM–EDX study. Admittedly, the mapped area is very small in size, but it was previously demonstrated that the chemical zoning in LT chlorites is restricted to small distances. Applied on chlorites from Glarus (Switzerland), Bourdelle et al. [81] exhibited the first quantitative redox maps at the nanoscale as evidence of the compositional zoning of LT chlorites. To prove the existence of this nanoscale zoning, to date not detected with classical methods such as EMPand SEM–EDX, the combination of this kind of map with TEM–EDX data should allow us to obtain all compositional data useful for applications of the Inoue et al. [28] and Vidal et al. [50,53] thermometers.

5. Perspectives of New Analytical and Thermometric Developments

The question of ferric iron in LT chlorites remains extensively studied. New ferric end-members could be envisaged to complete thermodynamic models, provided that their standard-state thermodynamic properties can be determined. Cationic substitutions where Fe^{3+} is implicated can also be redefined. Masci et al. [16] investigated a wide range of chlorite compositions by XANES and EMP, and claimed that the DT substitution implicating Fe^{3+} is an artefact. This substitution is usually considered as the result of the combination of the Al-DT substitution and AF exchange but without the intervention of Al. If the $Al^{3+} = Fe^{3+}$ exchange is still considered by the authors for low-Fe systems, a new exchange vector, implying deprotonation, is envisaged for Fe-rich chlorites. Here, the structural formula is not calculated on a 14-oxygen basis but on a 15-oxygen basis, as $R^{2+} + H^{+} = Fe^{3+}$, leading to a reduction of octahedral vacancies. Masci et al. [16] campaigned for the consideration of an oxychlorite member (or "oxidized chlorite"), close to tri-trioctahedral chlorite, and for a new cation site assignment as a part of a chlorite solid-solution accounting for proton loss in M1 and M2 sites. If the deprotonation becomes a verified process, the published data on Fe^{3+}-rich chlorites with a high vacancy content will have to be reappraised, and the implications for chlorite thermometry are far-reaching.

Aja [56] proposed a new approach for solid-solution modeling by defining a molecular solid-solution model instead of the atomic site-mixing models conventionally used by chlorite geothermometers, such as in [9]. The author starts from the observation that regular-solution site-mixing models are, at least in part, speculative because they deduce a putative structural chemistry from the general chlorite stoichiometry and do not account for the true atomic site occupancy of each investigated sample. To circumvent this issue, a molecular model was developed in which each end-member contribution (mole fraction) is depicted by a single aggregate parameter, presuming that local charge remains balanced in the chlorite structure. The author advances that the molecular solid solution brings out insufficiencies in the thermodynamic database available for chlorite, and suggests new ternary solid-solution models, opening up new thermometric perspectives.

Currently, geothermometers are developed for the most common chlorites, i.e., Al-, Fe-, or Mg-rich chlorites. However, it could be worth considering extending these models to rarer—but characteristic of specific systems—chlorites, like Mn-rich chlorites. As an example, Sussenberger et al. [82] suggest that the Mn content in chlorite could be a proxy for chemo-stratigraphic conditions in a depositional environment. For their part, Bobos et al. [83] established a link between Mn-chlorite and wolframite, with the Mn content in chlorite becoming an indicator of W–Mo mineralization. In this way, geothermometers should include an Mn end-member like pennantite [84] (similar to clinochlore and daphnite). However, the question of the structural formula calculation, similarly to ferric iron, will arise without quantification of the Mn^{3+}/Mn_{total} ratio. An STXM–XANES protocol investigating the Mn $L_{2,3}$-edges (630–660 eV) could be contemplated for nanoscale mapping of the Mn redox, analogously to the Fe redox [81].

Other perspectives of development could be envisaged, like the extension of thermodynamic models toward Al-free chlorite, the comprehension of chloritization mechanisms according to the system and the precursor which, although very widely studied, may need to be reevaluated with new analysis techniques and new experiments, or the coupling of STEM–EDX maps (scanning transmission electron microscopy) with STXM–XANES maps. This latter development could facilitate the detection of compositional variations within chlorite crystals, knowing that it is currently difficult to understand whether the zoning is a progressive zoning from crystal core to rim resulting from a continuous evolution, or is the consequence of a clear juxtaposition of different independent compositions in the same chlorite.

6. Conclusions

Chlorite geothermometry has already shown its relevance and contribution in numerous studies of low-grade metamorphic, diagenetic, and hydrothermal contexts, especially when coupled with other thermometric tools, but it should only be used when also considering the recommendations of thermometer developers, choosing a suitable analysis scale/techniques, and being aware of possible deficiencies. A review of the chlorite geothermometers for LT contexts is inevitably at the center of an old debate on their relevance. With the intention of demonstrating the strong link between the evolution of thermometer concepts and development of new analytical tools, this study takes the view that chlorite geothermometry is appropriate for estimating the P-T-X paleo-conditions prevailing in low-temperature–low-pressure contexts. However, the debate, especially on the thermodynamic status of intermediate phases involved in chloritization processes, cannot be settled here and will require new data and new analytical or experimental advances, as in [64]. Beyond this debate, the highlighting of chemical zoning inside LT chlorites seems to lead in the direction of the local equilibrium model, justifying a thermodynamic/thermometric approach.

Funding: This research received no external funding.

Acknowledgments: The discussions and comments of the associate editor and of reviewers are gratefully acknowledged.

Conflicts of Interest: The author declares no conflict of interest.

References

1. Okuyama-Kusunose, Y.; Itaya, T. Metamorphism of Carbonaceous Material in the Tono Contact Aureole, Kitakami Mountains, Japan. *J. Metamorph. Geol.* **1987**, *5*, 121–139. [CrossRef]

2. Beyssac, O.; Pattison, D.R.M.; Bourdelle, F. Contrasting Degrees of Recrystallization of Carbonaceous Material in the Nelson Aureole, British Columbia and Ballachulish Aureole, Scotland, with Implications for Thermometry Based on Raman Spectroscopy of Carbonaceous Material. *J. Metamorph. Geol.* **2019**, *37*, 71–95. [CrossRef]
3. Beyssac, O.; Rouzaud, J.N.; Goffe, B.; Brunet, F.; Chopin, C. Graphitization in a High-Pressure, Low-Temperature Metamorphic Gradient: A Raman Microspectroscopy and HRTEM Study. *Contrib. Mineral. Petrol.* **2002**, *143*, 19–31. [CrossRef]
4. Lahfid, A.; Beyssac, O.; Deville, E.; Negro, F.; Chopin, C.; Goffe, B. Evolution of the Raman Spectrum of Carbonaceous Material in Low-Grade Metasediments of the Glarus Alps (Switzerland). *Terra Nova* **2010**, *22*, 354–360. [CrossRef]
5. Dubois, M.; Ayt Ougougdal, M.; Meere, P.; Royer, J.-J.; Boiron, M.-C.; Cathelineau, M. Temperature of Paleo- to Modern Self-Sealing within a Continental Rift Basin; the Fluid Inclusion Data (Soultz-Sous-Forets, Rhine Graben, France). *Eur. J. Mineral.* **1996**, *8*, 1065–1080. [CrossRef]
6. Roedder, E. *Fluid Inclusions: An Introduction to Studies of All Types of Fluid Inclusions, Gas, Liquid, or Melt, Trapped in Materials from Earth and Space, and Their Application to the Understanding of Geologic Processes*; Reviews in Mineralogy; Mineral Soc. of America: Washington, DC, USA, 1984; ISBN 978-0-939950-16-4.
7. Samson, I. (Ed.) *Fluid Inclusions: Analysis and Interpretation*; Short Course Series; Mineralogical Association: Ottawa, ON, Canada, 2003; ISBN 978-0-921294-32-0.
8. Burkhard, M.; Kerrich, R.; Maas, R.; Fyfe, W.S. Stable and Sr-Isotope Evidence for Fluid Advection during Thrusting of the Glarus Nappe (Swiss Alps). *Contrib. Mineral. Petrol.* **1992**, *112*, 293–311. [CrossRef]
9. Vidal, O.; Parra, T.; Trotet, F. A Thermodynamic Model for Fe-Mg Aluminous Chlorite Using Data from Phase Equilibrium Experiments and Natural Pelitic Assemblages in the 100 Degrees to 600 Degrees C, 1 to 25 Kb Range. *Am. J. Sci.* **2001**, *301*, 557–592. [CrossRef]
10. Bailey, S.W. (Ed.) Chlorites: Structures and Crystal Chemistry. In *Hydrous Phyllosilicates*; De Gruyter: Berlin, Germany; Boston, MA, USA, 1988; ISBN 978-1-5015-0899-8.
11. Walshe, J.L. A Six-Component Chlorite Solid Solution Model and the Conditions of Chlorite Formation in Hydrothermal and Geothermal Systems. *Econ. Geol.* **1986**, *81*, 681–703. [CrossRef]
12. Cathelineau, M.; Nieva, D. A Chlorite Solid Solution Geothermometer the Los Azufres (Mexico) Geothermal System. *Contrib. Mineral. Petrol.* **1985**, *91*, 235–244. [CrossRef]
13. De Caritat, P.; Hutcheon, I.; Walshe, J. Chlorite Geothermometry—A Review. *Clays Clay Miner.* **1993**, *41*, 219–239. [CrossRef]
14. Essene, E.J.; Peacor, D.R. Clay Mineral Thermometry—A Critical Perspective. *Clays Clay Miner.* **1995**, *43*, 540–553. [CrossRef]
15. Beaufort, D.; Rigault, C.; Billon, S.; Billault, V.; Inoue, A.; Inoue, S.; Patrier, P. Chlorite and Chloritization Processes through Mixed-Layer Mineral Series in Low Temperature Geological Systems—A Review. *Clay Miner.* **2015**, *50*, 497–523. [CrossRef]
16. Masci, L.; Dubacq, B.; Verlaguet, A.; Chopin, C.; De Andrade, V.; Herviou, C. A XANES and EPMA Study of Fe^{3+} in Chlorite: Importance of Oxychlorite and Implications for Cation Site Distribution and Thermobarometry. *Am. Mineral.* **2019**, *104*, 403–417. [CrossRef]
17. Wiewiora, A.; Weiss, Z. Crystallochemical Classifications of Phyllosilicates Based on the Unified System of Projection of Chemical Composition; II, The Chlorite Group. *Clay Miner.* **1990**, *25*, 83–92. [CrossRef]
18. Meunier, A. *Clays*; Springer: Berlin/Heidelberg, Germany; New York, NY, USA, 2005; ISBN 978-3-540-21667-4.
19. Holland, T.; Baker, J.; Powell, R. Mixing Properties and Activity-Composition Relationships of Chlorites in the System MgO-FeO-Al_2O_3-SiO_2-H_2O. *Eur. J. Mineral.* **1998**, *10*, 395–406. [CrossRef]
20. Helgeson, H.C.; Aagaard, P. Activity/Composition Relations among Silicates and Aqueous Solutions; I, Thermodynamics of Intrasite Mixing and Substitutional Order/Disorder in Minerals. *Am. J. Sci.* **1985**, *285*, 769–844. [CrossRef]
21. Stoessell, R.K. Regular Solution Site-Mixing Model for Chlorites. *Clays Clay Miner.* **1984**, *32*, 205–212. [CrossRef]
22. Vidal, O.; Parra, T. Exhumation Paths of High-Pressure Metapelites Obtained from Local Equilibria for Chlorite-Phengite Assemblages. *Geol. J.* **2000**, *35*, 139–161. [CrossRef]
23. Bourdelle, F.; Cathelineau, M. Low-Temperature Chlorite Geothermometry: A Graphical Representation Based on a T-R^{2+}-Si Diagram. *Eur. J. Mineral.* **2015**, *27*, 617–626. [CrossRef]
24. Hillier, S.; Velde, B. Octahedral Occupancy and the Chemical-Composition of Diagenetic (Low-Temperature) Chlorites. *Clay Miner.* **1991**, *26*, 149–168. [CrossRef]
25. Cathelineau, M. Cation Site Occupancy in Chlorites and Illites as a Function of Temperature. *Clay Miner.* **1988**, *23*, 471–485. [CrossRef]
26. Velde, B.; Medhioub, M. Approach to Chemical Equilibrium in Diagenetic Chlorites. *Contrib. Mineral. Petrol.* **1988**, *98*, 122–127. [CrossRef]
27. Jahren, J.S. Compositional Variations in Diagenetic Chlorites and Illites, and Relationships with Formation-Water Chemistry. *Clay Miner.* **1989**, *24*, 157–170. [CrossRef]
28. Inoue, A.; Meunier, A.; Patrier-Mas, P.; Rigault, C.; Beaufort, D.; Vieillard, P. Application of Chemical Geothermometry to Low-Temperature Trioctahedral Chlorites. *Clays Clay Miner.* **2009**, *57*, 371–382. [CrossRef]
29. Bourdelle, F.; Parra, T.; Beyssac, O.; Chopin, C.; Vidal, O. Clay Minerals as Geo-Thermometer: A Comparative Study Based on High Spatial Resolution Analyses of Illite and Chlorite in Gulf Coast Sandstones (Texas, USA). *Am. Mineral.* **2013**, *98*, 914–926. [CrossRef]

30. McDowell, S.D.; Elders, W.A. Authigenic Layer Silicate Minerals in Borehole Elmore 1, Salton Sea Geothermal Field, California, USA. *Contrib. Mineral. Petrol.* **1980**, *74*, 293–310. [CrossRef]
31. Bevins, R.; Robinson, D.; Rowbotham, G. Compositional Variations in Mafic Phyllosilicates from Regional Low-Grade Metabasites and Application of the Chlorite Geothermometer. *J. Metamorph. Geol.* **1991**, *9*, 711–721. [CrossRef]
32. Jahren, J.; Aagaard, P. Diagenetic Illite-Chlorite Assemblages in Arenites. 1. Chemical Evolution. *Clays Clay Miner.* **1992**, *40*, 540–546. [CrossRef]
33. Rahn, M.; Mullis, J.; Erdelbrock, K.; Frey, M. Very Low-Grade Metamorphism of the Taveyanne Greywacke, Glarus Alps, Switzerland. *J. Metamorph. Geol.* **1994**, *12*, 625–641. [CrossRef]
34. Schmidt, D.; Livi, K.J.T.; Frey, M. Reaction Progress in Chloritic Material: An Electron Microbeam Study of the Taveyanne Greywacke, Switzerland. *J. Metamorph. Geol.* **1999**, *17*, 229–241. [CrossRef]
35. Mas, A.; Guisseau, D.; Mas, P.P.; Beaufort, D.; Genter, A.; Sanjuan, B.; Girard, J.P. Clay Minerals Related to the Hydrothermal Activity of the Bouillante Geothermal Field (Guadeloupe). *J. Volcanol. Geotherm. Res.* **2006**, *158*, 380–400. [CrossRef]
36. Koroknai, B.; Arkai, P.; Horvath, P.; Balogh, K. Anatomy of a Transitional Brittle-Ductile Shear Zone Developed in a Low-T Meta-Andesite Tuff: A Microstructural, Petrological and Geochronological Case Study from the Bukk Mts. (NE Hungary). *J. Struct. Geol.* **2008**, *30*, 159–176. [CrossRef]
37. Beaufort, D.; Patrier, P.; Meunier, A.; Ottaviani, M. Chemical Variations in Assemblages Including Epidote and or Chlorite in the Fossil Hydrothermal System of Saint Martin (Lesser Antilles). *J. Volcanol. Geotherm. Res.* **1992**, *51*, 95–114. [CrossRef]
38. Xie, X.G.; Byerly, G.R.; Ferrell, R.E. IIb Trioctahedral Chlorite from the Barberton Greenstone Belt: Crystal Structure and Rock Composition Constraints with Implications to Geothermometry. *Contrib. Mineral. Petrol.* **1997**, *126*, 275–291. [CrossRef]
39. Lopez-Munguira, A.; Nieto, F.; Morata, D. Chlorite Composition and Geothermometry: A Comparative HRTEM/AEM-EMPA-XRD Study of Cambrian Basic Lavas from the Ossa Morena Zone, SW Spain. *Clay Miner.* **2002**, *37*, 267–281. [CrossRef]
40. Potel, S. Very Low-Grade Metamorphic Study in the Pre-Late Cretaceous Terranes of New Caledonia (Southwest Pacific Ocean). *Isl. Arc* **2007**, *16*, 291–305. [CrossRef]
41. Percival, J.B.; Kodama, H. Sudoite from Cigar Lake, Saskatchewan. *Can. Mineral.* **1989**, *27*, 633–641.
42. Hutcheon, I. Clay Carbonate Reactions in the Venture Area, Scotian Shelf, Nova Scotia, Canada. *Geochem. Soc.* **1990**, *2*, 199–212.
43. Bourdelle, F.; Beyssac, O.; Parra, T.; Chopin, C. Nanoscale Chemical Zoning of Chlorite and Implications for Low-Temperature Thermometry: Application to the Glarus Alps (Switzerland). *Lithos* **2018**, *314*, 551–561. [CrossRef]
44. Hutcheon, I.; Oldershaw, A.; Ghent, E.D. Diagenesis of Cretaceous Sandstones of the Kootenay Formation at Elk Valley (Southeastern British Columbia) and Mt Allan (Southwestern Alberta). *Geochim. Cosmochim. Acta* **1980**, *44*, 1425–1435. [CrossRef]
45. Zang, W.; Fyfe, W. Chloritization of the Hydrothermally Altered Bedrock at the Igarape-Bahia Gold Deposit, Carajas, Brazil. *Miner. Depos.* **1995**, *30*, 30–38. [CrossRef]
46. Waldie, C.J.; Jowett, E.C.; Swinden, H.S. The Crescent Lake Copper Deposit, Central Newfoundland: Deep Levels of a Volcanogenic Hydrothermal System? *Atl. Geol.* **1991**, *27*. [CrossRef]
47. Kranidiotis, P.; MacLean, W.H. Systematics of Chlorite Alteration at the Phelps Dodge Massive Sulfide Deposit, Matagami, Quebec. *Econ. Geol.* **1987**, *82*, 1898–1911. [CrossRef]
48. Jowett, E.C. Fitting Iron and Magnesium into the Hydrothermal Chlorite Geothermometer. In Proceedings of the GAC/MAC/SEG Joint Annual Meeting, Toronto, ON, Canada, 27–29 May 1991.
49. Vidal, O.; Parra, T.; Vieillard, P. Thermodynamic Properties of the Tschermak Solid Solution in Fe-Chlorite: Application to Natural Examples and Possible Role of Oxidation. *Am. Mineral.* **2005**, *90*, 347–358. [CrossRef]
50. Vidal, O.; De Andrade, V.; Lewin, E.; Munoz, M.; Parra, T.; Pascarelli, S. P-T-Deformation-Fe^{3+}/Fe^{2+} Mapping at the Thin Section Scale and Comparison with XANES Mapping: Application to a Garnet-Bearing Metapelite from the Sambagawa Metamorphic Belt (Japan). *J. Metamorph. Geol.* **2006**, *24*, 669–683. [CrossRef]
51. Lanari, P.; Wagner, T.; Vidal, O. A Thermodynamic Model for Di-Trioctahedral Chlorite from Experimental and Natural Data in the System MgO-FeO-Al2O3- SiO2-H2O: Applications to P-T Sections and Geothermometry. *Contrib. Mineral. Petrol.* **2014**, *167*, 968. [CrossRef]
52. Parra, T.; Vidal, O.; Theye, T. Experimental Data on the Tschermak Substitution in Fe-Chlorite. *Am. Mineral.* **2005**, *90*, 359–370. [CrossRef]
53. Vidal, O.; Lanari, P.; Munoz, M.; Bourdelle, F.; De Andrade, V. Deciphering Temperature, Pressure and Oxygen-Activity Conditions of Chlorite Formation. *Clay Miner.* **2016**, *51*, 615–633. [CrossRef]
54. Trincal, V.; Lanari, P. Al-Free Di-Trioctahedral Substitution in Chlorite and a Ferri-Sudoite End-Member. *Clay Miner.* **2016**, *51*, 675–689. [CrossRef]
55. Bourdelle, F.; Parra, T.; Chopin, C.; Beyssac, O. A New Chlorite Geothermometer for Diagenetic to Low-Grade Metamorphic Conditions. *Contrib. Mineral. Petrol.* **2013**, *165*, 723–735. [CrossRef]
56. Aja, S. Excess Functions of Chlorite Solid Solutions and Neoformation of Fe-Chlorites: Some Implications of Recent Thermochemical Measurements. *Am. Mineral.* **2019**, *104*, 232–243. [CrossRef]
57. Inoue, A.; Kurokawa, K.; Hatta, T. Application of Chlorite Geothermometry to Hydrothermal Alteration in Toyoha Geothermal System, Southwestern Hokkaido, Japan. *Resour. Geol.* **2010**, *60*, 52–70. [CrossRef]
58. Harbi, H.M.; Surour, A.A.; Davidson, G.J. Genesis of Neoproterozoic Au-Bearing Volcanogenic Sulfides and Quartz Veins in the Ar Rjum Goldfield, Saudi Arabia. *Ore Geol. Rev.* **2014**, *58*, 110–125. [CrossRef]

59. Mamadou, M.M.; Cathelineau, M.; Bourdelle, F.; Boiron, M.-C.; Elmaleh, A.; Brouand, M. Hot Fluid Flows Around a Major Fault Identified by Paleothermometric Studies (Tim Mersoi Basin, Niger). *J. Sediment. Res.* **2016**, *86*, 914–928. [CrossRef]
60. Perez-Caceres, I.; Martinez Poyatos, D.J.; Vidal, O.; Beyssac, O.; Nieto, F.; Simancas, J.F.; Azor, A.; Bourdelle, F. Deciphering the Metamorphic Evolution of the Pulo Do Lobo Metasedimentary Domain (SW Iberian Variscides). *Solid Earth* **2020**, *11*, 469–488. [CrossRef]
61. Vazquez, M.; Nieto, F.; Morata, D.; Droguett, B.; Carrillo-Rosua, F.J.; Morales, S. Evolution of Clay Mineral Assemblages in the Tinguiririca Geothermal Field, Andean Cordillera of Central Chile: An XRD and HRTEM-AEM Study. *J. Volcanol. Geotherm. Res.* **2014**, *282*, 43–59. [CrossRef]
62. Dolores Rodriguez-Ruiz, M.; Abad, I.; Bentabol, M.J. Permo-Triassic Clastic Rocks from the Ghomaride Complex and Federico Units (Rif Cordillera, N Morocco): An Example of Diagenetic-Metamorphic Transition. *Minerals* **2019**, *9*, 738. [CrossRef]
63. Dowey, P.J.; Hodgson, D.M.; Worden, R.H. Pre-Requisites, Processes, and Prediction of Chlorite Grain Coatings in Petroleum Reservoirs: A Review of Subsurface Examples. *Mar. Pet. Geol.* **2012**, *32*, 63–75. [CrossRef]
64. Mosser-Ruck, R.; Pignatelli, I.; Bourdelle, F.; Abdelmoula, M.; Barres, O.; Guillaume, D.; Charpentier, D.; Rousset, D.; Cathelineau, M.; Michau, N. Contribution of Long-Term Hydrothermal Experiments for Understanding the Smectite-to-Chlorite Conversion in Geological Environments. *Contrib. Mineral. Petrol.* **2016**, *171*, 97. [CrossRef]
65. Jiang, W.; Peacor, D.; Buseck, P. Chlorite Geothermometry—Contamination and Apparent Octahedral Vacancies. *Clays Clay Miner.* **1994**, *42*, 593–605. [CrossRef]
66. Hillier, S.; Velde, B. Chlorite Interstratified with a 7-a Mineral—An Example from Offshore Norway and Possible Implications for the Interpretation of the Composition of Diagenetic Chlorites. *Clay Miner.* **1992**, *27*, 475–486. [CrossRef]
67. Curtis, C.D.; Ireland, B.J.; Whiteman, J.A.; Mulvaney, R.; Whittle, C.K. Authigenic Chlorites: Problems with Chemical Analysis and Structural Formula Calculations. *Clay Miner.* **1984**, *19*, 471–481. [CrossRef]
68. Putnis, A. Mineral Replacement Reactions. In *Thermodynamics and Kinetics of Water-Rock Interaction*; Oelkers, E.H., Schott, J., Eds.; De Gruyter: Berlin, Germany, 2009; Volume 70, pp. 87–124.
69. Putnis, C.V.; Geisler, T.; Schmid-Beurmann, P.; Stephan, T.; Giampaolo, C. An Experimental Study of the Replacement of Leucite by Analcime. *Am. Mineral.* **2007**, *92*, 19–26. [CrossRef]
70. Trincal, V.; Lanari, P.; Buatier, M.; Lacroix, B.; Charpentier, D.; Labaume, P.; Munoz, M. Temperature Micro-Mapping in Oscillatory-Zoned Chlorite: Application to Study of a Green-Schist Facies Fault Zone in the Pyrenean Axial Zone (Spain). *Am. Mineral.* **2015**, *100*, 2468–2483. [CrossRef]
71. Ganne, J.; De Andrade, V.; Weinberg, R.F.; Vidal, O.; Dubacq, B.; Kagambega, N.; Naba, S.; Baratoux, L.; Jessell, M.; Allibon, J. Modern-Style Plate Subduction Preserved in the Palaeoproterozoic West African Craton. *Nat. Geosci.* **2012**, *5*, 60–65. [CrossRef]
72. Vidal, O.; Baldeyrou, A.; Beaufort, D.; Fritz, B.; Geoffroy, N.; Lanson, B. Experimental Study of the Stability and Phase Relations of Clays at High Temperature in a Thermal Gradient. *Clays Clay Miner.* **2012**, *60*, 200–225. [CrossRef]
73. Jahren, J. Evidence of Ostwald Ripening Related Recrystallization of Diagenetic Chlorites from Reservoir Rocks Offshore Norway. *Clay Miner.* **1991**, *26*, 169–178. [CrossRef]
74. Bourdelle, F.; Parra, T.; Beyssac, O.; Chopin, C.; Moreau, F. Ultrathin Section Preparation of Phyllosilicates by Focused Ion Beam Milling for Quantitative Analysis by TEM-EDX. *Appl. Clay Sci.* **2012**, *59–60*, 121–130. [CrossRef]
75. Wirth, R. Focused Ion Beam (FIB): A Novel Technology for Advanced Application of Micro- and Nanoanalysis in Geosciences and Applied Mineralogy. *Eur. J. Mineral.* **2004**, *16*, 863–876. [CrossRef]
76. Munoz, M.; De Andrade, V.; Vidal, O.; Lewin, E.; Pascarelli, S.; Susini, J. Redox and Speciation Micromapping Using Dispersive X-Ray Absorption Spectroscopy: Application to Iron Chlorite Mineral of a Metamorphic Rock Thin Section. *Geochem. Geophys. Geosystems* **2006**, *7*, Q11020. [CrossRef]
77. Rat, J.; Mouthereau, F.; Brichau, S.; Cremades, A.; Bernet, M.; Balvay, M.; Ganne, J.; Lahfid, A.; Gautheron, C. Tectonothermal Evolution of the Cameros Basin: Implications for Tectonics of North Iberia. *Tectonics* **2019**, *38*, 440–469. [CrossRef]
78. Grosch, E.G.; Vidal, O.; Abu-Alam, T.; McLoughlin, N. P-T Constraints on the Metamorphic Evolution of the Paleoarchean Kromberg Type-Section, Barberton Greenstone Belt, South Africa. *J. Petrol.* **2012**, *53*, 513–545. [CrossRef]
79. Grosch, E.G.; McLoughlin, N.; Lanari, P.; Erambert, M.; Vidal, O. Microscale Mapping of Alteration Conditions and Potential Biosignatures in Basaltic-Ultramafic Rocks on Early Earth and Beyond. *Astrobiology* **2014**, *14*, 216–228. [CrossRef] [PubMed]
80. Inoue, A.; Inoue, S.; Utada, M. Application of Chlorite Thermometry to Estimation of Formation Temperature and Redox Conditions. *Clay Miner.* **2018**, *53*, 143–158. [CrossRef]
81. Bourdelle, F.; Benzerara, K.; Beyssac, O.; Cosmidis, J.; Neuville, D.R.; Brown, G.E.; Paineau, E. Quantification of the Ferric/Ferrous Iron Ratio in Silicates by Scanning Transmission X-Ray Microscopy at the Fe L-2,L-3 Edges. *Contrib. Mineral. Petrol.* **2013**, *166*, 423–434. [CrossRef]
82. Sussenberger, A.; Pospiech, S.; Schmidt, S.T. [MnO Vertical Bar SiO_2, Al_2O_3, FeO, MgO] Balanced Log-Ratio in Chlorites: A Tool for Chemo-Stratigraphic Mapping and Proxy for the Depositional Environment. *Clay Miner.* **2018**, *53*, 351–375. [CrossRef]
83. Bobos, I.; Noronha, F.; Mateus, A. Fe-, Fe,Mn- and Fe,Mg-Chlorite: A Genetic Linkage to W, (Cu, Mo) Mineralization in the Magmatic-Hydrothermal System at Borralha, Northern Portugal. *Mineral. Mag.* **2018**, *82*, S259–S279. [CrossRef]
84. Smith, W.C.; Bannister, F.A.; Hey, M.H. Pennantite, a New Manganese-Rich Chlorite from Benallt Mine, Rhiw, Carnarvonshire. *Mineral. Mag.* **1946**, *27*, 217–220.

minerals

MDPI

Article

Mineral Chemistry of Low-Temperature Phyllosilicates in Early Paleozoic Metaclastic Rocks, Eastern Tauride Belt, Türkiye

Ömer Bozkaya [1,*] and Hüseyin Yalçın [2]

[1] Department of Geological Engineering, Pamukkale University, 20160 Denizli, Turkey
[2] Department of Geological Engineering, Sivas Cumhuriyet University, 58140 Sivas, Turkey
* Correspondence: obozkaya@pau.edu.tr

Abstract: The mineral chemistry of illite/mica and chlorites, together with the evaluation of textural data of low-temperature metaclastic rocks, plays an important role in determining their origin and metamorphic grade. This study aimed to investigate the chemical properties of phyllosilicates in early Paleozoic metaclastic rocks in the Eastern Tauride Belt, Türkiye. The textural (electron microscopy) and chemical (mineral chemistry analysis) analyses were performed on the samples representing different grades of metamorphism. The illites/micas and chlorites are observed as detrital (chlorite–mica stacks) and neoformation origin. Trioctahedral chlorites (chamosite) exhibit different chemistry for detrital and neoformed origin as well as the metamorphic grade. Tetrahedral Al and octahedral Fe + Mg increase, whereas octahedral Al decreases together with the increasing grade of metamorphism. The detrital chlorites have higher tetrahedral Al and Fe contents than their neoformed counterparts. Chlorite geothermometry data (detrital: 241–$\geq$350 °C; neoformed: 201–268 °C) are compatible with the texture and illite Kübler index data. Illite/white-mica compositions display muscovite and Na-K mica. Tetrahedral Al and interlayer K + Na contents of illites/micas increase with metamorphic grade. Na-K mica and paragonite are observed as replacement-type developments within the detrital CMS. The obtained data indicate that phyllosilicate chemistry can be used effectively for determining the geological evolution of low-grade metamorphic sequences.

Keywords: phyllosilicate; mineral chemistry; metaclastic rocks; low-grade metamorphism

1. Introduction

Phyllosilicates in diagenetic to very low-grade metamorphic sedimentary units exhibit textural and mineralogical features based on their occurrence in different tectonic and burial settings; they have the potential, therefore, to be used as indicators of time and temperature history [1,2]. The cleavage development is a common indicator for increasing temperature and pressure for extensional and compressional settings, whereas the observation of chlorite–mica stacks (CMS: [3,4]) or intergrown phyllosilicate grains [5], related to chloritization of detrital biotites, is a characteristic indicator of extensional basin activity (very-low grade metamorphism) [2].

Mineralogic (mineral assemblages, illite Kübler index, polytypes of K-micas and chlorite and *b* cell dimension of illite/white K-micas) studies of diagenetic to very low-grade metamorphic argillaceous sedimentary rocks have been used to reconstruct the thermal histories and the structural evolution of sedimentary basins and orogenic belts (i.e., [2,6–9]). In addition to mineralogical data, the chemistry of detrital and neoformed (authigenic) illite/mica and chlorites, forming the main components of metaclastic rocks, plays an important role in determining their origins and diagenetic–metamorphic grades, respectively. During the progressive transformation of illite to white K-mica, the interlayer K and tetrahedral charge increase due to Al^{3+} substitution for Si^{4+}, and the octahedral charge increases due to Fe^{2+} and Mg^{2+} substitution for Al^{3+} [10–12]. Chlorite composition is also sensitive to the diagenetic/metamorphic grade. The tetrahedral Al (Al^{IV}) and octahedral R^{2+} (Mg and

Fe^{2+}) contents of chlorite increase with increasing grade [13–20]. Thus, different chemical compositions for detrital (metamorphic) and authigenic (diagenetic) chlorites have been reported in many studies (e.g., [17,21–25]). During the last 10 years, semi-empirical methods using ratios of end-member activities to link chlorite composition to temperature through the equilibrium constant K [26–30] have been preferred for temperature estimations from diagenetic to very low-grade metamorphic chlorite compositions. Additionally, the new T–$R2^{+}$–Si diagram was also provided as a practical tool for thermometric, in particular for diagenesis and low-grade metamorphism [31], based on the chlorite solid-solution model.

The Tauride–Anatolide Platform or the Tauride–Anatolide Composite Terrain (TACT [32]), represents an Alpine continental microplate (Figure 1a). Surrounded by the northern and southern branches of the Neotethys, the TACT experienced crustal thickening related to the Latest Cretaceous Alpine closure of the Neotethyan ocean branches and subsequent collision with the surrounding continental microplates (e.g., [33]). The commonly accepted scenario for the evolution of the TACT is that nappes or tectonostratigraphic units at the northern margin of the platform were sliced and emplaced towards the south onto a relatively autochthonous central part (Geyik Dağı Unit, [34]). In contrast to the allochthonous units of the Taurides, the relatively autochthonous Geyik Dağı Unit contains better-preserved successions and a wide range of grades (diagenesis to epizone), in which Precambrian and Lower Paleozoic sedimentary sequences [35–37].

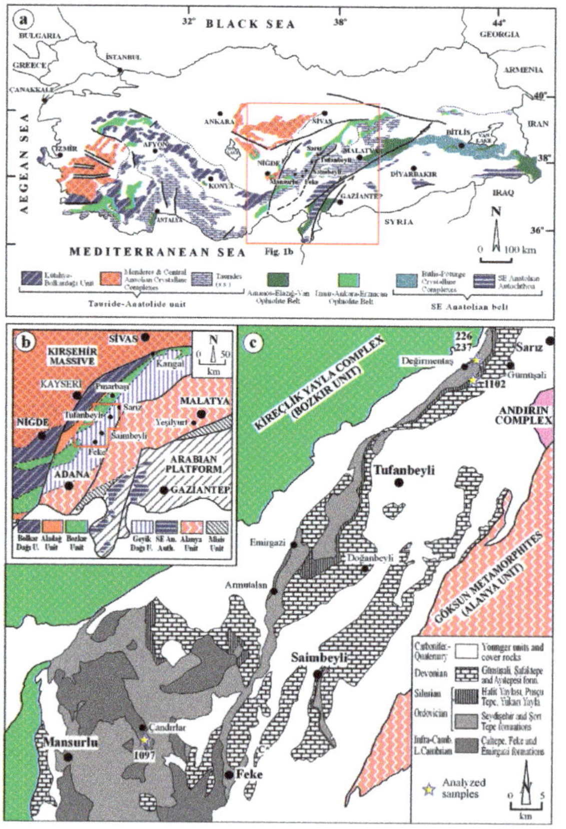

Figure 1. (**a**) Tectonic map of southern Turkey (Modified from [32]). (**b**) Tectono-stratigraphic units of the Eastern Tauride belt [34]. (**c**) Geologic map of the Geyik Dağı Unit between Sarız and Feke (simplified from [38]) and sample locations.

This study aimed to investigate the mineral chemistry of illites/micas and chlorites from the anchizonal to epizonal grade of the Precambrian–Early Paleozoic metaclastic sequence of the Geyik Dağı Unit. In this context, the obtained temperature conditions of neoformed and detrital phyllosilicate minerals correlate with the relationship between phyllosilicate chemistry and crystal–chemical data of phyllosilicates (illite Kübler index, polytype of illite/mica) which have been previously investigated by [36].

2. Geological Setting and Stratigraphy

Tectono-stratigraphic units of the TACT and their paleo-geographic setting relative to the autochthonous Geyik Dağı Unit are distinguished and differentiated by their stratigraphic and structural characteristics [34]. Based on the palinspastic reconstruction and interpretation of Özgül [34], the other units, called "allochthonous units", are palinspastically restored from north to south as the Bozkır Unit, Bolkar Dağı Unit, Aladağ Unit, and Alanya Unit in the Eastern Tauride Belt (Figure 1a,b).

The study area is located between Kayseri and Adana in the western part of the Eastern Taurides (Figure 1a,b). The Eastern Taurus Autochthonous Unit [38], or the eastern part of the Geyik Dağı Unit [34], shows a relatively well-exposed section in the Sarız–Tufanbeyli–Saimbeyli area (Figure 1c). In this area, the Geyik Dağı Unit is overthrusted by the Kireçlik Yayla Complex [39], Andırın Complex [38], and Göksun Metamorphites [40]. The Lower Paleozoic parts of the autochthonous units, which were investigated in this study, consist of Emirgazi, Feke, Çaltepe, Seydisehir, Sort Tepe, Halit Yaylası, Pusçu Tepe, Yukarı Yayla, and Ayı Tepesi formations from bottom to top.

The Emirgazi formation is composed of phyllitic slates and/or phyllites, alternating with metasiltstone, metasandstone, and rarely, recrystallized limestone (metalimestone). The Feke formation is mainly made up of purplish red colored metasandstones with yellow-green colored slate laminations. It overlies the Emirgazi formation as a parallel unconformity [41] and overlying by the Çaltepe formation, which is characterized, from bottom to top, by white-gray dolomite, dark gray recrystallized limestone, and green-pink nodular metalimestone of Upper Cambrian age. A transitional boundary exists to the overlying Seydişehir formation, which contains slate–metasiltstone alternations, intercalated with bands of recrystallized nodular limestone at the lower part and shales alternating with siltstones at the middle to the upper part.

The lower parts of the Seydişehir formation (Tremadocian parts) are brighter and more cleaved than the overlying Arenigian parts, and they show a distinct mineral orientation. After a stratigraphic gap including the middle Ordovician and probably the lower part of the Upper Ordovician, the shales of the Ashgillian Şort Tepe formation unconformably overly the Seydisehir formation. The late Ashgillian Halit Yaylası formation paraconformably overlies the Şort Tepe formation and is made up of glacial conglomerates [42] and sandstones, including silty shale alternations. The lower Silurian Pusçu Tepe formation contains shales and black shale–siltstone alternations. The Yukarı Yayla formation includes shales with limestone interlayers and was deposited during mid-Silurian. After a stratigraphic gap that may cover the Upper Silurian, the lower Devonian Ayı Tepesi formation unconformably covers the older units. It is followed by the middle Devonian Şafaktepe formation, represented by dolomitic limestone and limestone with sandstone–shale alternations. It is unconformably overlain by Gümüşali (upper Devonian) and Ziyarettepe (lower Carboniferous) formations, respectively.

3. Materials and Methods

Four rock samples from the Precambrian Emirgazi and Ordovician Seydişehir and Şort Tepe formations of the Eastern Tauride Belt Unit (Figure 1b,c) were selected for electron microscopy and mineral–chemistry analysis. Mineralogical–petrographic investigations using X-ray diffraction (XRD, Rigaku DMAX IIIC, Applied Rigaku Technologies, Austin, TX, USA) and optical microscopy (OM, Nikon, Tokyo, Japan), were previously carried out by [36] at the Geological Engineering Laboratory, Cumhuriyet University, Sivas, Turkey.

In this study, in addition to additional optical microscopic study, the electron microscopic mineral image (scanning electron microscope—secondary electrons and backscattered electrons/SEM-SE and SEM-BSE) and chemistry (energy dispersion spectrometry—EDS and electron microprobe-EMP) studies were performed on the polished and carbon coated surfaces on the samples.

SEM-SE and SEM-BSE images and EDS analyses were conducted at the Georgia Electron Microscopy (GEM) laboratory at the University of Georgia, Athens, GA, USA. The SEM investigations were performed using a Zeiss 1450EP instrument equipped with an Oxford INCA EDS system. This has a variable pressure sample chamber which allows imaging from high vacuum mode to 2600 Pa. The samples were coated with carbon and prepared as both freshly crushed grains (for SEM-SE images) and polished rock slabs (for SEM-BSE images). Thin-section slides were prepared using standard mounting techniques and polished to a thickness of 30 μm.

EMP analysis of minerals on carbon-coated polished rock slabs was performed with a JEOL 8600 electron microprobe using a 15 kV accelerating voltage and 15 nA beam current. Mineral grains were identified qualitatively using a Bruker 5010 Silicon Drift Detector (SDD) controlled by a Bruker Quantax energy dispersive analysis system. Quantitative analyses were performed with wavelength dispersive spectrometers (WDS) automated with Advanced Microbeam, Inc. EMP software, using 10 s counting times and natural and synthetic mineral standards. Analyses were calculated using Armstrong's Phi-Rho-Z matrix correction model [43,44]. The BSE and SE images and X-ray maps were acquired using the imaging software of the Quantax analysis system.

A total of 286 mineral chemistry analyses were conducted on micas/illites (117 EDS, 39 EMP) and chlorites (81 EDS, 49 EMP). Both EDS and EMP analyses of illite/mica and chlorites were performed on the same samples so that semi-quantitative EDS data could be calibrated with EMP data [25]. Thus, the EDS element oxide values were converted to EMP values, and these values could be used as EMP data.

4. Results

4.1. Texture

The main mineralogical–petrographical and textural characteristics of the Precambrian–Ordovician metaclastic rocks are given in Table 1. The grade of phyllosilicate orientation and cleavage developments in the matrix increases toward older units, i.e., from late Ordovician Şort Tepe formation to Precambrian Emirgazi formation (Figure 2).

Table 1. Optical mineralogical properties of the studied samples (Qz = Quartz, Pl = Plagioclase, Chl = Chlorite, Ms = Muscovite, Bt = Biotite, Tur = Tourmaline, Zrn = Zircon, Om = Opaque mineral.

Sample/ Formation	Rock Type/ Texture	Optical Mineralogy	Some Textural Properties
TFK-1097 Precambrian	Metasiltstone Blastopelitic	Qz + Pl + Ms + Chl + Om	Distinct orientation, roughly developed crenulation cleavage, scarce CMS are rich in muscovite
TTB-226 E.Ordovician	Chlorite slate Blastopelitic	Qz + Pl + Chl + Ms + Bt + Tur + Zrn + Om	Microlamination, distinct orientation, well-developed slaty cleavage, abundant CMS are rich in chlorite
TTB-237 E.Ordovician	Metasandstone Blastopsammitic	Qz + Pl + Chl + Ms + Tur + Zrn + Om	Abundant CMS are very rich in chlorite, distinct orientation, poorly-developed slaty cleavage
TTB-1102 L.Ordovician	Silty slate Blastopelitic	Qz + Fel + Ms + Om	Abundant CMS are rich in chlorite, less developed orientation formed from detrital CMS, mica and chlorite

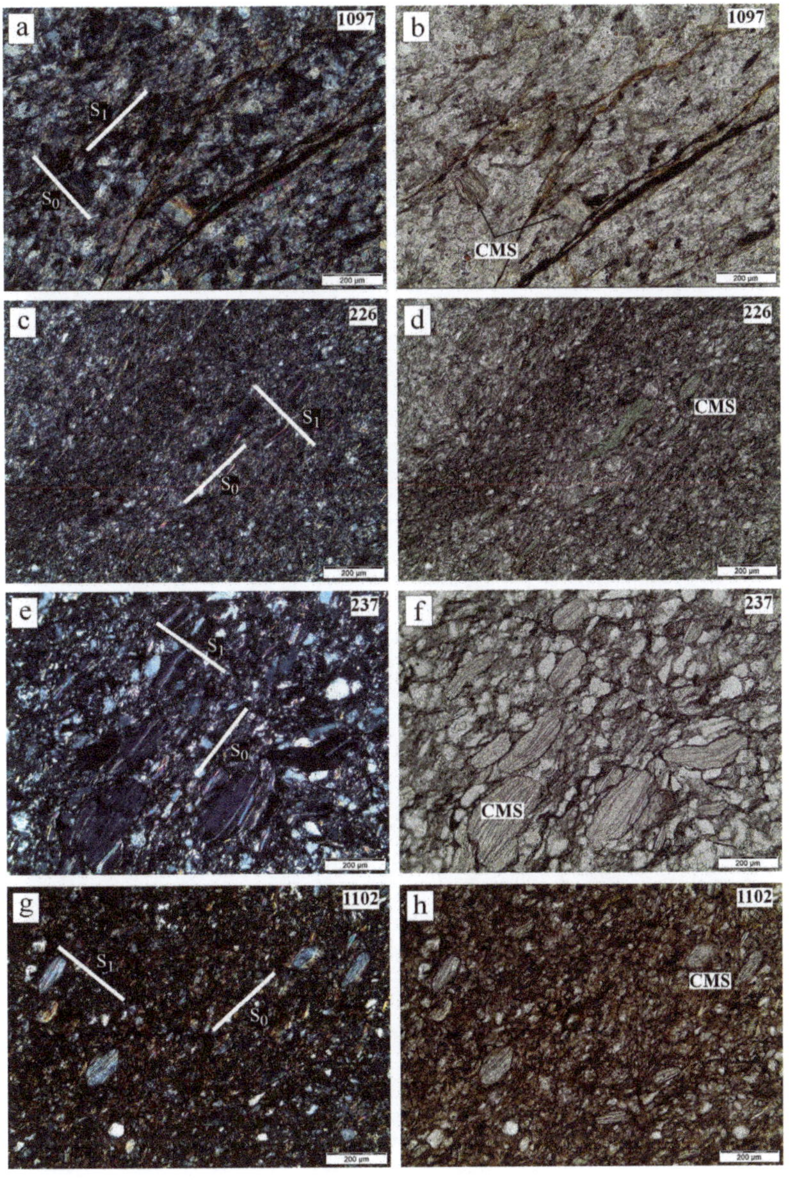

Figure 2. Photomicrographs of the Early Paleozoic siliciclastic rocks from the Geyik Dağı Unit in the Eastern Tauride Belt. Photos on the left were taken under crossed nicols, and on the right were taken under plane-polarized light. (**a**,**b**) Detrital CMS parallel to bedding planes (S_0) and perpendicular to slaty cleavage planes (S_1) in metasiltstone from Precambrian Emirgazi formation. (**c**,**d**) Chlorite-rich CMS; their long axes are parallel to S_0 and perpendicular to S_1 in chlorite slate from lower parts of early Ordovician Seydişehir formation, (**e**,**f**) Abundant chlorite-rich CMS which is parallel to S_0 in the anchimetamorphic sandstone from the lower–middle part of the Seydişehir formation, (**g**,**h**) White K-mica-rich CMS in silty slate from the late Ordovician Şort Tepe formation.

Optical microscopic investigations of the Precambrian and Ordovician siliciclastic rocks revealed distinctive bedding orientations (S_0), formed by chloritized and illitized groundmass and coarse detrital mica flakes, and poorly developed cleavage planes (S_1) which are nearly perpendicular to S_0 (Figure 2a–h). The schistosity and cleavage development in the samples increased from the late Ordovician to the Precambrian, together with an increasing degree of diagenesis/metamorphism. Intense chloritization of biotite and lozenge-shaped CMS are encountered in all samples.

Detrital mineral (quartz and plagioclase) boundaries are sutured with a groundmass of fine-grained micas (sericite) and chlorites in the Precambrian phyllitic slate sample (Figure 2a,b). Chloritization of coarse-grained biotite and the presence of chlorite-bearing mica stacks suggest that detrital biotite was altered under very low-grade metamorphic conditions [45,46]. Similar features for early Paleozoic units were also noted in the western and central Taurides [25,47]. CMS in the Precambrian lower parts of the Seydişehir formation are rich in chlorite components (Figure 2c–g), whereas CMS in the Emirgazi and Şort Tepe formation have relatively low chlorite content (Figure 2a,b,g,h) and some stacks in the Emirgazi formation consist almost entirely of white micas (Figure 2a,b).

The SEM-SE images of the phyllitic slate sample show that chlorite–muscovite–NaK mica stacks are highly deformed and their long axes were almost angularly oriented to the cleavage and/or shear zones (Figure 3a,b). The groundmass consists of relatively fine-grained flaky white K-mica and neoformed chlorites. Neoformed illite and chlorites were developed in voids and pores as fine-grained aggregates (Figure 3c). Neoformed illites were also developed onto coarse-grained muscovite plates (Figure 3d).

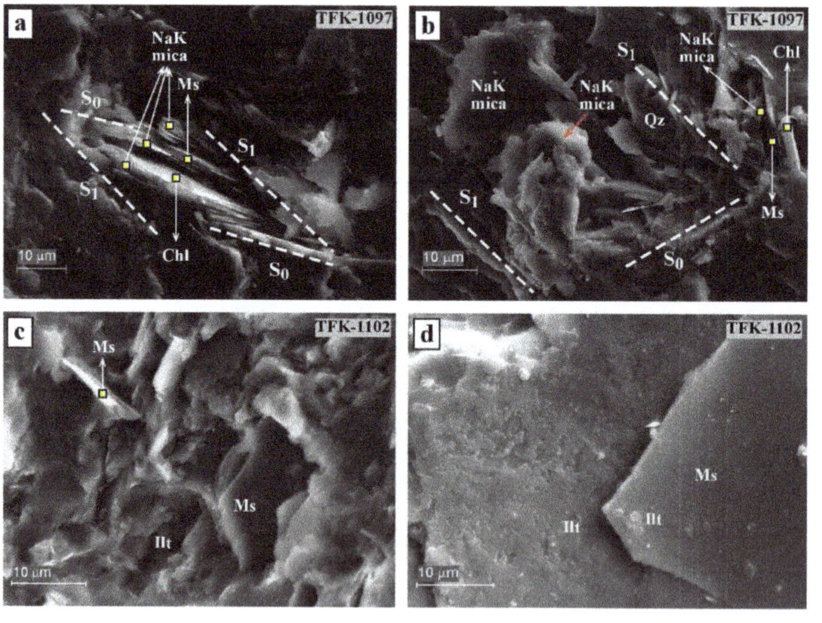

Figure 3. SEM-SE photomicrographs of the Precambrian and late Ordovician metapelitic rocks from the Eastern Tauride Autochthon. (**a**,**b**) Chlorite–muscovite–NaK mica stack showing its stacking planes are parallel to bedding plane (S_0) and high angle to cleavage (shear) planes (S_1). (**c**,**d**) The association of detrital micas and neoformed illites, which developed in voids and pores and onto the coarse-grained muscovite plates, as well.

The SEM-BSE images of the metasiltstone sample of Precambrian Emirgazi formation exhibit coarse-grained CMS. The stacking planes of chlorite-rich CMS were distinctly oriented as parallel to bedding planes (Figure 4a,b), whereas stacking planes of white mica-rich CMS, developed alongside the tectonically disturbed zones (syn-tectonic origin), were nearly perpendicular to bedding planes which were also parallel to shear planes (Figure 4c–f). White mica-rich and strongly disturbed CMS also contained the replacement-type NaK-micas within the stacks (Figure 4e,f), as demonstrated by the X-ray elementary map data (Figure 4g–j). Pore-filling neoformed illites and chlorites developed with a non-oriented texture, suggesting post-tectonic formation during a later (post-tectonic) phase. Additionally, the neoformed phyllosilicates were also developed on the coarse mica and chlorites less frequently.

The SEM-BSE photomicrographs of the slate sample of the early Ordovician part of the Seydişehir formation show that the long axes of CMS have a distinctly oriented parallel to the bedding planes (Figure 5a–f). CMS are generally chlorite-rich and coarse-grained (the lengths of the stacks reach 300 μm and their thickness reaches 100 μm) and tectonically disturbed in places (Figure 5a,c). The neoformed illites and chlorites were developed in the pores of the groundmass having partial orientation.

The SEM-BSE images of the metasandstone sample of the late–early Ordovician part of the Seydişehir formation exhibit coarse (>100 μm) CMS composed of almost chlorite components, including thin white K-mica layers in rare amounts (Figure 6a–f). Elongated and lozenge-shaped CMS have stacking planes nearly parallel to bedding planes and are partly disturbed due to cutting by slightly developed cleavage planes in places (Figure 6b,f).

The SEM-BSE photomicrographs of the late Ordovician silty slate display distinct orientation formed by CMS and detrital micas and chlorites (Figure 7a–f). Elongated and lozenge-shaped CMS are rich in chlorite components and relatively less deformed with respect to the other samples (Figure 7a–c,f). The groundmass is formed from fine-grained quartz grains and pore-filling neoformed phyllosilicates (illites and chlorites) without orientation.

4.2. Mineralogy

Bulk and phyllosilicate fraction and crystal–chemical data (illite Kübler index, $\Delta°2\theta$, [48]) polytype, and b cell dimension of illite/muscovite), which were previously studied by the current authors [36], are given in Figure 8. The main mineralogical composition is, in the order of abundance, phyllosilicates, quartz, alkali feldspar, plagioclase, and small amounts of calcite and dolomite. The amounts of quartz are increased in the Halit Yaylası formation, of feldspar in the Emirgazi and Feke formations, and of carbonates in the Çaltepe and Yukarı Yayla formations, depending on the main lithological types (metacarbonates and carbonate rocks are composed of calcite, dolomite, and detrital clasts such as quartz, plagioclase, and phyllosilicates). With respect to their relative amounts, phyllosilicates show a wide variety in mineral contents of illite, chlorite, mixed-layered chlorite–vermiculite (C-V), chlorite–smectite (C-S), illite–smectite (I-S), PM, and paragonite.

In terms of the vertical distribution of the phyllosilicates, illite/muscovite is found in all units. Chlorite is mostly detected in the Cambrian–Ordovician, while C-V occurs in the Silurian–Devonian and the Cambrian earliest Ordovician. C-S, PM, and paragonite are only detected in the Emirgazi formation, while I-S is only observed in the Ayı Tepesi formation (Figure 8). Increasing amounts of chlorite and illite but decreasing C-V and I-S towards depth is related to progressive evolution with stratigraphic depth, as stated by [36].

According to illite Kübler index data, Precambrian to early Ordovician units have epimetamorphic, early–middle Ordovician to middle Silurian, and late Silurian to early Devonian have anchimetamorphic and high-grade diagenetic degrees (Figure 8). The $d_{(060)}$ or b_0 values of illites and $2M_1$-contents ($2M_1/[2M_1 + 1M_d]$) of illite increase from Devonian to Ordovician formations.

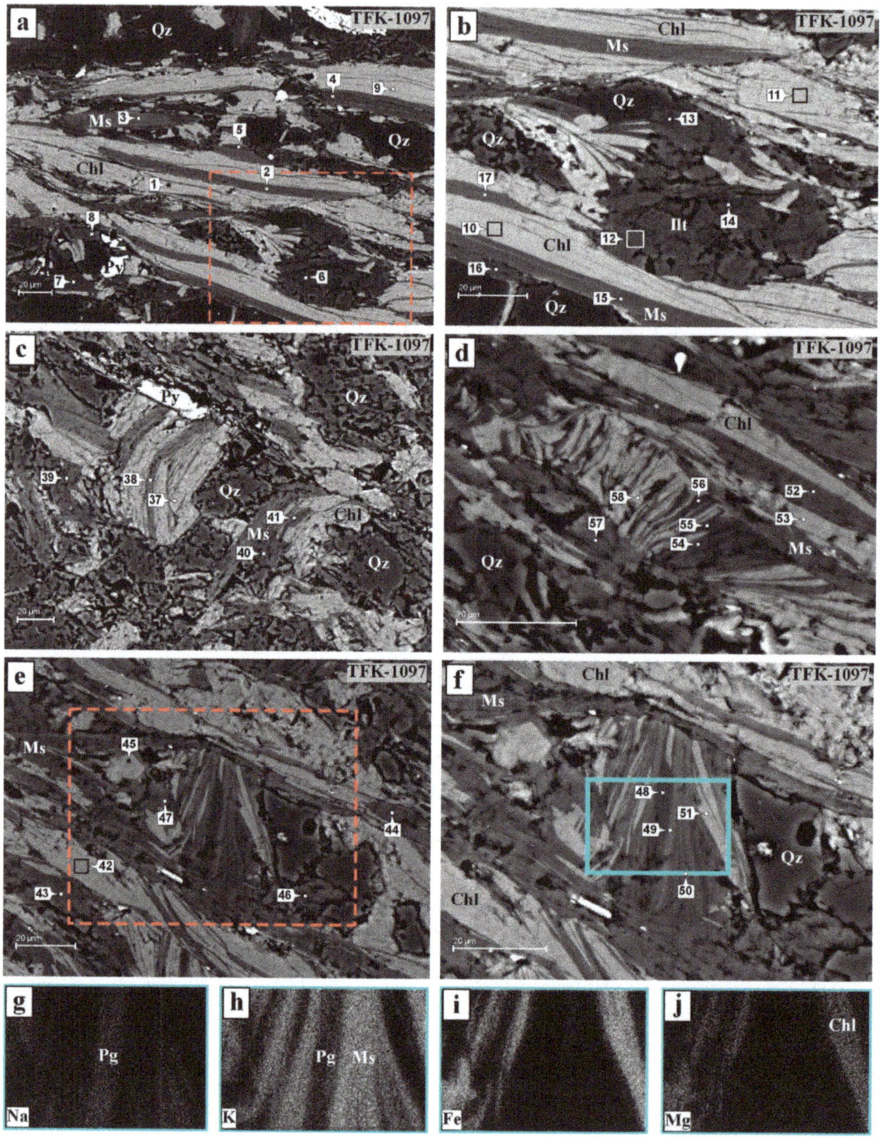

Figure 4. SEM-BSE photomicrographs of the Precambrian phyllitic slate with locations of some analyzed points and areas (black rectangulars). (**a**,**b**) Coarse-grained CMS oriented their long axes along the bedding planes and post-metamorphic illite occurrences, (**c**,**d**) The associations of CMS; both their stacking {001} planes are parallel to bedding planes and {001} are nearly perpendicular to bedding planes, i.e., along the schistosity or cleavage planes, (**e**,**f**) Highly disturbed white mica-rich CMS between shear zones in which {001} planes of CMS are nearly perpendicular to bedding plane and also to the shear band planes, (**g**–**j**) X-ray elementary maps (Na, K, Fe, Mg) of the CMS displaying the chemical differences of chlorite, white K- and Na-K mica/paragonite. Chl = Chlorite, Ilt = Illite, Ms = Muscovite, Pg = Paragonite, Py = Pyrite, Qz = Quartz.

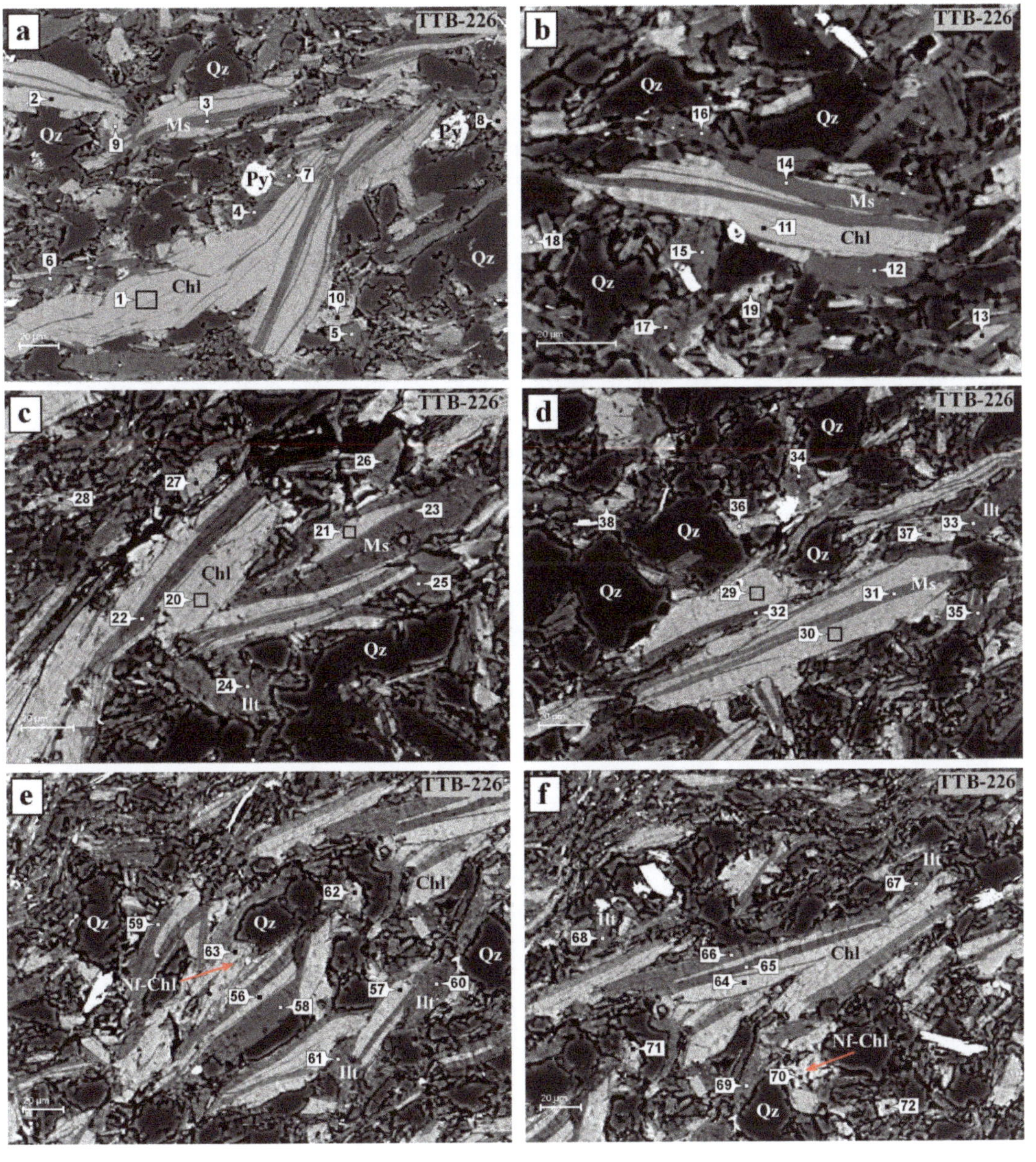

Figure 5. SEM-BSE photomicrographs of the early Ordovician slate with locations of some analyzed points and areas (black rectangulars). (**a**,**b**) Coarse-grained tectonically disturbed chlorite-rich CMS which parallel the bedding planes within the slightly oriented groundmass, (**c**,**d**) Chlorite-rich CMS within the quartz, neoformed illite (Ilt) and chlorite (Nf-Chl) bearing groundmass, (**e**,**f**) Muscovite-rich CMS with long axes were almost parallel to bedding planes and widespread occurrences of neoformed illites and chlorites in the pores of the groundmass. Chl = Chlorite, Nf-Chl: Neoformed chlorite, Ilt = Illite, Ms = Muscovite, Py = Pyrite, Qz = Quartz.

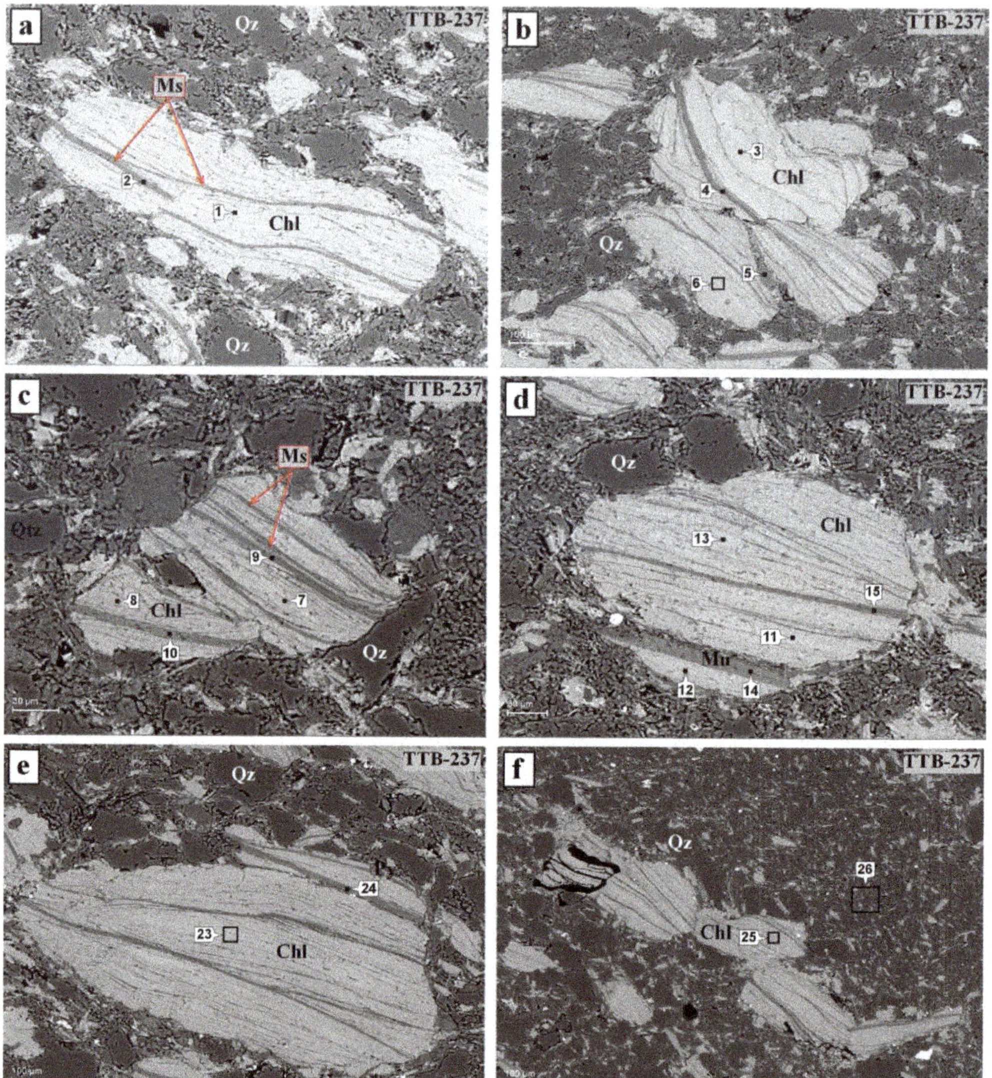

Figure 6. SEM-BSE photomicrographs of the early–middle Ordovician metasandstone with locations of some analyzed points and areas (black rectangulars). (**a**) Elongated chlorite-rich CMS which their stacking planes are nearly parallel to bedding planes and partly disturbed due to cutting by slightly developed cleavage planes, (**b**–**e**) Lozenge-shaped chlorite-rich CMS cutting by cleavage planes, (**f**) Elongated chlorite-rich CMS which multiple sliced by slightly developed cleavage planes within the groundmass composed of quartz-fine grained-white K-mica and chlorite.

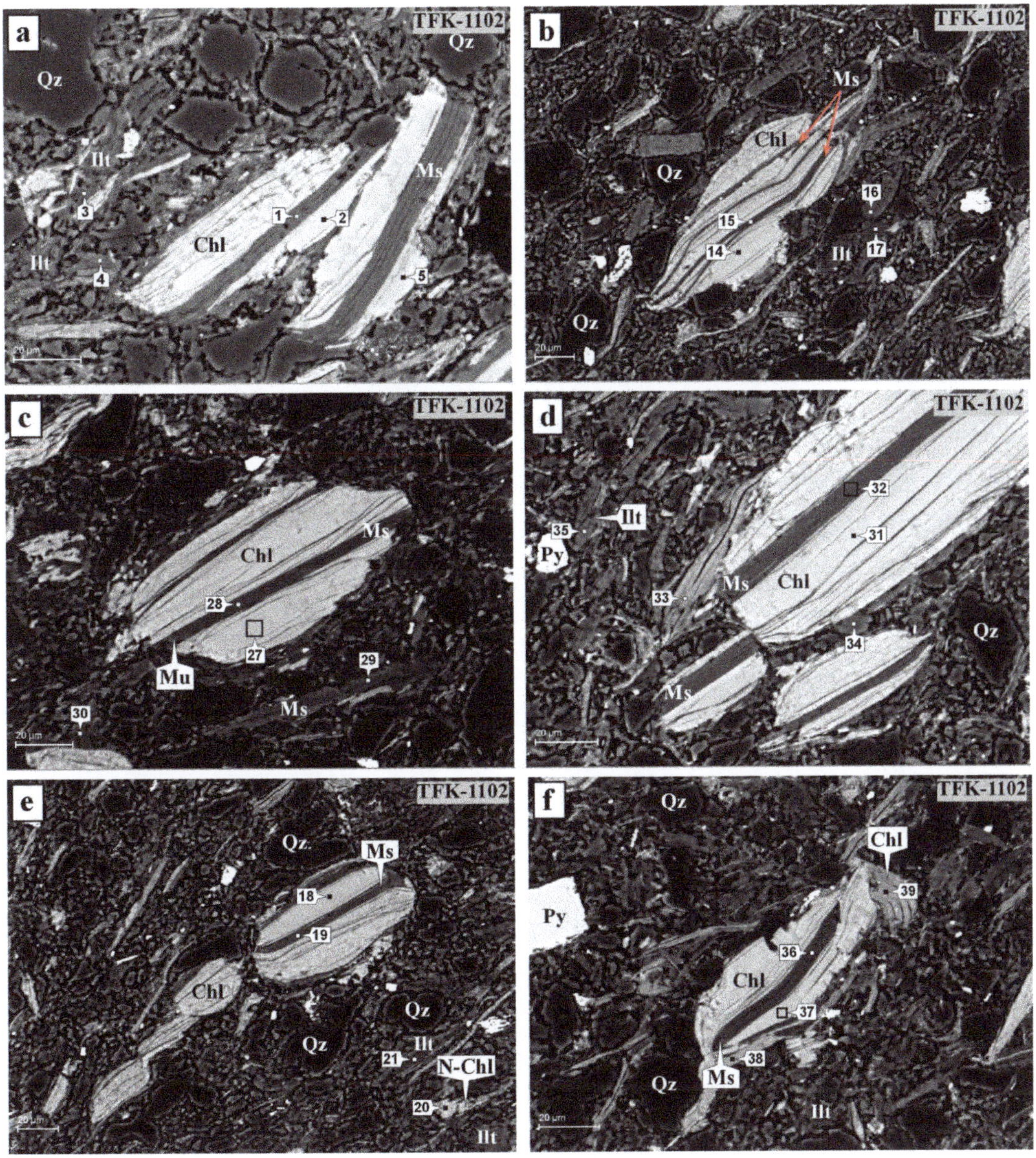

Figure 7. SEM-BSE photomicrographs of the late Ordovician silty slate with locations of some analyzed points and areas (black rectangulars). (**a**) Elongated chlorite-rich CMS which their stacking planes nearly parallel to bedding planes within the groundmass, including neoformed illites and chlorites, (**b**,**c**) Lozenge-shaped and curved chlorite-rich CMS, (**d**,**e**) Elongated chlorite-rich CMS, their long axes are parallel to bedding planes and sliced by crenulation type of cleavage planes in the quartz-fine grained-white K-mica and chlorite groundmass, (**f**) Elongated and curved chlorite-rich CMS within the groundmass containing pore filled neoformed illites and chlorites.

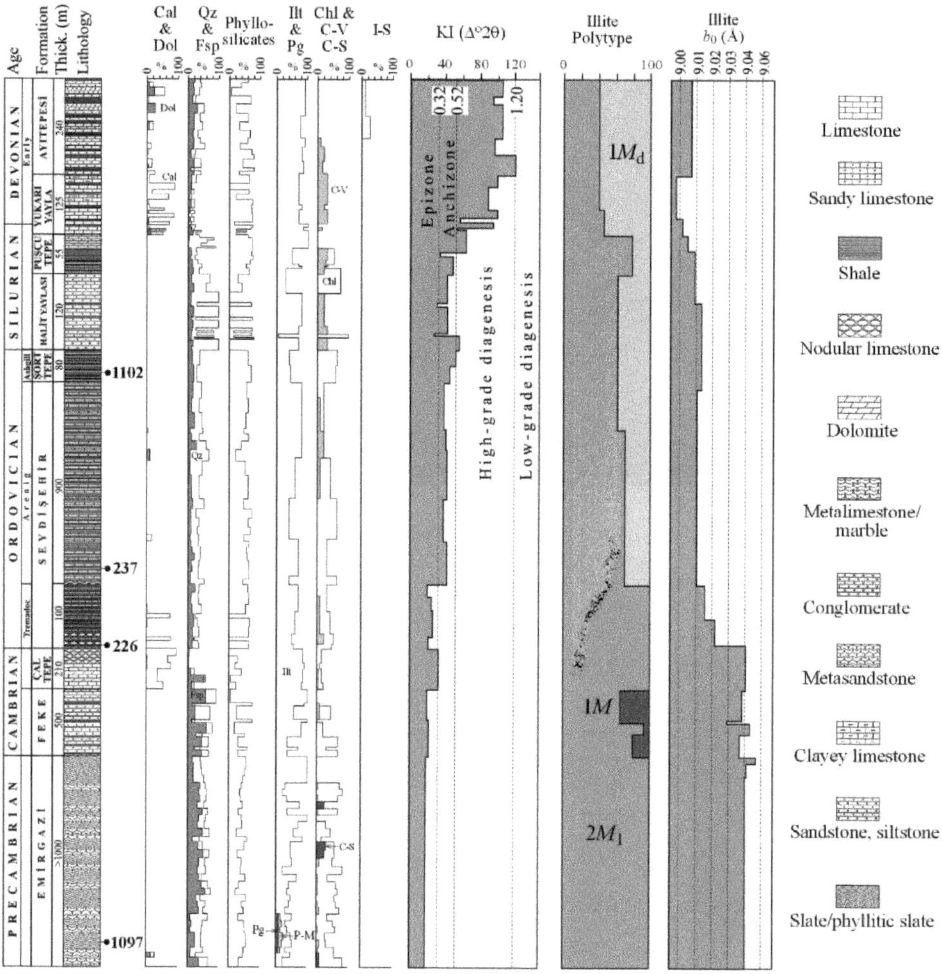

Figure 8. Vertical distribution of mineralogic data in the Lower Paleozoic units (Modified from [36]). Cal = Calcite, Chl = Chlorite, Dol = Dolomite, Fsp = Feldspar, Ilt = Illite, I-S = Mixed-layered illite-smectite, Ms = Muscovite, Pg = Paragonite, Qz = Quartz, C-S = Mixed-layered chlorite–smectite, C-V = Mixed-layered chlorite–vermiculite, KI = Illite Kübler index.

There is a positive correlation among % KI, $2M_1$%, and b_0 of illites/micas, indicating that the diagenetic–metamorphic characteristics were mainly caused by sedimentary burial at the initial site of deposition in the Lower Paleozoic part of the Eastern Taurus Autochthonous Unit [36]. The increasing $2M_1$-contents and b_0 values of the illites refer to increasing temperature and pressure conditions toward the lower part, respectively [2,6–8].

The XRD data of studied samples are made up mainly of quartz, feldspar (plagioclase), and phyllosilicates (Figure 9, Table 2 [36]). Phyllosilicate minerals of the samples are mainly represented by illite/muscovite, chlorite, and scarce mixed-layer chlorite–vermiculite (C-V).

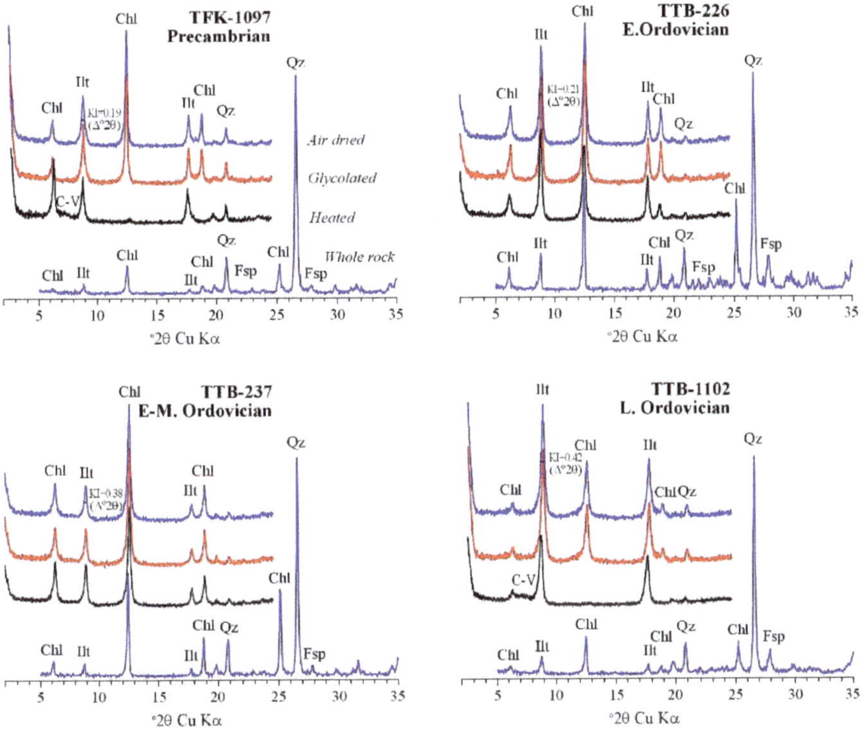

Figure 9. XRD patterns of whole rock and clay fraction of Precambrian–Ordovician metaclastic rocks. Chl = Chlorite, Fsp = Feldspar, Ilt = Illite, Ms = Muscovite, Qz = Quartz, C-V = Mixed-layer chlorite–vermiculite, KI = Illite Kübler index.

Table 2. Bulk and clay/phyllosilicate mineral composition and crystal–chemical data of illites of studied samples.

Sample/ Formation	Whole Rock			Clay Fraction				Crystal–Chemical Data			
	Qz	Fsp	Phl	Ilt	Chl	C-V	Pg-PM	KI	b_0	$2M_1$	$1M_d$
TFK-1097 Precambrian	40	8	52	35	50	5	10	0.19		100	
TTB-226 E.Ordovician	20	15	65	55	45			0.21	9.041	100	
TTB-237 E.Ordovician	27	6	67	30	70			0.38	9.023		
TTB-1102 L.Ordovician	29	22	49	55	35	10		0.42		90	10

Qz = Quartz, Fsp = Feldspar, Phl = Phyllosilicate, Ilt = Illite, Chl = Chlorite, C-V = Mixed-layered chlorite–vermiculite, Pg = Paragonite, PM = NaK mica, KI = Illite Kübler index, $2M_1$ = 2 layers monoclinic polytype, $1M_d$ = Disordered 1 layer monoclinic polytype.

KI (Δ°2θ) data of the samples correspond to epizone (TFK-1097, TTB-226), high- (TTB-237) and low-anchizone (TTB-1102) grades (Table 2). *b*0 (Å) values of illites/micas indicate the moderate to high pressure facies series [49,50]. The dominance of $2M_1$ polytype illite/mica contents corresponds to anchizonal-epizonal grades [6,7]. The phyllosilicate associations and crystal–chemical data exhibit reasonable compatibility for increasing diagenetic/metamorphic grades, together with increasing sedimentary burial, from late Ordovician to Precambrian.

4.3. Mineral Chemistry

Average major element compositions obtained from EMPA and EDS data and structural formulae of phyllosilicates are listed in Tables 3 and 4 for chlorites and illites/micas, respectively. Unit-cell compositions were calculated on the basis of 11 oxygen equivalents for I-S and illite–mica and 14 oxygen equivalents for chlorite [51].

Table 3. Average chemical composition and structural formulae of chlorites of the early Paleozoic rocks from the Eastern Tauride Belt (CMS: Chlorite–mica stacks, NFC: Neoformed chlorites).

	EMP				EDS						
Sample Origin (*n*)	**TFK-1097 CMS (5)**	**TTB-226 CMS (12)**	**TTB-237 CMS (6)**	**TTB-1102 CMS (5)**	**TFK-1097 CMS (16)**	**TTB-226 CMS (15)**	**TTB-226 NFC (18)**	**TTB-237 CMS (13)**	**TTB-237 NFC (2)**	**TTB-1102 CMS (11)**	**TTB-1102 NFC (3)**
SiO_2	22.47	24.46	23.89	22.38	26.27	28.52	29.40	26.90	29.24	26.01	30.08
TiO_2	0.02	0.04	0.29	0.02	0.00	0.00	0.00	0.00	0.00	0.00	0.00
Al_2O_3	24.97	21.75	24.96	24.28	28.34	25.73	25.22	28.90	27.58	27.65	25.35
FeO_{tot}	26.26	27.03	26.44	31.50	33.80	31.36	30.48	32.19	31.37	38.11	31.67
MnO	0.17	0.15	0.14	0.09	0.00	0.00	0.00	0.00	0.00	0.00	0.00
MgO	11.65	12.07	9.74	7.21	11.60	14.39	14.89	12.01	11.80	8.22	12.71
CaO	0.01	0.01	0.00	0.01	0.00	0.00	0.00	0.00	0.00	0.00	0.00
Na_2O	0.01	0.01	0.02	0.02	0.00	0.00	0.00	0.00	0.00	0.00	0.00
K_2O	0.00	0.00	0.01	0.00	0.00	0.00	0.00	0.00	0.00	0.00	0.18
Total	85.56	85.51	85.49	85.50	100.00	100.00	100.00	100.00	100.00	100	100
Si	2.45	2.67	2.59	2.51	2.48	2.66	2.72	2.52	2.71	2.51	2.80
Al^{IV}	1.55	1.33	1.41	1.49	1.52	1.34	1.28	1.48	1.29	1.49	1.20
Al^{VI}	1.65	1.47	1.78	1.72	1.64	1.49	1.48	1.70	1.72	1.66	1.58
Ti	0.00	0.00	0.02	0.00	0.00	0.00	0.00	0.00	0.00	0.00	0.00
Fe^{2+}	2.39	2.47	2.40	2.95	2.67	2.44	2.36	2.52	2.43	3.08	2.46
Mn	0.01	0.01	0.01	0.01	0.00	0.00	0.00	0.00	0.00	0.00	0.00
Mg	1.89	1.96	1.57	1.20	1.63	2.00	2.06	1.67	1.63	1.18	1.76
OC	1.54	1.32	1.40	1.48	1.52	1.34	1.28	1.48	1.29	1.49	1.18
TOC	5.94	5.92	5.79	5.88	5.94	5.93	5.90	5.89	5.78	0.00	0.00
Ca	0.00	0.00	0.00	0.00	0.00	0.00	0.00	0.00	0.00	0.00	0.00
Na	0.00	0.00	0.00	0.00	0.00	0.00	0.00	0.00	0.00	0.00	0.00
K	0.00	0.00	0.00	0.00	0.00	0.00	0.00	0.00	0.00	0.00	0.02
ILC	0.00	0.00	0.01	0.01	0.00	0.00	0.00	0.00	0.00	0.00	0.00
TLC	0.00	0.00	0.01	0.01	0.00	0.00	0.00	0.00	0.00	0.00	0.00
°C	>350	343	241	318	>350	343	268	321	204	>350	201

OC = Octahedral Charge, TOC = Total Octahedral Cation, ILC = Interlayer Charge, TLC = Total Layer Charge, *n* = number of analyses.

The mineral chemistry data of chlorites in the early Paleozoic metaclastic rocks (Table 3) show moderately narrow distributions of Si–R^{2+} and Al^{IV}–Fe/(Fe + Mg) diagrams (Figure 10). However, the compositions of chlorites in CMS and neoformed ones range among clinochlore/daphnite, amesite and sudoite end-members, almost places between clinochlore/daphnite and amesite (Figure 10a,b). Both neoformed and detrital (CMS) chlorites of the sample in early Ordovician (Tremadocian) parts exhibit reasonably different compositions than others because of the higher Si and R^{2+} contents. Neoformed chlorites have higher Si contents than those of detrital (CMS) chlorites. Although the chlorites are shown to have brunsvigite-ripidolite composition in the Al^{IV}–Fe/(Fe + Mg) diagram (Figure 10c,d), in accordance with AIPEA nomenclature [52], the Fe/(Fe + Mg) content of chlorites indicate the chamosite composition (Fe/(Fe + Mg > 0.5). The detrital chlorites have higher Al^{IV} and Fe contents than their neoformed counterparts.

Table 4. Average chemical composition and structural formulae of illite/micas of the early Paleozoic rocks from the Eastern Tauride Belt (Ms = Muscovite, Ilt = Illite, P-M = Paragonite-muscovite).

	EMP						EDS										
Sample	**1097**	**1097**	**226**	**237**	**1102**	**1102**	**1097**	**1097**	**1097**	**226**	**226**	**226**	**237**	**237**	**1102**	**1102**	**1102**
Origin	**Ms**	**Ilt**	**Ms**	**Ms**	**Ms**	**Ilt**	**Ms**	**Ilt**	**P-M**	**Ms**	**Ilt**	**P-M**	**Ms**	**Ilt**	**Ms**	**Ilt**	**P-M**
(*n*)	**(3)**	**(1)**	**(6)**	**(4)**	**(3)**	**(1)**	**(14)**	**(16)**	**(11)**	**(10)**	**(22)**	**(5)**	**(9)**	**(3)**	**(11)**	**(10)**	**(2)**
SiO_2	45.44	46.60	45.46	44.99	45.04	44.90	47.87	48.65	47.97	49.13	50.08	48.10	47.58	50.07	49.65	49.38	48.17
TiO_2	0.30	0.22	0.55	1.37	0.54	0.42	0.00	0.00	0.00	0.00	0.00	0.00	0.00	0.00	0.00	0.00	0.00
Al_2O_3	37.67	37.14	33.32	34.84	35.45	35.31	38.28	41.03	40.56	35.67	34.39	38.28	37.51	35.59	35.37	37.79	37.58
FeO_{tot}	0.93	0.81	2.49	3.07	2.93	0.83	2.43	0.68	2.48	2.82	3.62	1.79	3.14	2.51	2.75	1.75	3.00
MnO	0.03	0.01	0.07	0.05	0.06	0.01	0.00	0.00	0.00	0.00	0.00	0.00	0.00	0.00	0.00	0.00	0.00
MgO	0.60	0.68	1.55	1.15	0.92	1.82	1.36	0.55	1.06	1.91	1.93	1.46	1.68	1.79	2.09	1.43	1.69
CaO	0.01	0.02	0.02	0.01	0.05	0.05	0.00	0.00	0.00	0.00	0.00	0.00	0.00	0.00	0.00	0.00	0.00
Na_2O	0.80	1.23	0.50	0.33	0.64	0.44	0.95	2.23	3.91	0.58	0.27	1.28	0.60	0.35	0.45	0.32	1.49
K_2O	7.77	7.35	9.47	8.67	7.65	9.65	9.11	6.86	4.02	9.89	9.70	9.09	9.49	9.69	9.68	9.32	8.07
Total	93.55	94.07	93.44	94.50	93.28	93.43	100.00	100.00	100.00	100.00	100.00	100.00	100.00	100.00	100.00	100.00	100.00
Si	3.02	3.07	3.09	3.02	3.04	3.03	3.02	3.01	2.97	3.12	3.18	3.03	3.02	3.16	3.14	3.10	3.04
Al^{IV}	0.98	0.93	0.91	0.98	0.96	0.97	0.98	0.99	1.03	0.88	0.82	0.97	0.98	0.84	0.86	0.90	0.96
Al^{VI}	1.98	1.96	1.77	1.78	1.86	1.84	1.87	2.00	1.93	1.78	1.75	1.87	1.83	1.80	1.78	1.89	1.83
Ti	0.02	0.01	0.03	0.07	0.03	0.02	0.00	0.00	0.00	0.00	0.00	0.00	0.00	0.00	0.00	0.00	0.00
Fe^{2+}	0.05	0.04	0.14	0.17	0.17	0.05	0.13	0.04	0.13	0.15	0.19	0.09	0.17	0.13	0.15	0.09	0.16
Mn	0.00	0.00	0.00	0.00	0.00	0.00	0.00	0.00	0.00	0.00	0.00	0.00	0.00	0.00	0.00	0.00	0.00
Mg	0.06	0.07	0.16	0.12	0.09	0.18	0.13	0.05	0.10	0.18	0.18	0.14	0.16	0.17	0.20	0.13	0.16
OC	0.21	0.15	0.02	0.19	0.21	0.07	0.13	0.18	0.24	0.01	0.00	0.08	0.14	0.02	0.02	0.12	0.13
TOC	2.10	2.08	2.10	2.14	2.15	2.09	2.13	2.09	2.16	2.11	2.13	2.10	2.15	2.11	2.12	2.11	2.15
Ca	0.00	0.00	0.00	0.00	0.00	0.00	0.00	0.00	0.00	0.00	0.00	0.00	0.00	0.00	0.00	0.00	0.00
Na	0.10	0.16	0.07	0.04	0.08	0.06	0.12	0.27	0.47	0.07	0.03	0.16	0.07	0.04	0.06	0.04	0.18
K	0.66	0.62	0.82	0.74	0.66	0.83	0.73	0.54	0.32	0.80	0.79	0.73	0.77	0.78	0.78	0.75	0.65
ILC	0.76	0.78	0.89	0.79	0.75	0.90	0.85	0.81	0.79	0.87	0.82	0.89	0.84	0.82	0.84	0.79	0.83
TLC	0.76	0.78	0.89	0.79	0.75	0.90	0.85	0.81	0.79	0.87	0.82	0.89	0.84	0.82	0.84	0.79	0.83

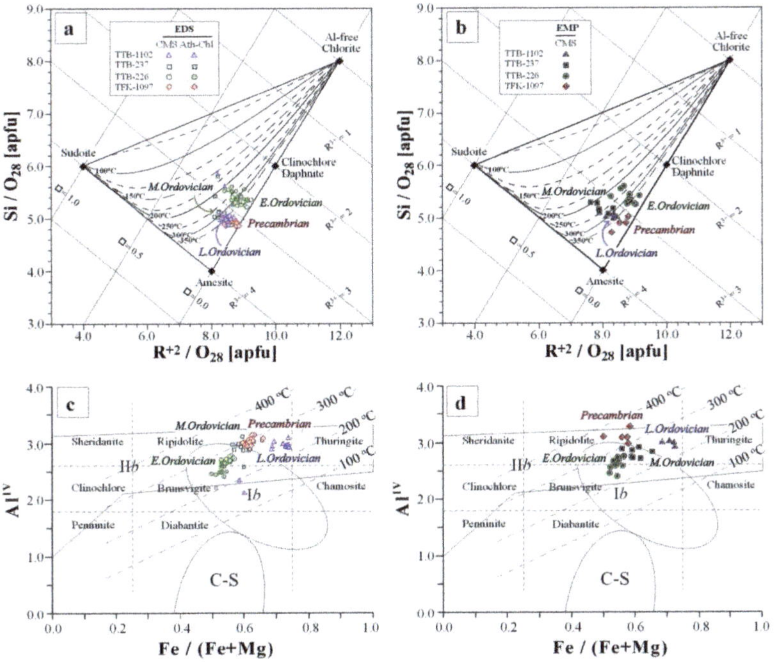

Figure 10. Distribution of chlorite compositions on (**a**,**b**) Si vs. R^{2+} (simplified from [53]), and (**c**,**d**) Al^{IV} vs. Fe/(Fe + Mg) diagram (modified from [54]; IIb chlorite area and compositional limits from [55]; Ib and swelling chlorites (C-S) areas from [56]. (**a**,**c**) and (**b**,**d**) represent the analytical results by EDS and EMP, respectively. Temperature contours on Si − R^{2+} and Al^{IV} − Fe/(Fe + Mg) diagrams are from [30] and [52], respectively.

The tetrahedral and octahedral Al data of the chlorites in the sample of early Ordovician (Tremadocian) parts and of the chlorites from Precambrian, early–middle, and late Ordovician units assembled in two different areas in the Al^{VI} + 2Ti vs. Al^{IV} diagram (Figure 11a,b). Neoformed chlorites are distributed towards the left corner of the Al^{VI} + 2Ti vs. Al^{IV} diagram (Figure 11a), indicating an increase in the dioctahedral substitution (AM, Al^{3+} for Mg^{2+}). Tschermak substitution (TM), e.g., Al^{IV} for Si and Al^{VI} for Mg, indicates higher octahedral substitution. The octhedral-element distribution of chlorites in the Al+□-Mg–Fe diagram refers to completely trioctahedral (Type I) chlorites (Figure 11c,d). Chlorites within the phyllosilicate stacks have slightly greater Fe related to originating from detrital biotites. The chlorites of early Ordovician (Tremadocian) parts are associated with metapelitic rocks, whereas chlorites of the other units are accompanied by both felsic and metapelitic rocks.

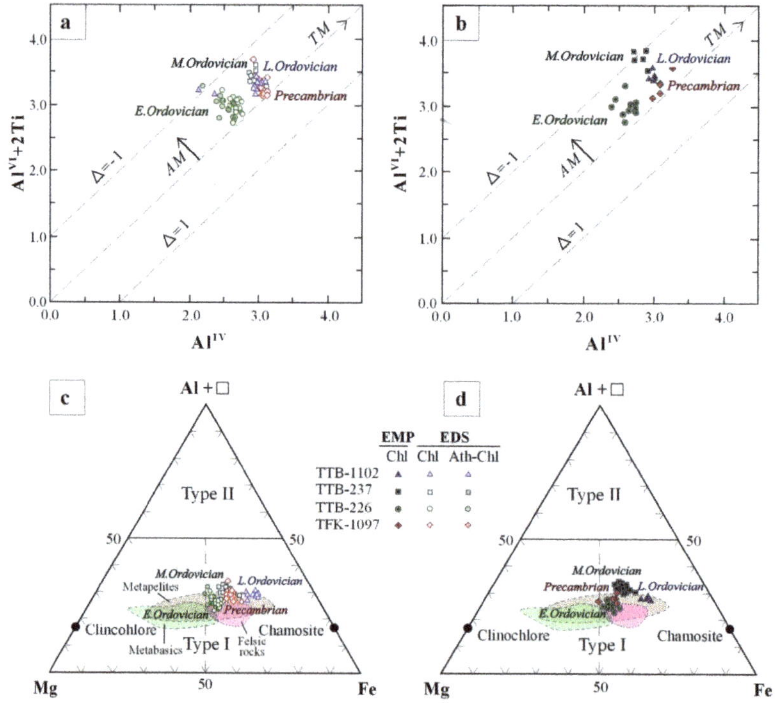

Figure 11. Distribution of chlorite compositions on classification diagrams. (**a**,**b**) Al^{VI} + Ti vs. Al^{IV} [57]; (**c**,**d**) octahedral Al + □–Mg–Fe [58]. (**a**,**c**) and (**b**,**d**) represent the analytical results by EDS and EMP, respectively. $\Delta = Al^{IV} - Al^{VI} + 2Ti$, □ = Octahedral vacancy, Type I = $X_{Mg} + X_{Fe} \geq X_{Al} + X_{□}$, Type II = $X_{Mg} + X_{Fe} < X_{Al} + X_{□}$. The TM and AM arrows point towards increasing Tschermak and dioctahedral substitutions, respectively.

The chlorite thermometry method of [30] was used for temperature estimation (Table 3). Temperature values from chlorite compositions were also obtained from the temperature contours in Si–R^{2+} [31] and Al^{IV}–Fe/(Fe + Mg) diagrams [19] (Figure 10a,b). The temperature estimations of chlorites from these diagrams show a wide range of temperatures from ~150 °C to ≥350 °C (Figure 10a–c). The average temperature data of detrital (CMS) and neoformed chlorites were calculated as 241 °C to ≥350 °C and 201 °C to 268 °C, respectively (Table 3, Figure 12). The temperature conditions of detrital chlorites display differences for individual samples, i.e., detrital chlorites in the samples from the lower parts reveal higher temperature conditions than those of the upper parts. Similarly, the temperature condition of the neoformed chlorite in the epimetamorphic slate was higher than the low-grade anchimetamorphic slate sample (Figure 12).

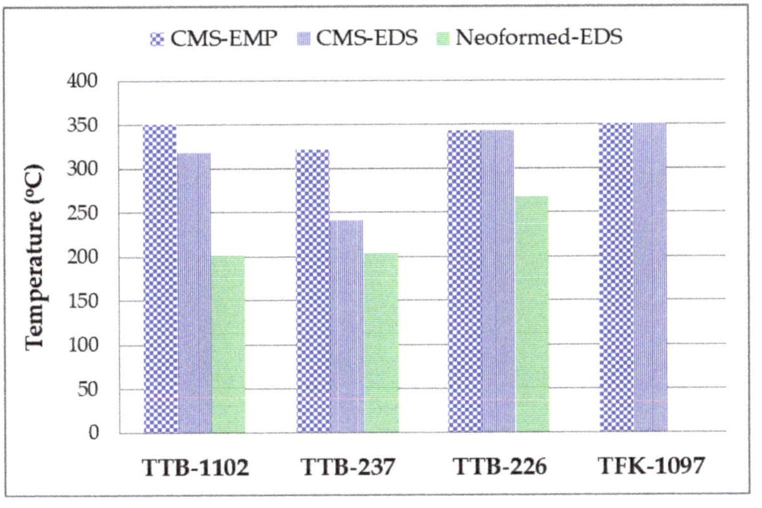

Figure 12. Average temperature distributions of detrital (chlorite–mica stacks-CMS) and neoformed chlorites obtained from EMP and EDS data based on the thermometry method of [28]. Metamorphic grade and stratigraphic depth increase toward the right from low-grade anchizone/late Ordovician (TTB-1102) to epizone/Precambrian (TFK-1097).

The compositional variations of white mica and illite are shown in Table 4, Figure 13. Mineral chemistry studies performed with SEM-BSE images revealed the presence of NaK mica and paragonites, which could not be detected by XRD, in two samples. K-white micas often plot in the compositional space between ideal muscovite and phengite, whereas neoformed illite minerals are projected in the phengitic to celadonitic compositional space in the M^{+}-4Si-R^{2+} diagram (Figure 13a,b). In the Fe + Mg-Al^{IV}–Al^{VI} mica diagram (Figure 13c,d), detrital white micas have muscovitic compositions, but neoformed illite compositions shift to the Fe + Mg corner towards phengite and ferrimuscovite compositions. Tetrahedral Si and total interlayer cation occupancy of detrital K-white micas, NaK micas, and neoformed illites are clearly distinguished in the Si−(Na + K) diagram (Figure 13e,f).

Illites have relatively higher Si-, Fe-, and Mg- and lower K-content than K-white micas. The distributions of octahedral Fe, Mg, and interlayer Na content in neoformed illites with respect to white micas in the Precambrian unit are remarkably different from other units. Contrary to detrital white micas, neoformed illites in the Precambrian unit have lower Fe- and Mg- content but higher interlayer Na-content. NaK micas are distinguished by lower Si-, Fe-, Mg- and higher Al- content than muscovite and illite, in addition to higher interlayer Na content (Figure 13a–f).

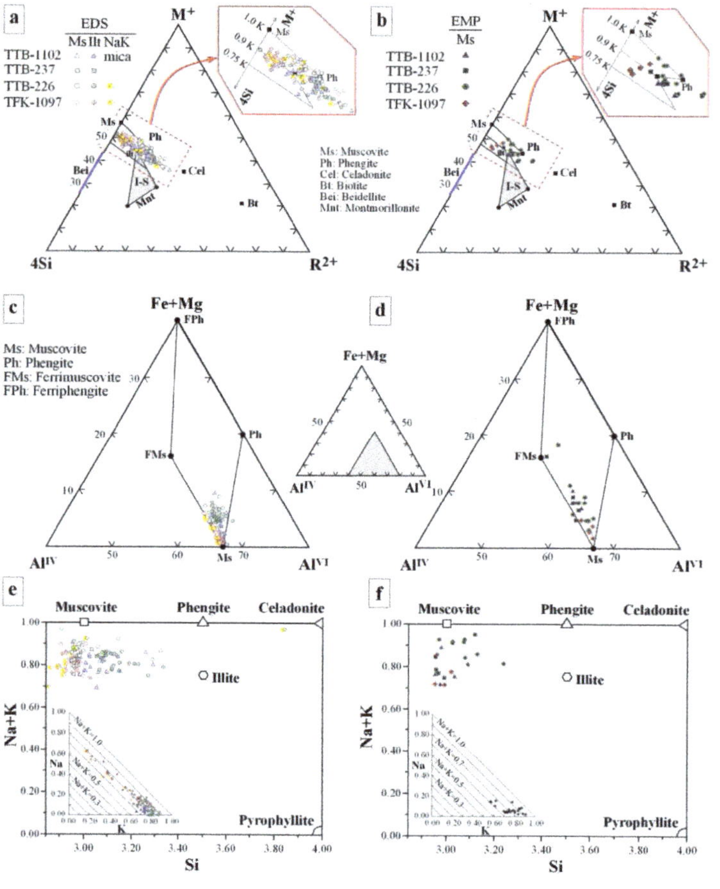

Figure 13. The compositional distributions of illites/micas. (**a**,**b**) M^{+}–4Si–R^{2+} triangular diagram [59], (**c**,**d**) Fe + Mg–Al^{IV}-Al^{VI} [60], (**e**,**f**) Na + K vs. Si diagram.

5. Discussion

5.1. Chlorite Chemistry

Chlorite composition can be sensitive to the diagenetic/metamorphic grade. The tetrahedral Al (Al^{IV}) and octahedral R^{2+} (Mg and Fe^{2+}) contents of chlorite increase with increasing grade [13–20]. Thus, different chemical compositions for detrital (metamorphic) and authigenic (diagenetic) chlorites have been reported in many studies (e.g., [17,21–24].

Chlorites in studied samples exhibit different compositions for the samples which have different metamorphic grades. Tetrahedral Al and octahedral Fe + Mg increase, whereas octahedral Al decreases together with the increasing grade of metamorphism. Chlorites metasiltstone and slate samples from Precambrian–Early Ordovician (Tremadocian) parts have ripidolite, whereas metasandstone and slate samples from the early–middle Ordovician–Silurian have brunsvigite compositions. It has been noticed that the chlorites in Tremadocian (early Ordovician) metaclastics have different compositions with respect to the others. It was evaluated that this difference is related not only to the metamorphic grade but also to different detrital origins. The chemistry data of chlorites indicating a different source for early Ordovician (see Figure 11b) also support this opinion.

Neoformed chlorites have higher Si-but lower Al^{IV}- and Fe-contents than those detrital (CMS) chlorites. The greater Fe content of chlorites in CMS seems to be related to originating from biotites. During the alteration of biotites to chlorite, the released excess Fe ions [61–63] could be precipitated as iron oxides and/or sulfur (i.e., hematite, pyrite).

Chlorite geothermometry data for detrital and neoformed origin (Figure 12) are well-matched with KI data (Figure 14), as well as with the textural (slaty cleavage) and other mineralogical (phyllosilicate mineral associations, polytype) maturation data. The different temperature data for detrital (higher temperature) and neoformed (lower temperature) chlorites indicate that the detrital chlorites preserve their metamorphic nature even in anchizonal conditions.

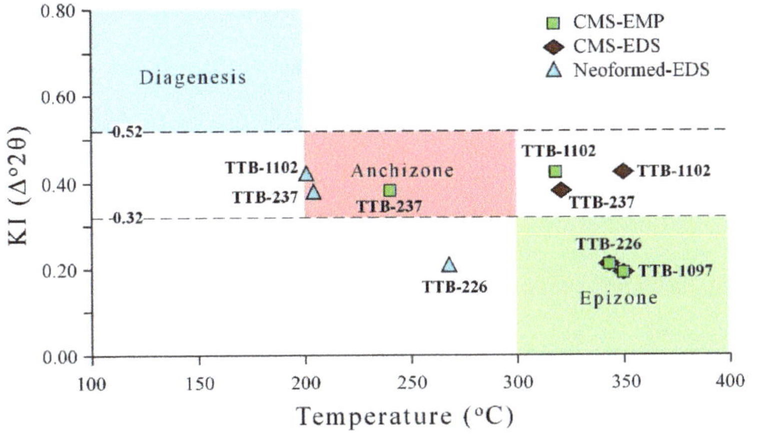

Figure 14. Chlorite thermometry data versus illite Kübler index data of the samples. The Diagenesis–Anchizone–Epizone regions are drawn based on assumed temperature limits for low and upper limits of anchizone [2,7–9].

5.2. Illite/White Mica Chemistry

Illite/mica compositions display the ranges among muscovite–phengite–celadonite. Illites with neoformation origins represent phengitic, whereas micas have commonly muscovitic and partly paragonitic and celadonitic compositions. Tetrahedral Al and interlayer K + Na contents of illite/micas increase together with the increasing grade of diagenesis-metamorphism and change the compositions from illite and phengite to ideal muscovite. Increasing the interlayer K^+ content and layer charge (especially tetrahedral charge) during the transformation from illite to muscovite, together with the increasing diagenesis/metamorphism, are known in the literature [11,25].

In compensation for tetrahedral-charge change, octahedral-charge increases because of the Fe^{2+} and Mg substitutions for Al^{VI}. An increase in Fe^{2+} and Mg (or phengitic) content is mainly related to temperature-related diagenetic-metamorphic grade [10–12,64] but may also be associated with changes in the composition of diagenetic/metamorphic solutions [15,60]. The higher Fe and Mg contents of neoformed illite should be related to Mg- and Fe-rich diagenetic solutions, derived from the conversion of biotite to chlorite and then C-V, as stated for the Fe-rich (chamositic) chlorites.

NaK mica and paragonite-bearing samples in the Precambrian and Lower Ordovician units are restricted to high anchizonal–epizonal grades. Their occurrences are characteristic for the extensional basins [46,65], where the temperature is dominant, which are observed as replacement type developments within the detrital CMS [25]. Na,K-mica were developed within the CMS, amongst the {001} layers of muscovite (see Figure 4e,f). This occurrence may be related to the replacement of muscovite with Na-rich solutions. Retrograde or second metamorphic recrystallization occurrences of paragonite and/or mixed Na,K-mica were documented in the early Paleozoic sedimentary unit [25,66,67].

6. Conclusions

The Precambrian–Early Paleozoic metaclastic rocks of the Eastern Tauride Belt were metamorphosed under low temperature (low anchizone to epizone) conditions. In addition to textural and mineralogical changes, the mineral chemistry of phyllosilicates also affected the metamorphism. Under low-temperature metamorphic conditions, detrital biotite–muscovite stacks altered to CMS, in which the chlorites were derived from biotites, and therefore their composition is rich in Fe (chamosite). In addition to the transformation of clastic phyllosilicates, phyllosilicate neoformation processes have also taken place. The neoformed illites and chlorites were formed in the matrix, whereas replacement type NaK mica and paragonite minerals are also developed within the {001} planes of muscovites in CMS. Neoformed chlorites have higher Si but lower Al^{IV} and Fe contents than detrital (CMS) chlorites. The chlorite geothermometry data indicate that the calculated temperature values are compatible with the metamorphic grades. The increasing degree of metamorphism was reflected in the chlorite geothermometer, in other words, in the chlorite chemistry. In terms of the effect of metamorphism grade on the illite/mica composition, tetrahedral Al and interlayer K + Na contents of illite/micas increase together with the increasing grade of diagenesis-metamorphism and change the compositions from illite and phengite to ideal muscovite. The variations of phyllosilicate mineral chemistry data in accordance with the textural and mineralogical data indicate that the mineral chemistry data of phyllosilicate can be used as a useful tool for low-temperature metamorphic petrology.

Author Contributions: Investigation, Ö.B., H.Y.; formal analysis, Ö.B.; writing—original draft preparation, Ö.B.; resources, Ö.B., H.Y. All authors have read and agreed to the published version of the manuscript.

Funding: This research was funded by the Research Fund of Cumhuriyet University, part of Project Number M-469.

Acknowledgments: The authors also acknowledge the contributions of Paul Schroeder, Doug Crowe, Chris Fleischer, and John P. Schields from the University of Georgia (Athens, Georgia, USA) in terms of the electron microscope and electron microprobe analyses. The authors are also grateful to the editorial office and two anonymous referees for their constructive comments leading to important improvements in the manuscript.

Conflicts of Interest: The authors declare no conflict of interest.

References

1. Robinson, D. Transition from diagenesis to metamorphism in extensional and collision settings. *Geology* **1987**, *15*, 866–869. [CrossRef]
2. Merriman, R.J.; Frey, M. Patterns of very low grade metamorphism in metapelitic rocks. In *Low-Grade Metamorphism*; Frey, M., Robinson, D., Eds.; Blackwell Science: Oxford, UK, 1999; pp. 61–107.
3. Voll, G. New work on petrofabrics. *Liverp. Manch. Geol. J.* **1960**, *2*, 503–567. [CrossRef]
4. Craig, J.; Fitches, W.R.; Maltman, A.J. Chlorite-mica stacks in low-strain rocks from Central Wales. *Geol. Mag.* **1982**, *119*, 243–256. [CrossRef]
5. Giorgetti, G.; Memmi, F.; Nieto, F. Microstructures of intergrown phyllosilicate grains from Verrucano metasediments (northern Apennines, Italy). *Contrib. Mineral. Petrol.* **1997**, *128*, 127–138. [CrossRef]
6. Kisch, H.J. Mineralogy and petrology of burial diagenesis (burial metamorphism) and incipient metamorphism in clastic rocks. In *Diagenesis in Sediments and Sedimentary Rocks*; Larsen, G., Chilingar, G.V., Eds.; Elsevier: Amsterdam, The Netherlands, 1983; Volume 2, pp. 289–493; 513–541.
7. Frey, M. Very low-grade metamorphism of clastic sedimentary rocks. In *Low Temperature Metamorphism*; Frey, M., Ed.; Blackie & Son: Glasgow, UK, 1987; pp. 9–58.
8. Merriman, R.J.; Peacor, D.R. Very low-grade metapelites: Mineralogy, microfabrics and measuring reaction progress. In *Low-Grade Metamorphism*; Frey, M., Robinson, D., Eds.; Blackwell Science: Oxford, UK, 1999; pp. 10–60.
9. Ferreiro-Mählmann, R.; Bozkaya, Ö.; Potel, S.; Le Bayon, R.; Nieto, F. The pioneer work of Bernard Kübler and Martin Frey in very low-grade metamorphic terranes: Paleo-geothermal potential of variation in Kübler-Index/organic matter reflectance correlations. A review. *Swiss J. Geosci.* **2012**, *105*, 121–152. [CrossRef]
10. Ireland, B.J.; Curtis, C.D.; Whiteman, J.A. Compositional variation within some glauconites and illites and implications for their stability and origins. *Sedimentology* **1983**, *30*, 769–786. [CrossRef]

11. Hunziker, J.C.; Frey, M.; Clauer, N.; Dallmeyer, R.D.; Fredrichsen, H.; Flehmig, W.; Hochstrasser, K.; Roggviler, P.; Schwander, H. The evolution of illite to muscovite: Mineralogical and isotopic data from the Glarus Alps. Switzerland. *Contrib. Mineral. Petrol.* **1986**, *92*, 157–180. [CrossRef]
12. Brill, B.A. Illite crystallinity, b0 and Si content of K-white mica as indicators of metamorphic conditions in low-grade metamorphic rocks at Cobar, New South Wales. *Aust. J. Earth Sci.* **1988**, *35*, 295–302. [CrossRef]
13. Weaver, C.E. *Shale Slate Metamorphism in Southern Appalachians*; Elsevier: Amsterdam, The Netherlands, 1984.
14. Cathelineau, M.; Nieva, D.A. Chlorite solid solution geothermometer, the Los Azufres geothermal system (Mexico). *Contrib. Mineral. Petrol.* **1985**, *91*, 235–244. [CrossRef]
15. Cathelineau, M. Cation site occupancy in chlorites and illites as a function of temperature. *Clay Miner.* **1988**, *23*, 471–485. [CrossRef]
16. Velde, B.; Medhioub, M. Approach to chemical equilibrium in diagenetic chlorites. *Contrib. Mineral. Petrol.* **1988**, *98*, 122–127. [CrossRef]
17. Hillier, S.; Velde, B. Octahedral occupancy and the chemical composition of diagenetic (low temperature) chlorites. *Clay Miner.* **1991**, *26*, 149–168. [CrossRef]
18. Jahren, J.S.; Aagard, P. Diagenetic illite-chlorite assemblages in arenites. I. Chemical evolution. *Clays Clay Miner.* **1992**, *40*, 540–546. [CrossRef]
19. Xie, X.G.; Byerly, G.R.; Ferrell, R.E. IIb trioctahedral chlorite from the Barberton greenstone belt: Crystal structure and rock composition constraints with implications to geothermometry. *Contrib. Mineral. Petrol.* **1997**, *126*, 275–291. [CrossRef]
20. Potel, S.; Ferreiro-Mählmann, R.; Stern, W.B.; Mullis, J.; Frey, M. Very low-grade metamorphic evolution of pelitic rocks under high-pressure/low-temperature condition, NW New Caledonia (SW Pacific). *J. Petrol.* **2006**, *47*, 991–1015. [CrossRef]
21. Hillier, S.; Velde, B. Chlorite interstratified with a 7 Å mineral: An example from offshore Norway and possible implications for the interpretation of the composition of diagenetic chlorites. *Clay Miner.* **1992**, *27*, 475–486. [CrossRef]
22. Hillier, S. Origin, diagenesis, and mineralogy of chlorite minerals in Devonian lacustrine mudrocks, Orcadian Basin, Scotland. *Clays Clay Miner.* **1993**, *41*, 240–259. [CrossRef]
23. Jiang, W.T.; Peacor, D.R. Formation of corrensite, chlorite and chlorite-mica stacks by replacement of detrital biotite in low-grade pelitic rocks. *J. Metamorph. Geol.* **1994**, *12*, 867–884. [CrossRef]
24. Bartier, D.; Buatier, M.; Lopez, M.; Potdevin, J.L.; Chamley, H.; Arostegui, J. Lithological control on the occurrence of chlorite in the diagenetic Wealden complex of the Bilbao anticlinorium (Basco-Cantabrian Basin, Northern Spain). *Clay Miner.* **1998**, *33*, 317–332. [CrossRef]
25. Bozkaya, Ö.; Yalçın, H.; Schroeder, P.A. Two-step mode of clay formation in the extensional basins: Cambrian–Ordovician clastic rocks of the Antalya unit, SW Turkey. *Clay Miner.* **2017**, *52*, 365–389. [CrossRef]
26. Inoue, A.; Meunier, A.; Patrier-Mas, P.; Rigault, C.; Beaufort, D.; Vieillard, P. Application of chemical geothermometry to low-temperature trioctahedral chlorites. *Clays Clay Miner.* **2009**, *57*, 371–382. [CrossRef]
27. Inoue, A.; Inoué, S.; Utada, M. Application of chlorite thermometry to estimation of formation temperature and redox conditions. *Clay Miner.* **2018**, *53*, 143–158. [CrossRef]
28. Bourdelle, F.; Parra, T.; Chopin, C.; Beyssac, O. A new chlorite geothermometer for diagenetic to low-grade metamorphic conditions. *Contrib. Mineral. Petrol.* **2013**, *165*, 723–735. [CrossRef]
29. Lanari, P.; Wagner, T.; Vidal, O. A thermodynamic model for di-trioctahedral chlorite from experimental and natural data in the system MgO-FeO-Al_2O_3- SiO_2-H_2O: Applications to P-T sections and geothermometry. *Contrib. Mineral. Petrol.* **2014**, *167*, 968. [CrossRef]
30. Bourdelle, F. Low-temperature chlorite geothermometry and related recent analytical advances: A review. *Minerals* **2021**, *11*, 130. [CrossRef]
31. Bourdelle, F.; Cathelineau, M. Low-temperature chlorite geothermometry: A graphical representation based on a T-R^{2+}-Si diagram. *Eur. J. Mineral.* **2015**, *27*, 617–626. [CrossRef]
32. Göncüoglu, M.C.; Dirik, K.; Kozlu, H. General characteristics of pre-Alpine and Alpine Terranes in Turkey: Explanatory notes to the terrane map of Turkey. Annales Geologique de Pays Hellenique. *Geol. Soc. Greece* **1997**, *37*, 515–536.
33. Şengör, A.M.C.; Yılmaz, Y. Tethyan evolution of Turkey: A plate tectonic approach. *Tectonophysics* **1981**, *75*, 181–241. [CrossRef]
34. Özgül, N. Some geological aspects of the Taurus orogenic belt (Turkey). *Bull. Geol. Soc. Turk.* **1976**, *19*, 65–78. (In Turkish with English Abstract)
35. Özgül, N.; Metin, S.; Dean, W.T. Lower Paleozoic stratigraphy and faunas of the Eastern Taurus mountains in the Tufanbeyli region, southern Turkey. *Bull. Miner. Res. Explor.* **1973**, *79*, 9–20.
36. Bozkaya, Ö.; Yalçın, H.; Göncüoglu, M.C. Mineralogic and organic responses to the stratigraphic irregularities: An example from the lower Paleozoic very low-grade metamorphic units of the eastern Taurus Autochthon, Turkey. *Schweiz. Mineral. Petrogr. Mitt.* **2002**, *82*, 355–373.
37. Göncüoğlu, M.C.; Göncüoğlu, Y.; Kozur, H.W.; Kozlu, H. Paleozoic stratigraphy of the Geyik Dağı unit in the eastern Taurides (Turkey): New age data and implications for Gondwanan evolution. *Geol. Carpath.* **2004**, *55*, 433–447.
38. Metin, S.; Ayhan, A.; Papak, I. Geology of the western part of the eastern Taurus (SSE Turkey). *Bull. Miner. Res. Explor.* **1987**, *107*, 1–12. (In Turkish)

39. Erkan, E.; Özer, S.; Sümengen, M.; Terlemez, I. *Basic Geology of the Area Between Sarız-Şarkısla-Gemerek-Tomarza*; M.T.A. Report No. 6546; Institute of Mineral Research and Exploration: Ankara, Turkey, 1978. (In Turkish)
40. Demirtaşlı, E.; Metin, S.; Ayhan, A. Stratigraphy of the eastern Taurus Autochthon. In Proceedings of the International Symposium on the Geology of the Taurus Belt, Ankara, Turkey, 26–29 September 1983; pp. 7–16.
41. Kozlu, H.; Göncüoglu, M.C. Stratigraphy of the Infracambrian rock-units in the Eastern Taurides and their correlation with similar units in southern Turkey. In *Lower Palaeozoic Evolution in Northwest Gondwana*; Göncüoğlu, M.C., Derman, A.S., Eds.; Turkish Association Petroleum Geologists Special Publication: Ankara, Turkey, 1997; Volume 3, pp. 50–61.
42. Monod, O.; Kozlu, H.; Ghienne, J.-F.; Dean, W.T.; Günay, Y.; Le Hérissé, A.; Paris, F.; Robardet, M. Late Ordovician glaciation in southern Turkey. *Terra Nova* **2003**, *15*, 249–257. [CrossRef]
43. Armstrong, J.T. Quantitative analysis of silicate and oxide materials: Comparison of Monte Carlo, ZAF, and phi-rho-z procedures. In *Microbeam Analysis*; Newbury, D.E., Ed.; San Francisco Press: San Francisco, CA, USA, 1988; pp. 239–246.
44. Armstrong, J.T. CITZAF: A package of correction programs for the quantitative electron microbeam X-ray analysis of thick polished materials, thin films, and particles. *Microbeam Anal.* **1995**, *4*, 177–200.
45. Kisch, H.J. Development of slaty cleavage and degree of very-low-grade metamorphism: A review. *J. Metamorph. Geol.* **1991**, *9*, 735–750. [CrossRef]
46. Merriman, R.J. Clay minerals and sedimentary basin history. *Eur. J. Mineral.* **2005**, *17*, 7–20. [CrossRef]
47. Bozkaya, Ö.; Gürsu, S.; Göncüoglu, M.C. Textural and mineralogical evidence for a Cadomian tectonothermal event in the eastern Mediterranean (Sandıklı-Afyon area, western Taurides, Turkey). *Gondwana Res.* **2006**, *10*, 301–315. [CrossRef]
48. Kübler, B. Evaluation quantitative du métamorphisme par la cristallinité de l'illite. *Bull. Centre Rech. Pau-SNPA* **1968**, *2*, 385–397.
49. Sassi, F.P.; Scolari, A. The b_0 value of the potassic white micas as a barometric indicator in low-grade metamorphism of pelitic schists. *Contrib. Mineral. Petrol.* **1974**, *45*, 143–152. [CrossRef]
50. Guidotti, C.V.; Sassi, F.P. Classification and correlation of metamorphic facies series by means of muscovite b0 data from low-grade metapelites. *Neues Jahrb. Mineralogie. Abh.* **1986**, *153*, 363–380.
51. Weaver, C.E.; Pollard, L.D. The chemistry of clay minerals. In *Developments in Sedimentology*, 15th ed.; Elsevier: Amsterdam, The Netherlands, 1973; p. 272.
52. Bailey, S.W. Summary of recommendations of AIPEA nomenclature committee on clay minerals. *Am. Mineral.* **1980**, *65*, 1–7. [CrossRef]
53. Wiewióra, A.; Weiss, Z. Crystallochemical classifications of phyllosilicates based on the unified system of projection of chemical composition: II. The chlorite group. *Clay Miner.* **1990**, *25*, 83–92. [CrossRef]
54. Hayes, J.B. Polytypism of chlorite in sedimentary rocks. *Clays Clay Miner.* **1970**, *18*, 285–306. [CrossRef]
55. Foster, M.D. Interpretation of the composition and a classification of the chlorites. *US Geol. Surv. Prof. Pap.* **1962**, *414*, 1–33.
56. Curtis, C.D.; Hughes, C.R.; Whiteman, J.A.; Whittle, C.K. Compositional variations within some sedimentary chlorites and some comments on their origin. *Mineral. Mag.* **1985**, *49*, 375–386. [CrossRef]
57. Zane, A.; Weiss, Z. A procedure for classification of rock-forming chlorites based on microprobe data. *Rend. Fis. Acc. Lincei.* **1998**, *9*, 51–56. [CrossRef]
58. Zane, A.; Sassi, R.; Guidotti, C.V. New data on metamorphic chlorite as a petrogenetic indicator mineral, with special regard to greenschist facies rocks. *Canad. Mineral.* **1998**, *36*, 713–726.
59. Meunier, A.; Velde, B. *Illite. Origins, Evolution and Metamorphism*; Springer: Berlin, Germany, 2004; p. 286.
60. Guidotti, C.V. Micas in metapelitic rocks. In *Micas. Reviews in Mineralogy 13*; Bailey, S.W., Ed.; Mineralogical Society of America: Washington, DC, USA, 1984; pp. 357–467.
61. Parry, W.T.; Downey, L.M. Geochemistry of hydrothermal chlorite replacing igneous biotite. *Clays Clay Miner.* **1982**, *30*, 81–90. [CrossRef]
62. Veblen, D.R.; Ferry, J.M. A TEM study of the biotite-chlorite reaction and comparison with petrologic observations. *Am. Mineral.* **1983**, *68*, 1160–1168.
63. Eggleton, R.A.; Banfield, J.F. The alteration of granitic biotite to chlorite. *Am. Mineral.* **1985**, *70*, 902–910.
64. Robinson, D.; Nicholls, R.A.; Thomas, L.J. Clay mineral evidence for low-grade Caledonian and Variscan metamorphism in south-western Dyfed, South Wales. *Mineral. Mag.* **1980**, *43*, 857–863. [CrossRef]
65. Merriman, R.J. Clay mineral assemblages in British Lower Paleozoic mudrocks. *Clay Miner.* **2006**, *41*, 473–512. [CrossRef]
66. Árkai, P. Very low- and low-grade Alpine regional metamorphism of the Paleozoic and Mesozoic formations of the Bükkium, NE-Hungary. *Acta Geol. Hung.* **1983**, *26*, 83–101.
67. Árkai, P.; Abad, I.; Nieto, F.; Németh, T.; Horváth, P.; Kis, V.K.; Judik, K.; Jiménez-Millán, J. Retrograde alterations of phyllosilicates in low-grade metapelite: A case study from the Szendrő Paleozoic, NE-Hungary. *Swiss J. Geosci.* **2012**, *105*, 263–282. [CrossRef]

minerals

MDPI

Article

Kaolinite-to-Chlorite Conversion from Si,Al-Rich Fluid-Origin Veins/Fe-Rich Carboniferous Shale Interaction

Franck Bourdelle [1,*], Michel Dubois [1], Emily Lloret [1], Cyril Durand [1], Ahmed Addad [2], Schéhérazade Bounoua [1], Sandra Ventalon [3] and Philippe Recourt [3]

[1] Univ. Lille, IMT Lille Douai, Univ. Artois, Yncrea Hauts-de-France, ULR 4515-LGCgE, Laboratoire de Génie Civil et géo-Environnement, F-59000 Lille, France; michel.dubois@univ-lille.fr (M.D.); emily.lloret@univ-lille.fr (E.L.); cyril.durand@univ-lille.fr (C.D.); cheche59@sfr.fr (S.B.)

[2] Univ. Lille, CNRS, INRAE, Centrale Lille, UMR 8207-UMET, Unité Matériaux et Transformations, F-59000 Lille, France; ahmed.addad@univ-lille.fr

[3] Univ. Lille, Univ. Littoral Côte d'Opale, CNRS, UMR 8187-LOG, Laboratoire d'Océanologie et de Geosciences, F-59000 Lille, France; sandra.ventalon@univ-lille.fr (S.V.); philippe.recourt@univ-lille.fr (P.R.)

* Correspondence: franck.bourdelle@univ-lille.fr

Abstract: The kaolinite-to-chlorite conversion is one of the chloritization processes that occurs in low temperature diagenetic and hydrothermal systems. The mechanism of this mineralogical transformation is still under discussion, since direct transformation, conversion via berthierine as intermediate phase or direct formation of berthierine/chlorite mix, either by dissolution-crystallization or by solid state transformation (or a combination of both), are all hypotheses put forward. In this context, each description of a kaolinite-to-chlorite conversion occurrence becomes an opportunity to shed new light and to renew this debate. Studying Carboniferous shale–crosscut by large quartz-kaolinite veins–from the mining basin of the North of France, we report therefore an uncommon kaolinite-Fe-rich chlorite assemblage. This assemblage appears as a chlorite fringe 20 μm wide along the interfaces between the shale and the quartz-kaolinite veins. All petrographical, mineralogical and chemical data suggest that the Fe-chlorite results from the interaction between the shale, providing the Fe,Mg supply, and the Si,Al-rich veins, leading to the chloritization of the kaolinite at a small scale via at least one dissolution-recrystallisation step. High-resolution observations highlight that neoformed Fe-rich chlorite contains some 7Å isochemical layers, as relict of berthierine. Therefore, we advance that the conversion takes place either through the precipitation of berthierine following by a second step involving solid state berthierine-chlorite conversion, or through the direct precipitation of a chlorite-rich/berthierine-poor mix driven by the Fe/(Fe + Mg) ratio, at low temperature and in reducing conditions. The comparison of our data with the recent literature allows to prefer the second hypothesis.

Keywords: Fe-rich chlorite; kaolinite-to-chlorite conversion; berthierine; Carboniferous shale

1. Introduction

Chlorite, a 2:1:1 phyllosilicate crystallizing from 80 °C [1] up to 600 °C [2], is ubiquitous in diagenetic, hydrothermal and metamorphic rocks. Considering that the chlorite composition depends in part on the pressure-temperature conditions [3], chlorite is widely used as an indicator mineral and provides the basis of many geothermobarometers (e.g., [4–12]). However, the chlorite thermometry application in diagenetic and hydrothermal contexts is debated, especially because low-T chlorites (i.e., T < 350 °C) are often considered metastable, or contaminated by metastable phases (e.g., [13–16]). In fact, chlorite formation, and more widely clay mineral transformation, may depend on kinetic factors. This point is moreover coupled to the complexity of the conversion pathways of a precursor into chlorite, and more widely, of the chlorite formation processes.

Whether in hydrothermal or diagenetic systems, chlorite can be formed by: (i) direct precipitation (DP) from a supersaturated solution-more frequent in hydrothermal context- or by (ii) conversion of a precursor, often 1:1 or 2:1 phyllosilicates, either via a solid-state transformation (SST), either via dissolution-crystallization process (DC) ([15] and references therein). Mineral conversions driving to the chlorite formation regularly highlighted are the kaolinite–AlMg-chlorite conversion (1:1 dioctaedral conversion; e.g., [17]) via DC, the berthierine–Fe-chlorite (1:1 tricoctaedral conversion; e.g., [18]) mainly via SST, the saponite–Mg-chlorite (2:1 trioctaedral conversion; e.g., [16]) via SST or/and DC, and the montmorillonite–chlorite (2:1 dioctaedral conversion; e.g., [19]) via SST or/and DC. The diversity of precursors and reaction schemes (chloritization processes or direct precipitation) which can lead to chlorite formation, is one of the key elements of the low-T chlorite thermometry applicability. Thus, it is crucial to understand the conditions and mechanisms of the precursor-to-chlorite transformation.

The formation of chlorite from serpentine or smectite is already covered by the literature, but the kaolinite-to-chlorite conversion is less documented. Few studies report two conversion schemes at low-T: the kaolinite-to-sudoite–a di-trioctahedral structure containing a large amount of vacancies (until 1 apfu)–conversion (e.g., [17,20]) mainly in Mg systems, and the kaolinite-to-Fe-rich chlorite conversion, often via berthierine (e.g., [21]). Concerning the latter, several hypotheses have been put forward, like the direct formation of chlorite from kaolinite, a conversion of kaolinite into berthierine followed by the conversion of berthierine into chlorite via DC and/or SST, and the precipitation of a berthierine/chlorite mix from kaolinite via DC. In this context, each occurrence of the kaolinite-to-Fe-rich chlorite conversion must be studied in order to provide new information likely to tip the balance in favor of one or the other of hypotheses, or to demonstrate that all hypotheses can coexist, depending on the P-T-X conditions.

Here, we describe the formation of Fe-rich chlorite with kaolinite as primary precursor, at the interface between clay and oxide-rich system and quartz-kaolinite veins in a Carboniferous black shale from the North of France. This spectacular kaolinite-chlorite assemblage is characterized through a petrological and geochemical study, while the conversion process is discussed on the basis of crystallochemical data, from SEM-EDX, FIB/TEM-EDX and FIB/STXM-XANES analysis. Fluid inclusion thermometry is also applied on quartz from veins, to constrain the temperature of the primary fluid circulation.

2. Materials and Methods

2.1. Geological Setting and Sampling

The studied samples are Carboniferous "black shales" in a generic sense, or a "black schist" in an industrial sense, from the mining basin of the Northern France. The geological unit including these "shales" was put in place during the Westphalian (Pennsylvanian) epoch and corresponds to an alternation of black shales, black sandy shales, black siltstone and shaly sandstones, with several thin coal seams (Figure 1). After a long period of burial, deformation, fracturing and erosion, the Carboniferous unit was covered in unconformity by Cretaceous and Paleogene sediments (marl, limestone and sands), and currently no crops out. However, the past mining activity, related to the coal extraction during the 20th century, led to bring shale samples to the surface, piled up as spoil tips. Studied samples were taken from the spoil tip located at Ostricourt (Figure 1) resulting from the mining via a well drilled to 480 m (*fosse 7-7bis*), and correspond to the most abundant of the rock component of the coal-waste heap in the Coal Basin of the Northern France (Figure 2; [22]).

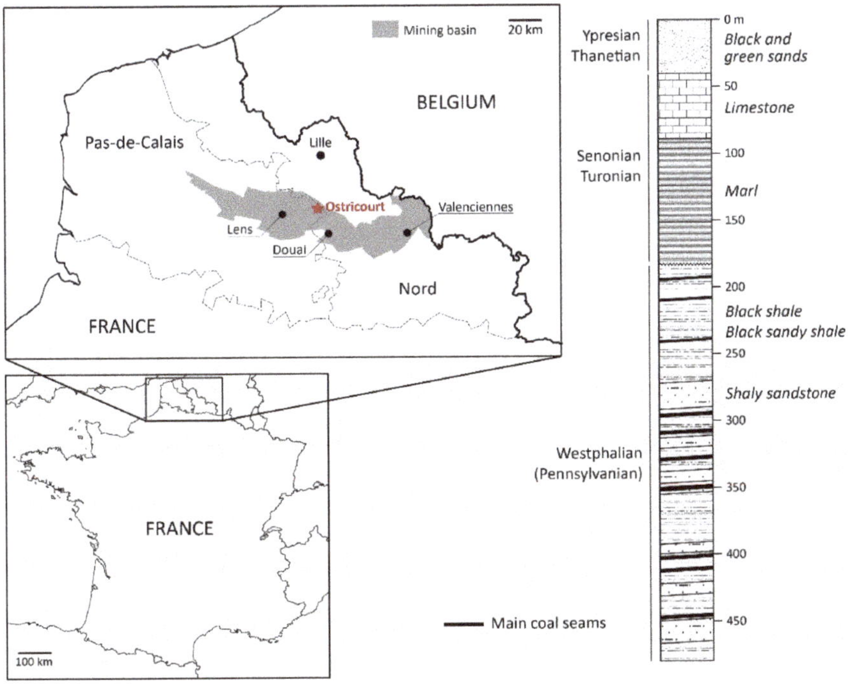

Figure 1. Location of the mining basin of the Northern France, and of the Ostricourt well (*fosse 7-7bis*) and spoil stip. Stratigraphic log, established from H.N. Groupes d'Oignies drilling report.

2.2. Chlorite Analysis

Scanning electron microscopy (SEM) images and Energy Dispersive X-ray spectroscopy (EDX) analysis were carried out on thin section cut across the shale, using a QUANTA 200 SEM instrument (FEI, Hillsboro, OR, USA; LOG laboratory, Université de Lille, Lille, France) operating at 15 kV with a 1.5 nA current intensity, and coupled to a calibrated-on-standards EDX probe.

Chemical analyses of chlorite were carried out with a CAMECA SX 100 electron microprobe (EMP, Cameca, Gennevilliers, France) equipped with wavelength dispersive spectrometers (WDS). The equipment belongs to the Chevreul Institute (Michel-Eugène Chevreul Institute IMET; CNRS Research Federation-FR2638) in Lille (France). The analytical conditions were the following: 15 nA, 15 kV, spot size of 1 μm, counting time of 20 s/element. The analysed elements were Si, Al, Fe, Mg, Mn, Ti, Ca, Na and K. The system was calibrated with albite for Na, orthoclase for Al and K, wollastonite for Ca and Si, MgO for Mg, $MnTiO_3$ for Ti and Mn, and Fe_2O_3 for Fe. Prior to analyse, thin sections were coated with carbon. Only analyses with an oxide weight % summing between 84.00 and 89.00 were considered as belonging to the chlorite group.

The area of interest, i.e., through kaolinite/chlorite interface, was cut with a focused ion beam (FIB). The FIB foils were taken using the FEI strata DualBeam 235 instrument (FEI, Hillsboro, OR, USA) of the Institut d'Électronique, de Microélectronique et de Nanotechnologie (IEMN, Lille, France), and cut out with a size of approximately 15 μm by 5 μm and a thickness around 100 nm. To check the preservation of the mineral crystalline structure after FIB-milling, lattice-fringe imaging was systematically carried out with a transmission electron microscope (TEM), following the protocol of Bourdelle et al. [23].

Observations and analyses were performed on a Tecnai G2 20 model transmission electronic microscope (FEI, Hillsboro, OR, USA; Plateforme de Microscopie Electronique (PMEL)–IMET, Lille, France) operating at 200 kV and equipped with a CCD camera and an EDX spectrometer (analysis with a counting time of 30 s for a dead time less than 10% and a beam defocused at 50 nm).

X-ray absorption near-edge structure spectroscopy (XANES) data were acquired on the PolLux beamline (beamline characteristics in [24]) at the Swiss Light Source (Paul Scherrer Institute–SLS, Villigen, Switzerland). During data collection, the SLS synchrotron storage ring operated at 2.4 GeV and 400 mA current in a top-up mode. The XANES spectroscopy was carried out with a scanning transmission X-ray microscope (STXM), allowing to collect spectrum on each pixel of a "sample image" and to evaluate the variation in the Fe^{3+}/Fe_{total} ratio with an uncertainty of around 5% (probably less for well-defined spectra). Spectra and stacks were obtained over the 700–730 eV energy range (Fe $L_{2,3}$-edges), using a circularly polarized light and a dwell time per pixel and energy point of 10 ms. Data are processed using the aXis2000 software [25]. The procedure for the Fe^{3+}/Fe_{total} ratio determination is detailed in Bourdelle et al. [26].

2.3. Fluid Inclusions and Microthermometry

Fluid inclusions (FIs) are relics of paleo-fluids trapped inside minerals during their growth. The study of FIs allows characterising the paleo-fluid by determining its composition and the minimum pressure-temperature conditions during its entrapment. The validity of fluid inclusion analyses relies on three assumptions: (i) the inclusions trapped a single, homogeneous phase; (ii) the volume of the inclusions has not changed since entrapment; (iii) the composition of the inclusions remains unchanged since it was trapped [27]. In this study, fluid inclusion analyses were performed at the Laboratory of Civil Engineering and geo-Environment (LGCgE, University of Lille, Lille, France), using a Fluid, Inc. microthermometric USGS-type stage (Fluid, Inc., Oakland, CA, USA). The calibration was done by measuring the melting of ice in a pure water bath (0.0 °C) and the melting temperature of CO_2 in synthetic aquo-carbonic fluid inclusions (−56.6 °C). An additional standard was used for low-temperature phase measurements (H_2O-KCl eutectic temperature, −10.7 °C). The maximum uncertainty above room temperature was 0.1 °C.

The "Fluid Inclusion Assemblage" (FIA) approach was used [28]. This method consists in gathering FIs all formed contemporaneously, by using only petrographic criteria. According to Goldstein and Reynolds [28], all the inclusions of a given FIA must have the same composition, density and homogenization temperatures (Th). A FIA is typically composed of 15 to 20 inclusions. This statistical approach allows to eliminate data that diverge more than 10–15 °C from the average value. Salinity based on the temperature of ice melting (Tmi) were calculated using the equation of Bodnar [29].

3. Results

3.1. Petrographical Observations

The petrological study is based on optical microscope and SEM observations, presented in Figures 2–7.

The black shale consists of quartz, mica, smectite, kaolinite, chlorite, iron oxides, titanium oxides, feldspar as major phases. Petrographically it can be defined as a siltstone rich in organic matter. The mineralogy mainly includes rounded or blunted well-sorted quartz grains (granulometry 10 to 50 μm) encased in by a clayey matrix. Pyrites, iron and titanium oxides are identified in the form of framboidal clusters or covering large areas. All these mineral phases are similar to those previously identified by X-ray diffraction (XRD) by Masalehdani et al. [22].

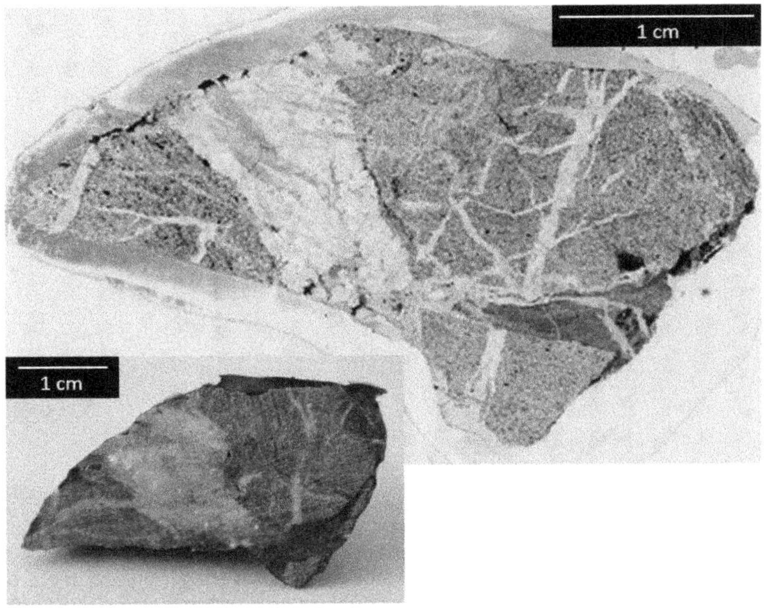

Figure 2. Main petrographic features of the studied sample. General view of the shale sample and of the thin section, on which the veins can be seen in white through the shale in grey (natural light).

The rock is crosscut by numerous fractures of several centimeters in length and 1 to 5 mm wide, without preferential orientation (Figures 2–4). Fractures generate disjunction of fragments in a jigsaw puzzle pattern. Veinlets are filled by a simple mineralogy composed of quartz, calcite and minerals of the kaolinite group. Two veinlet types can be distinguished as regards to the veinlet filling. The first type (type 1) is mainly composed of quartz. Moreover, at least two quartz generations can be distinguished inside. A first generation is composed of anhedral quartz defining a comb texture. This texture (Figure 3A) is characterized by the preferential alignment of parallel to sub-parallel elongated crystals either perpendicular to the veinlet wall [30,31], or with variable orientation relative to the wall. Crystals terminate in open space towards the vein center allowing for development of rhombohedral terminations. Crystals have a uniform size and zonation of crystals is common and attributed to variations in inclusion density, particularly fluid inclusions (Figure 3B). Quartz crystals present an undulose extension, which could testify of a slight deformation. The open space inside veinlets is occupied by a second quartz generation composed of euhedral quartz, micrometric to millimetric in size. Finally, the residual porosity is filled by carbonates (identified as ankerite) and minerals of the kaolinite group (Figure 3C,D, Figures 5 and 6; [32]). These latter form 4 to 10 μm-long vermicular booklets by stacking of thin pseudo-hexagonal plates, which is one of the typical morphology of kaolinite polymorph, and not of dickite [32].

The second veinlet type (type 2) clearly crosscuts both the host rock and the type 1 veinlet network and is composed either of kaolinite with minor euhedral quartz (Figures 4 and 6), or carbonates (ankerite). At the intersection between type 1 and type 2 veinlets a shift of type 1 veinlets is observable (Figure 3D).

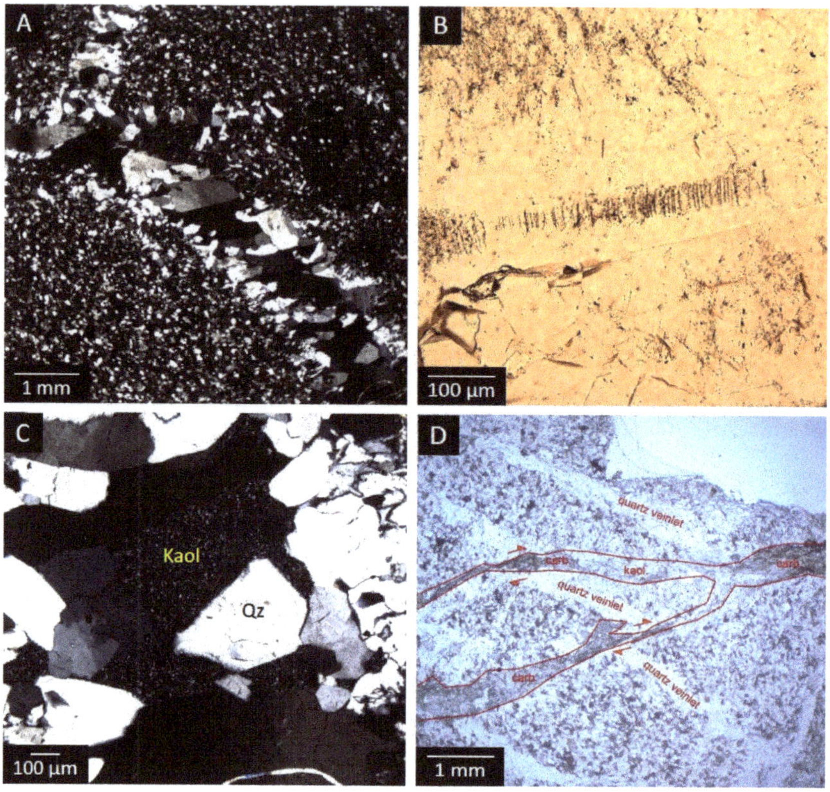

Figure 3. (**A**): general view of the romb quartz texture (crossed polarized light). (**B**): quartz zonation presenting a crystal core with numerous inclusions, particularly FIs (polarized light). (**C**): euhedral quartz crystal (type 1) enrobed with kaolinite in the middle of the vein. (**D**): distribution of the two veinlet types with criteria of relative chronology and displacement (microphotograph).

At the interface between the shale and the kaolinite areas of veins (type 1 and type 2), pleochroic very-light chlorites are observed, with irregular extinction under crossed-polarized light suggesting polycrystalline structures formed by very thin micrometric crystallites (Figures 4–7). These chlorites appear in the form of hair-like bundles, sometimes sparses (Figure 5; start of the chloritization process) along the interface, sometimes contiguous until a fringe 20 µm wide is formed (Figures 4 and 6). However, chlorites are almost exclusively present at the kaolinite/shale contacts (Figures 5A and 6A), rarely when the interface corresponds to euhedral quartz/shale contacts (mainly type 1 veins) or when a kaolinite pocket is isolated from shale by quartz crystals (Figures 3C and 6A). Sometimes chlorite forms intergrowths with kaolinite at the micrometric scale (Figure 6C), but chlorite and kaolinite are often not mixed (and never mixed at the nanometric scale, see below). Rare pockets of kaolinite are also observable into the shale, and partially replaced by chlorite with the same characteristics than the one at the shale/vein interface.

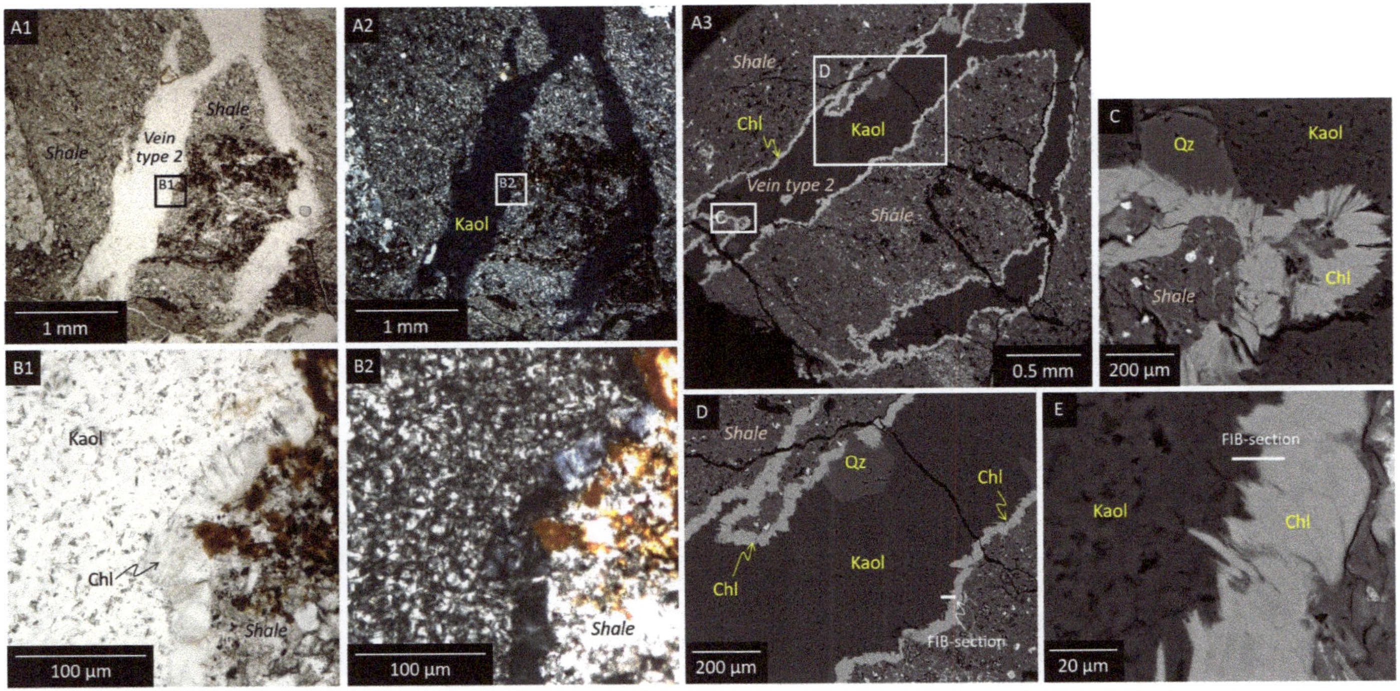

Figure 4. Optical microscope and SEM images of a carboniferous shale sample. (**A**,**B**): images of the large view of the veins/shale contact by plain light microscopy (**A1**,**B1**), crossed-polarized light microscopy (**A2**,**B2**) and SEM (backscattered-electrons) (**A3**), showing kaolinite vein (type 2) running through the shale. A: images highlight the wide expanse of the chlorite fringe along the kaolinite/shale interface. All kaolinite areas are surrounded by chlorite. (**C**,**D**) images are located. (**B**): plain light (**B1**) and crossed-polarized light (**B2**) microscopy images of the vein/shale contact, highlighting the chlorite fringe at the kaolinite/shale interface. (**C**): hairy chlorite seems to grow from the shale to kaolinite. (**D**): chlorite fringe on either side of a kaolinite vein. FIB section is located. (**E**): zoom of (**D**) image detailing the kaolinite/chlorite contact. Qz: quartz; Kaol: kaolinite; Chl: chlorite.

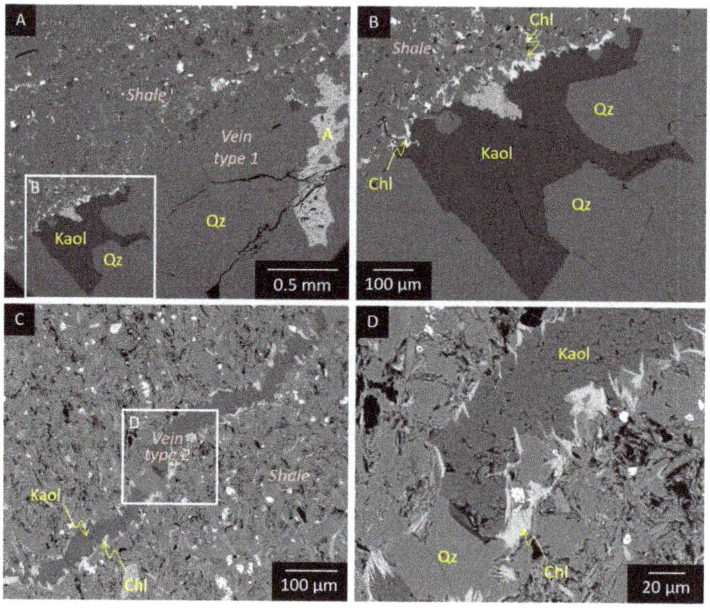

Figure 5. SEM images (backscattered-electrons) of a vein/shale contact (vein type 1). (**A**): large view of a vein/shale contact, showing the millimetric size of euhedral quartz in vein (type 1) and the absence of chlorite fringe at the quartz/shale interface. (**B**) image is located. (**B**): vein/shale contact, focused on the kaolinite/shale interface, underlining the presence of chlorite along this interface, and the occupation by kaolinite of the porosity left by quartz growth in the vein (type 1). (**C**): large view of a vein (type 2)/shale contact. (**D**) image is located. (**D**): vein/shale contact, focused on the start of the chloritization process, with sparse hairy chlorites which do not yet form a fringe. A: ankerite; Qz: quartz; Kaol: kaolinite; Chl: chlorite.

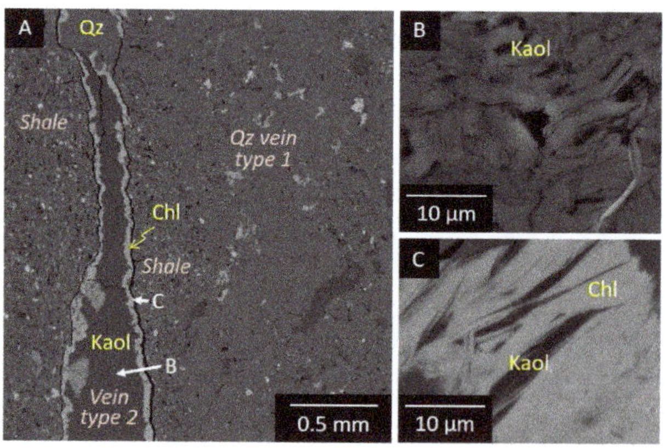

Figure 6. SEM images (backscattered-electrons) of veins/shale contacts. (**A**): On the left of the image, a vein mainly filled by kaolinite–type 2, and surrounded by a chlorite fringe. On the right, a vein mainly filled by quartz–type 1, not presenting chlorite fringe at the contact of the shale in absence of kaolinite/shale contacts. (**B**,**C**) images are located. B: platy crystals of kaolinite. C: kaolinite/chlorite intergrowths. Qz: quartz; Kaol: kaolinite; Chl: chlorite.

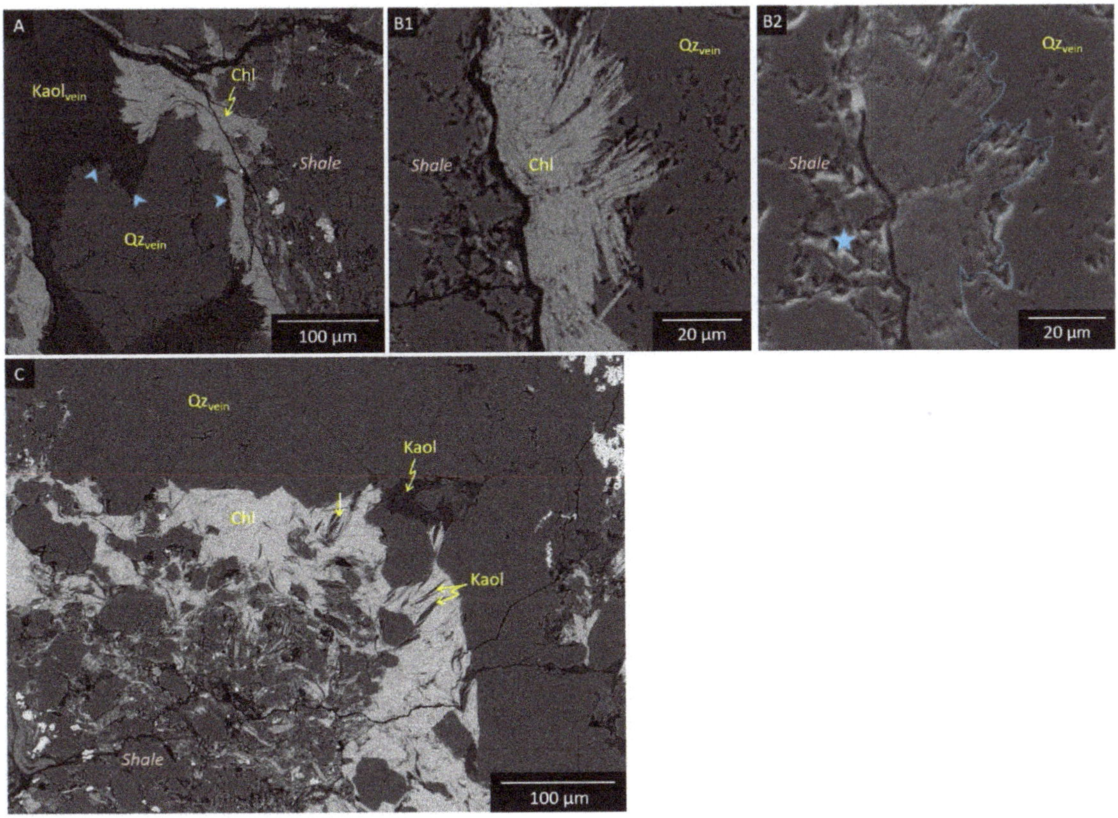

Figure 7. SEM images of veins/shale contacts. (**A**): backscattered-electron image of an euhedral quartz with cracked surfaces surrounded in part by a chlorite fringe. The cracks on the quartz crytsal surfaces are highlighted by blue arrows and are filled by kaolinite. The chlorite grows is stopped by the quartz, but progresses in the kaolinite. (**B1**,**B2**): backscattered and secondary electron images, respectively, of chlorite growing at the expense of the kaolinite nestled in quartz cracks. Kaolinite is no longer visible, but the cracks, whose limits are underlined by a blue dashed line on SE image, contain the neoformed chlorite. The blue star notifies in SE image an example of cracked area in quartz. (**C**): a rare–large–area where the chlorite has almost completely replaced kaolinite (noted residual kaolinite in chlorite fringe), surronding, bypassing or following the euhedral quartz. Qz: quartz; Kaol: kaolinite; Chl: chlorite.

In addition to Figures 4–6, Figure 7 allows one to confirm the anteriority of the quartz-kaolinite precipitation over the chlorite formation. In Figure 7A, the chlorite clearly bypasses the euhedral quartz of the vein, taking therefore locally a direction parallel to the vein/shale interface, to significantly grow trough the kaolinite (top of the image). The chlorite follows the cracked surfaces of the quartz, but textural insights do not allow to consider that the quartz is posterior to the chlorite, or encloses chlorite hairs. Sometimes, it might appear that the chlorite is "encapsulated" in quartz, but a focus on the quartz/chlorite interface (Figure 7B1,B2) indicates that the chlorite only replaces kaolinite which is nestled in quartz cracks. Observing the rare areas where the chlorite is developed beyond a 20 µm fringe, and almost completely replaced kaolinite (Figure 7C), we can identify: (i) residual kaolinite in the chlorite-the reverse being never found, (ii) a chlorite growth delimited by quartz faces, i.e., chlorite surronding, bypassing or following the euhedral quartz. Moreover, chlorite was found along the two types of veins, whereas they have different mineralogical assemblages (two generations of quartz, with one as large euhedral quartz and few kaolinite areas for veins type 1, and mainly kaolinite with few small quartz for veins type 2). By adding the Figure 5A observations where chlorite formation is restricted to the kaolinite areas of the veins (and not in front of quartz), all microstructures are in favor of a kaolinite → chlorite sequence, and not of a late quartz-kaolinite infill of a chlorite-lined fractures.

3.2. Fluid Inclusions

Fluid inclusions were observed in comb quartz crystals of the type 1 veinlets. The crystal cores generally present numerous FIs defining a cloudy aspect, whereas quartz borders are limpid and inclusions are scarce (Figure 8A). Detail examination shows that in crystal cores, inclusions are distributed along parallel tight alignments (Figure 8B) or as an unoriented cloud (Figure 8A). As the FI distribution can be related to growth of the crystal (concentric distribution, alignments perpendicular to the growth direction thus parallel to faces), these features indicate a primary origin [33]. FIs are very tiny, less than 5 µm in size and their shape is irregular, complex, often presenting necking down; they are therefore defined as immature [27]. At a FIA scale, defined as cluster inside the crystal core, FIs are predominantly one-phase and filled by liquid water. Very few inclusions present a vapour bubble at room temperature, sometimes rapidly moving (Figure 8B). The sample was stored one month in a deep freezer in order to try to bring about the nucleation of a vapour bubble during cooling. Additionally, inclusions were cooled down to −30 °C (just above the temperature of metastable water solidification) in the microthermometric stage and kept a few minutes at this temperature in order to induce bubble nucleation. All attempts (deep freezer and cooling down to −30 °C) were however unsuccessful.

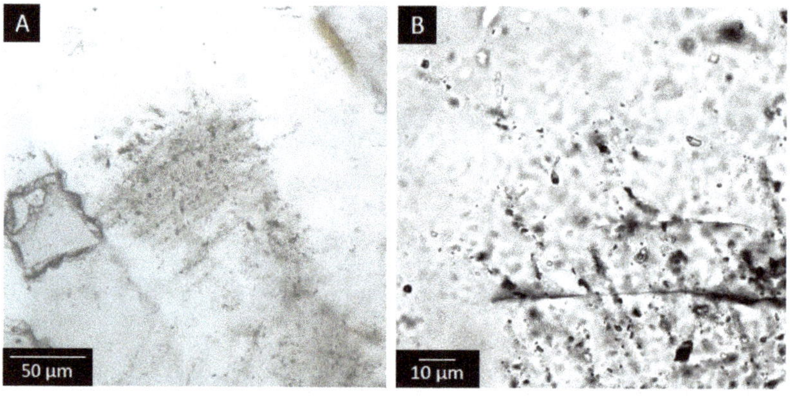

Figure 8. Fluid inclusion petrography. (**A**): Microphotograph of the cloudy core of a comb quartz crystal of vein type 1. (**B**): Typology of fluid inclusions inside a Fluid Inclusion Assemblage.

Experiments were performed on the thin sections used for petrographic observations. Indeed, the very small size of inclusions requires the use of thinner sections that the thick ones classically used in fluid inclusion studies. After precise location and mapping of Fluid Inclusion Assemblages, the thin section was cut using a putty knife into fragments of about 1 cm^2 in order to be introduced in the chamber of the microthermometric stage.

Ice melting temperatures (Tmi) were only measured in two-phase inclusions as Tmi can only be interpreted when at equilibrium with ice. Bubble nucleation in numerous inclusions was induced by inclusion freezing down to −50 °C, due to volume change and inclusion wall deformation during crystallization of ice. 17 values were kept in 2 distinct FIAs. All data lie in the range −1.6 to + 0.1 °C (Figure 9), with a mean value of −0.7 ± 0.5 °C. This corresponds to a value of salinity of 1.2 ± 0.8 wt% eq. NaCl.

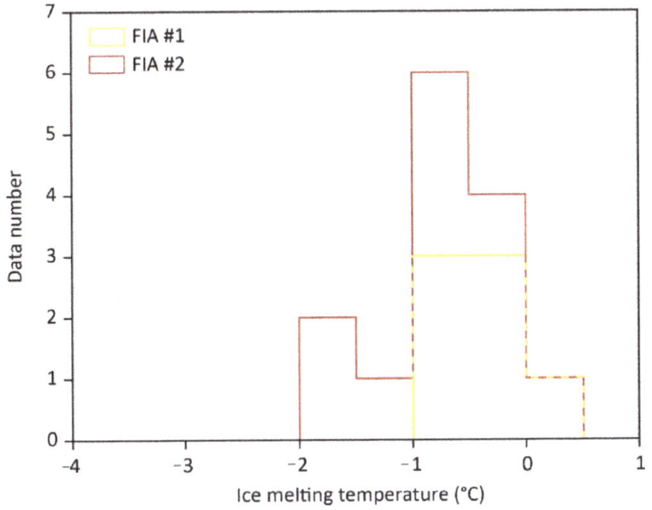

Figure 9. Frequency plot of ice melting temperature in 2 Fluid Inclusion Assemblage (FIAs) in quartz of type 1 veinlets.

Before any freezing run, the very rare two-phase inclusions (only 3) were heated in order to measure Th. Three values were obtained (58, 95 and 118 °C).

3.3. Chlorite Composition and Structure

3.3.1. SEM-EDX, EMPA and TEM-EDX Data

SEM-EDX, EMPA and TEM-EDX data were used to calculate the chlorite structural formulas on the basis of 14 O atoms per half formula unit (apfu). These data show the high homogeneity–at all scales–of the composition of the chlorite formed at the shale/vein interface (Table 1; Figures 10 and 11). Si is quite substituted by Al, i.e., Si ~2.7 atoms per formula unit, underlining the impact of the Tchermak substitution from a Si-pure end-member. This chlorite approaches tri-trioctaedral characteristics, with an R^{3+} content (Figures 10 and 11) around 30% of octahedral cations (~1.7 apfu), and consequently 70% of R^{2+} (~4 apfu), mainly–but not exclusively–due to the di-trioctahedral vector. The Fe/(Fe+Mg) ratio is high, ~0.7, bringing the composition of the considered chlorite between chamosite (daphnite) and Fe-amesite (Figures 10 and 11).

Table 1. EMPA, SEM-EDX and TEM-EDX analyses of Fe-chlorite$_{vein}$. Apfu; 14 oxygen basis. * analysis with 7Å layers.

	EMPA														
#	1	2	3	4	5	6	7	8	9	10	11	12	13	14	15
Si	2.68	2.59	2.71	2.75	2.69	2.71	2.70	2.76	2.68	2.71	2.65	2.70	2.69	2.71	2.73
Ti	0.00	0.00	0.00	0.00	0.00	0.00	0.00	0.00	0.00	0.00	0.00	0.00	0.00	0.00	0.00
Al	2.99	2.93	2.99	2.97	3.08	3.00	3.00	2.98	3.06	3.01	3.04	2.98	3.04	3.04	3.02
Fe^{2+}	2.89	2.88	2.88	2.91	2.84	2.90	2.88	2.76	2.90	2.83	2.95	2.87	2.87	2.85	2.81
Mn	0.06	0.34	0.01	0.01	0.00	0.01	0.01	0.02	0.00	0.01	0.01	0.01	0.01	0.01	0.01
Mg	1.19	1.20	1.19	1.12	1.15	1.17	1.19	1.20	1.14	1.22	1.19	1.25	1.17	1.15	1.18
Ca	0.00	0.01	0.00	0.00	0.00	0.00	0.00	0.01	0.01	0.00	0.01	0.00	0.00	0.00	0.00
Na	0.00	0.00	0.00	0.00	0.00	0.01	0.00	0.01	0.00	0.00	0.00	0.00	0.01	0.00	0.01
K	0.00	0.00	0.00	0.00	0.00	0.00	0.00	0.01	0.00	0.00	0.00	0.00	0.00	0.00	0.00
Fe/(Fe+Mg)	0.71	0.71	0.71	0.72	0.71	0.71	0.71	0.70	0.72	0.70	0.71	0.70	0.71	0.71	0.70
Σoct	5.81	5.94	5.78	5.76	5.76	5.79	5.78	5.72	5.78	5.78	5.84	5.81	5.78	5.76	5.75

Table 1. *Cont.*

	SEM-EDX							TEM-EDX							
#	16	17	18	19	20	21	22	23	24	25	26	27	28 *	29 *	30 *
Si	2.66	2.72	2.76	2.84	2.65	2.74	2.74	2.78	2.66	2.67	2.71	2.71	2.75	2.73	2.71
Ti	0.00	0.00	0.00	0.00	0.00	0.00	0.00	0.00	0.00	0.00	0.00	0.00	0.00	0.00	0.00
Al	3.07	2.98	3.03	3.12	3.11	3.01	3.09	3.09	3.04	3.02	3.02	3.05	3.00	3.03	2.99
Fe^{2+}	2.92	2.90	2.81	2.63	2.94	2.78	2.78	2.71	3.00	2.86	2.87	2.84	2.84	2.81	2.87
Mn	0.00	0.01	0.01	0.01	0.00	0.01	0.01	0.01	0.01	0.01	0.01	0.00	0.01	0.01	0.01
Mg	1.15	1.16	1.11	1.00	1.11	1.21	1.08	1.08	1.09	1.25	1.17	1.16	1.15	1.18	1.19
Ca	0.00	0.01	0.01	0.00	0.00	0.01	0.01	0.00	0.01	0.00	0.00	0.00	0.00	0.00	0.00
Na	0.00	0.01	0.01	0.00	0.00	0.00	0.00	0.00	0.00	0.00	0.00	0.00	0.00	0.00	0.00
K	0.00	0.00	0.00	0.00	0.00	0.00	0.00	0.00	0.00	0.00	0.00	0.00	0.00	0.00	0.00
Fe/(Fe+Mg)	0.72	0.71	0.72	0.72	0.73	0.7	0.72	0.72	0.73	0.70	0.71	0.71	0.71	0.70	0.71
Σoct	5.8	5.77	5.72	5.6	5.81	5.75	5.7	5.67	5.8	5.81	5.78	5.76	5.75	5.76	5.77

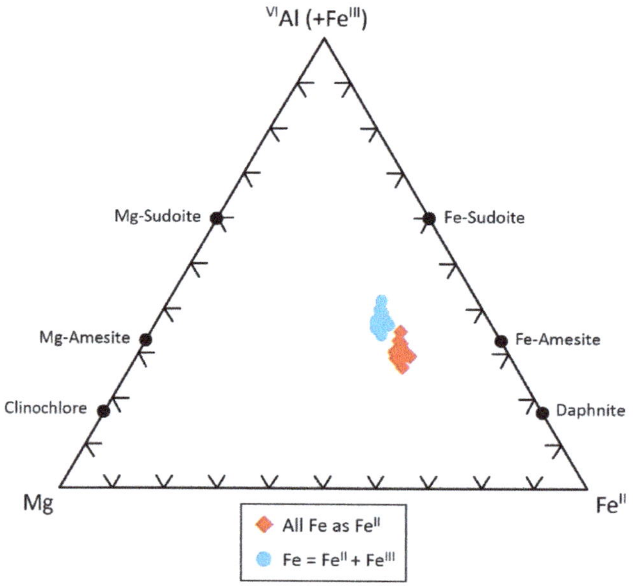

Figure 10. Compositions of analysed chlorites (SEM-EDX, EMPA and TEM-EDX data) in a ^{VI}Al-Mg-Fe^{II} ternary diagram. Chlorite end-members are indicated.

3.3.2. XFe^{3+} Quantification by STXM-XANES on FIB-Section

Estimate of the chlorite Fe^{3+} content, i.e., the Fe^{3+}/Fe_{total} ratio, was performed via STXM-XANES analysis, at the Fe $L_{2,3}$-edges (Figure 12). STXM-XANES were carried out on a FIB section cutting across the kaolinite/chlorite interface (Figure 12B; location on Figure 4D,E). Fe $L_{2,3}$-edge spectra display four major peaks, two for the L_3-edge and noted L_3-a (~708.7 eV) and L_3-b (~710.2 eV), and two for the L_2-edge and noted L_2-a (~720 eV) and L_2-b (~723.5 eV). L_3-a and L_2-a peaks are mainly affected by the Fe^{2+} content, while the L_3-b and L_2-b peaks are by the Fe^{3+} content (Figure 12A). Here, the estimation of Fe^{3+}/Fe_{total} ratio is based on the calibrated L_3-b/L_3-a peak intensity ratio. Spectra were acquired on several areas of the FIB section, into the chlorite, symbolized by squares on the Figure 12C. Rescaled spectra are compared in the Figure 12A, and Fe^{3+}/Fetotal ratio were calculated following the Bourdelle et al. [26] protocol. All spectra have a similar shape: a high L_3-a peak and a small L_3-b peak, indicating a ferrous phase.

All Fe^{3+}/Fe_{total} ratio estimates are around 14% +/−1, showing the homogeneity of the chlorite composition even at the nanoscale, from the contact with kaolinite to the contact with the shale. Combined with SEM-EDX, EMPA and TEM-EDX data, the mean structural formula for this chlorite is:

$$(Fe^{II}_{2.39}Mg_{1.14})(Fe^{III}_{0.42}Al_{1.65})(Si_{2.67}Al_{1.33})O_{10}(OH)_8.$$

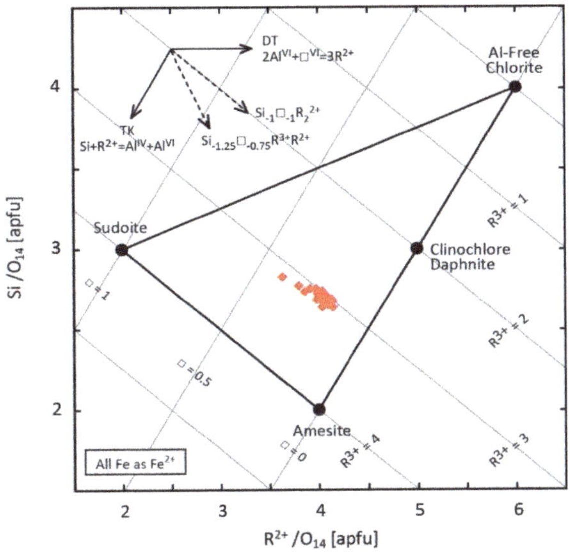

Figure 11. Chlorite chemical compositions (chlorite of the veins/shale interface) in R^{2+}–Si diagram [34]. R^{3+} refers to trivalent cations (Al and Fe^{3+}), □ to octahedral vacancies and R^{2+} to divalent cations (Fe^{2+} and Mg). TK and DT refer to the Tschermak (Si + R^{2+} = ^{IV}Al + ^{VI}Al) and the di-trioctahedral (2 ^{VI}Al + ^{VI}□ = 3 R^{2+}) exchanges, respectively.

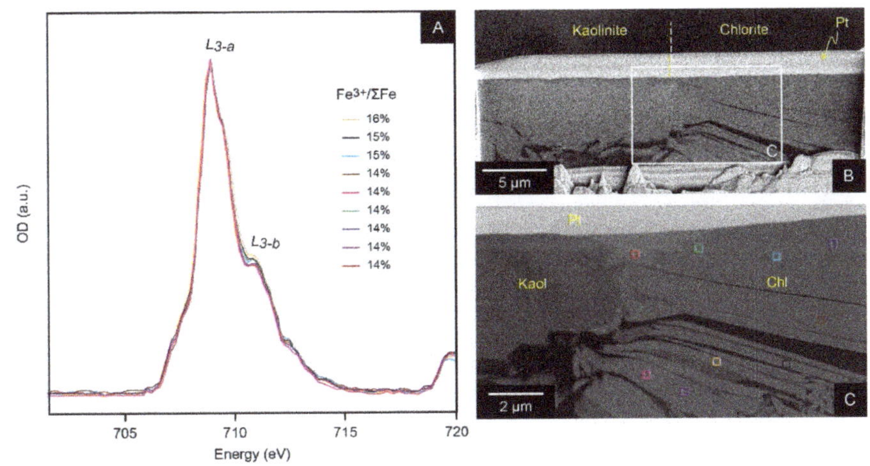

Figure 12. (**A**): Representative X-ray absorption spectra of chlorites, measured in transmission mode (STXM, SLS) at the Fe L_3-edge. Based on the protocol of Bourdelle et al. [26]. The calculated average Fe^{3+}/Fe_{tot} ratio is ~14% for all analysed chlorites. (**B**): SEM image of the FIB-section cut through the kaolinite/chlorite contact. (**C**) image is located. Pt: platinum, added before cutting the FIB-section to protect it. (**C**): location of the STXM-XANES analyses into chlorite, each square referring to one spectrum.

3.3.3. TEM Observations on FIB-Section

The nature of the phase was verified using TEM on the FIB section, by high-resolution images and electron diffraction on selected area. The chlorite fringe was specifically observed to ensure a 14 Å structure, and not a 7 Å structure. Therefore, a 14 Å layer structure is clearly identified, as well as by high-resolution observation than by electron diffraction (Figure 13A,B). The layered structure is regular, with very rare 7 Å layers which present the same composition (in term of Si, Al, Fe, Mg contents and Fe^{3+}/Fe_{total} ratio) as the 14 Å layers (Figure 13C). These berthierine-like layers visually represent less than 10% of the structure. In the same way, the platy crystallites of kaolinite have a regular, free from any defect, 7 Å layer structure (Figure 13C). Although kaolinite and chlorite sometimes form intergrowths at the micrometric scale (as on Figure 6C), the interface between the two phases at the nanoscale is clear, there is no continuity observed between kaolinite and chlorite layers whatever the orientation of the crystallites, and no chemical zonation on either side of the interface. The kaolinite composition was checked by EMPA and TEM-EDX, confirming a pure kaolinite without compositional variation (Table 2).

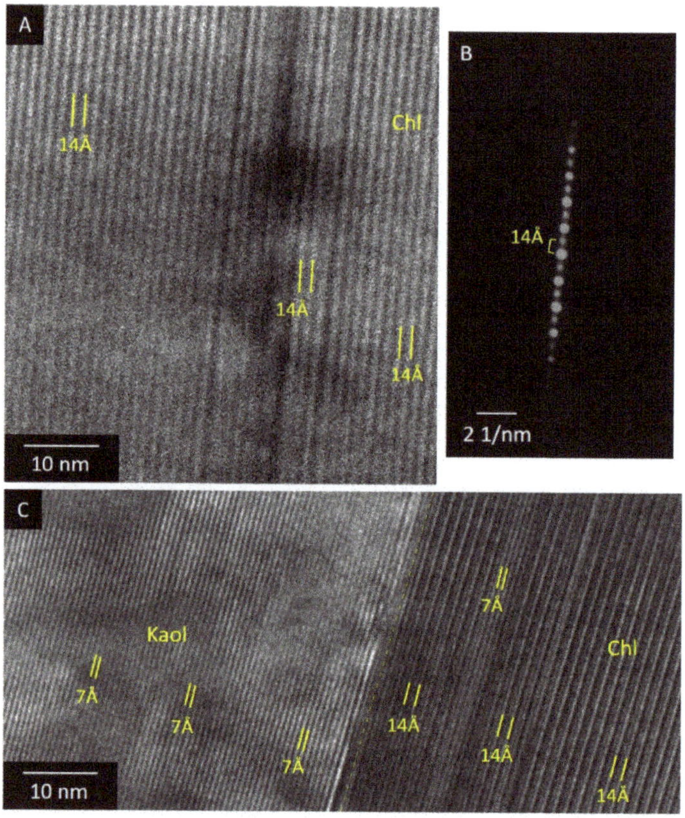

Figure 13. TEM images obtained on the FIB section. (**A**): high-resolution image of the chlorite formed at the vein/shale interface. (**B**): electron diffraction obtained on the area of A; 00*l* spots. (**C**): high-resolution of a kaolinite/chlorite contact. Kaol: kaolinite; Chl: chlorite.

Table 2. TEM-EDX analyses of kaolinite. Apfu; 7 oxygen basis.

#	1	2	3
Si	2.01	2.04	2.00
Ti	0.00	0.00	0.00
Al	1.98	1.96	1.99
Fe^{2+}	0.00	0.00	0.00
Mn	0.00	0.00	0.00
Mg	0.00	0.00	0.00
Ca	0.00	0.00	0.00
Na	0.00	0.00	0.00
K	0.00	0.00	0.00

4. Discussion

4.1. Kaolinite Precipitation

The presence of authigenic kaolinite in sandstones or shales is often the result of an early alteration of detrital material, as feldspar or mica (e.g., [35]), due to meteoric flushing in an open system, sometimes favored by acidic fluids (CO_2 or organic acid). In the present case, authigenic kaolinite is quite rare in the shale, but is one of the main phases filling the fractures–in association with quartz. The kaolinite morphology and location suggest direct precipitation from solutions [36], especially since these fractures constitute preferential paths for fluid circulation. In this way, the rock structure indicates that the veinlet formation is due to hydraulic fracturing (fluid-assisted brecciating). The chronological sequence was probably (i) a first quartz generation with a comb texture, (ii) a second quartz generation, represented by euhedral crystals, at the veinlet center, (iii) the precipitation of kaolinite (and locally carbonate like ankerite) in residual voids as well as in thin veinlet crosscutting all previous structures.

All these veins testify to the circulation of an aqueous fluid, confirmed by the presence of nearly pure water fluid inclusions. The fluid was probably charged in elements from the weathering and leaching of feldspar and/or mica at shallow depth/very low temperature before migrating to the depth. The feldspar and/or mica alteration provided the sufficient quantity of Si and Al, but also gave amounts of K and Na. The large fractures in the shale, in addition to facilitate the fluid circulation, led to local high water/rock ratio and fluid flow rates [37], sufficient to remove or to lower the concentration of K and Na, and to make Al mobile. In this way, the salinity determined from the Tmi for fluid inclusions is very low (1.2 ± 0.8 wt% eq. NaCl). Therefore, the system presented here favors the movement of the Si and Al ions to lead to the precipitation of quartz and kaolinite rather than illite.

Boles and Franks [38], also observing veins of coexisting kaolinite and quartz across sandstone and shale in the Eocene Wilcox formation (Texas Gulf Coast, TX, USA), estimate that the process of kaolinite formation occurs at $T < 170$ °C, beyond that, kaolinite is progressively less abundant. Here, fluid inclusions are mainly one-phase. The coexistence of one-phase and rare two-phase inclusions in a given FIA indicates a low formation temperature, probably less than 50 °C [28]. Indeed, one-phase inclusions are typical of low-temperature environments because metastability in this case prevents the nucleation of the vapor bubble. Moreover, fluid inclusions do not present any clues allowing to consider a subsequent significant increase in the temperature. Consequently, the observed kaolinite was formed at $T < 100$ °C, and probably $T < 50$ °C. This is in agreement with the literature reporting the kaolinite formation temperature in sedimentary rocks (e.g., [39]).

4.2. Mechanisms of Kaolinite-to-Chlorite Conversion

Two major mechanisms are involved in the conversion of a precursor into chlorite: the solid-state transformation (SST) and the dissolution-crystallization (DC). In the first case, the precursor is gradually replaced by chlorite from layers to layers in a topotactic transformation. This mechanism leads to a conservation of the volume, the shape and the size of the precursor crystallites. In the second case, the precursor is dissolved to a great extent, and chlorite nucleates and growths as epitaxial or separate crystallites [40]. In this way, the characteristics of the precursor plates, as shape or volume, are lost, and the reaction can involve important changes in chemistry and crystal texture. In our case, chlorite formed at the kaolinite/shale contact does not retain any property of the kaolinite precursor. Indeed, there is no continuity between chlorite and kaolinite crystallites, no relict of kaolinite layers in the chlorite structure, and the chlorite seems to grow as epitaxial crystallites on shale surface toward the kaolinite, at the expense of the latter. It is the same for the observed berthierine-like layers, which do not have the characteristics allowing to consider a transformation of kaolinite into berthierine by SST. So, all structural observations indicate that the kaolinite-to-chlorite conversion involves at least one dissolution-crystallization step.

In Al-rich systems, the kaolinite is transformed into sudoite (a di-trioctahedral MgAl-chlorite) via tosudite (a 1:1 regularly interstratified chlorite-smectite) as an intermediate phase of a conversion series [17]. This conversion may be through stepwise mineral reactions with DC as the main mechanism. In the present case, the kaolinite conversion occurs in a Fe-rich system, but similarly, kaolinite seems to be replaced by Fe-rich tri-trioctahedral-like chlorite via DC. This kind of kaolinite-Fe-rich chlorite assemblage was already observed in less ancient rocks, e.g., in Oligocene until recent sediments of Gulf Coast by Boles and Franks [38] and Burton et al. [21], or in Jurassic sandstones of the Moezian Platform (Northern Bulgaria) by Hrischeva and Gier [36], but indication for a kaolinite dissolution–chlorite replacement was not clear. Boles and Franks [38] highlighted a decrease in kaolinite amount with depth in the Gulf Coast basin in Texas and a counter-variation of the Fe-chlorite amount, and consequently suggested that kaolinite dissolves in favor of chlorite, unfortunately without petrographic evidence. Burton et al. [21], also studying Gulf Coast samples, confirms their hypothesis observing lamellar intermixture of kaolinite and chlorite, and proposed a kaolinite-to-chlorite conversion via berthierine as an intermediate phase. Here, several berthierine-like layers, or small stacks of 7 Å layers, were indeed found in the chlorite phase. Even if these types of chlorite isochemical 7 Å layers are in a very low number, they could be considered as relics of a berthierine-to-chlorite conversion which is almost complete. Berthierine was previously reported in hydrothermally altered shales [41], sandstones (e.g., [42]), arenites [43], or in Palaeogene and Triassic coal measures [44], but the coupling of kaolinite-to-berthierine and berthierine-to-chlorite conversions is in debate. Iljima and Matsumoto [44] argue that all the kaolinite-to-chlorite conversion process occurs by DC transformations: as a first step, kaolinite and siderite reacted to form berthierine, and as a second step, berthierine is converted into chlorite via DC since the two minerals present different chemical compositions. In the same way, Aagaard et al. [42], from the study of experimental and natural samples (in clastic quartz-rich sandstones), claim that chamosite formation from berthierine involved DC, with a chemical composition evolution between the two phases. On the other hand, SST transformation is the process privileged by many authors for various berthierine-to-chlorite conversion contexts (e.g., [45–52]). This hypothesis is often supported by the interpretation of polytype modifications or High-resolution TEM images showing intercalation of 7 Å layers and 14 Å layers. Especially, Banfield and Bailey [48], Xu and Veblen [51] and Ryan and Reynolds [49] suggest a SST mechanism of berthierine-to-chlorite conversion by inversion of half of the tetrahedral layers, i.e., from a TO TO to a TOT O structure. In the present study, the 7 Å layers are few in number, dispersed as small stacks in chlorite domains, and have compositions identical to the 14 Å layers, making two approaches conceivable:

(i) either the dissolution of the kaolinite and the crystallization of berthierine, followed by an advanced chloritization of the berthierine via SST,
(ii) or the dissolution of the kaolinite, and the direct crystallization of a chlorite containing isochemical 7 Å layers as defects.

Since the 14 Å layers are the predominant components, it does not seem possible to decide in favor of either of the hypotheses. However, Inoue and Kogure [53], studying chlorite crystallites from quartz veins of epithermal to xenothermal vein-type ore deposits (Japan), propose that the direct precipitation of chlorite containing 7 Å layers actually corresponds to the neoformation of a 7/14 Å interstratification phase enriched in 14 Å layers. In this case, the authors suggest that the proportion of 7 Å layers is driven by the Fe content and the Fe/(Fe + Mg) ratio. For Arakawa chlorites, which present the same chemical characteristics that the chlorite in the present study in term of Si content, Fe content, Fe/(Fe + Mg) ratio, chemical homogeneity of the 7/14 Å layers, and electron diffraction patterns, Inoue and Kogure [53] determined by HRTEM a proportion of 7 Å layers around 8%. TEM observations made in the present study are in agreement with those of [53], supporting the hypothesis of direct precipitation of chlorite in its most stable form, interlaying with infrequent metastable 7 Å layers observed only very locally.

4.3. Conditions of Chlorite Formation

The transformation of the kaolinite in a clay mineral of "higher" temperature like chlorite was probably triggered by a change in the P-T conditions of the system and/or an increase of Fe,Mg content in the pore-fluid. An increase of the temperature coupled with an increase of iron availability, either due to the burial (with contribution of pore-fluid) or by the circulation of fluids, can explain the dissolution of the kaolinite and the formation of a Fe-rich chlorite. From petrographic observations, this process was only made possible by the contribution of the shale, which provided Fe and Mg, and by the kaolinite as parental mineral. This explains why Fe-rich chlorite is rarely observed at the $quartz_{vein}$/shale contact or elsewhere in the rock excepted in some pockets of kaolinite/chlorite in the shale porosity. The phases in the shale likely to provide Fe and Mg supply are pyrite, oxides and detrital ferromagnesian minerals, by dissolution and/or transformation (as illitization of smectite). Therefore, the system represented by the kaolinite/shale contact has favored, beyond any effect of kinetics, as:

$$\text{kaolinite} + 5.7\ Fe^{2+}{}_{shale} + 2.3\ Mg_{shale} + 10\ H_2O = 2\ \text{chlorite} + 0.25\ SiO_2 + 16\ H^+$$

This type of reaction releases a small amount of silicon and consumes a considerable amount of water. However, the Fe and Mg-two elements deemed to be mobile-transfer seems to be limited, i.e., operates only on short distance, since the kaolinite is largely preserved and is often transformed only over a 20 µm wide fringe near the shale. It is permissible to consider that the chlorite, after needing Fe and Mg from the shale to nucleate and growth, formed a fringe wide and homogeneous enough to reduce the local permeability and to prevent Fe and Mg circulation to the kaolinite further in the vein, to keep kaolinite away from the shale and to stop the reaction. Sometimes, the element or fluid supply is not even enough to reach the fringe step, and only sparse chlorite hairs are formed. The mineral reaction and the new equilibrium are therefore very localized on the vein rims, testifying to a very small-scale system. The high homogeneity of chlorite composition (structural formula and Fe^{3+}/Fe_{total} ratio) despite the large extent of the reaction front, argues in favor of either a rapid mineralogical reaction, or of P-T-X conditions that have been maintained for a time, while the high Fe content, the high Fe/(Fe + Mg) ratio and the relatively low Fe^{3+}/Fe_{total} ratio (14%) of neoformed chlorite suggest reducing conditions. In fact, this chlorite formation only recorded one event of condition changes, since chlorite composition did not reequilibrated afterwards.

The mineral sequence from the literature is roughly kaolinite → berthierine → chlorite with increasing temperature. In the scenario of successive kaolinite-to-berthierine–berthierine-to-chlorite conversions, via DC and SST respectively, the first conversion probably occurs between 50 and 100 °C, according to the P-T-X context. Iljima and Matsumoto [44], studying diagenetic mineral reactions in coal measures, argue that kaolinite-siderite reaction leads to the formation of berthierine between 65 and 130 °C. In arenites, Velde et al. [43] suggest that the kaolinite-to-berthierine conversion takes place up to ~100 °C, similarly to Jahren and Aagaard [54] in clastic reservoir. The rest of the chloritization process, involving a berthierine-to-chlorite conversion via SST or DC, is often reported between 70 and 200 °C, as during burial diagenesis of shales and sands [38,42,44,46,54,55]. Considering a DC mechanism for this conversion, Iljima and Matsumoto [44] found chamosite from 160 °C, while Aagaard et al. [42] envisage that the transition is around 90–100 °C in natural environments.

In the scenario based on Inoue and Kogure [53], the chlorite-containing few 7 Å layers-precipitates directly from kaolinite dissolution fluid, at a temperature which can exceed 190 °C. Without invalidating the precursor character of berthierine in the chloritization process during burial diagenesis, Inoue and Kogure [53] suggest the possibility that berthierine is formed as a metastable phase from solutions at the same time as Fe-rich chlorite, which widens the range of possible temperatures of formation insofar as the berthierine content does not depend on it but on the Fe content. Although (i) the study of Inoue and Kogure [53] does not include low-temperature samples and (ii) the samples studied here do not come from hydrothermal ore deposits like those observed by Inoue and Kogure [53], chlorites described here also precipitate in quartz veins, and also show a high-homogeneous composition. Therefore, a similar direct DC conversion, but at T < 190 °C, of kaolinite into chlorite containing few metastable 7 Å layers is quite possible.

5. Conclusions

Here, we report the formation of Fe-rich chlorite as the result of an interaction between a black shale supplying Fe, Mg and large kaolinite areas formed from a Si,Al-rich fluid, like a small size system leading to a noteworthy mineralogical reaction front. The chloritization of the kaolinite begins with its dissolution, followed by the precipitation either of a chlorite-rich/berthierine-poor mix, or the precipitation of berthierine as intermediate phase. In this last case, a second step—berthierine-to-chlorite conversion—is required, probably through a solid state mechanism given the isochemistry of the observed 14 and 7 Å layers. The first hypothesis, suggests by Inoue and Kogure [53], signifies that the conversion is driven by the local P-T-X conditions, especially the Fe/(Fe+Mg) ratio and eventually the redox conditions, while the second hypothesis rather call to kinetic factors. Although we cannot totally exclude the second hypothesis, the first one is in our favor since we have found only few berthierine layers, and not clearly observed any signs of sheet-to-sheet transformation of berthierine into chlorite. This kaolinite-to-chlorite conversion occurs here at a very localized scale, under specific conditions such as a low-T, an interstitial fluid as transfer vector of Fe and Mg and reducing conditions.

Author Contributions: Conceptualization, F.B., M.D., E.L., and C.D.; analysis, F.B., M.D., E.L., C.D., A.A., S.B., S.V. and P.R.; writing—original draft preparation, F.B.; writing—review and editing, F.B., M.D., E.L. and C.D.; project administration, F.B.; funding acquisition, F.B. and E.L. All authors have read and agreed to the published version of the manuscript.

Funding: This research was funded by Centre National de la Recherche Scientifique (CNRS), France, INSU TelluS SYSTER program, and by Agence Nationale de la Recherche (ANR), France, JCJC ENTRESOL project, and LGCgE funding.

Data Availability Statement: On request from Franck Bourdelle.

Acknowledgments: We acknowledge four anonymous reviewers and the editor for their constructive comments. We are most grateful to the PSI SLS synchrotron, especially Benjamin Watts (PolLux beamline) for technical advice, and to UMET Laboratory (University of Lille). The TEM national facility at Institut Chevreul (Lille, France) is supported by the Conseil Régional du Nord-Pas de Calais, the European Regional Development Fund (ERDF), and the Institut National des Sciences de l'Univers (INSU, CNRS). Thanks are also extended to David Troadec and IEMN (Cité Scientifique, av. Poincaré CS 60069, 59652 Villeneuve d'Ascq Cedex). This study was carried out with the gracious assistance of the Département du Nord.

Conflicts of Interest: The authors declare no conflict of interest.

References

1. Hillier, S.; Velde, B. Octahedral Occupancy and the Chemical-Composition of Diagenetic (Low-Temperature) Chlorites. *Clay Miner.* **1991**, *26*, 149–168. [CrossRef]
2. Vidal, O.; Lanari, P.; Munoz, M.; Bourdelle, F.; De Andrade, V. Deciphering Temperature, Pressure and Oxygen-Activity Conditions of Chlorite Formation. *Clay Miner.* **2016**, *51*, 615–633. [CrossRef]
3. Bourdelle, F.; Cathelineau, M. Low-Temperature Chlorite Geothermometry: A Graphical Representation Based on a T-R2+-Si Diagram. *Eur. J. Mineral.* **2015**, *27*, 617–626. [CrossRef]
4. Cathelineau, M. Cation Site Occupancy in Chlorites and Illites as a Function of Temperature. *Clay Miner.* **1988**, *23*, 471–485. [CrossRef]
5. Vidal, O.; Parra, T.; Trotet, F. A Thermodynamic Model for Fe-Mg Aluminous Chlorite Using Data from Phase Equilibrium Experiments and Natural Pelitic Assemblages in the 100 Degrees to 600 Degrees C, 1 to 25 Kb Range. *Am. J. Sci.* **2001**, *301*, 557–592. [CrossRef]
6. Vidal, O.; Parra, T.; Vieillard, P. Thermodynamic Properties of the Tschermak Solid Solution in Fe-Chlorite: Application to Natural Examples and Possible Role of Oxidation. *Am. Mineral.* **2005**, *90*, 347–358. [CrossRef]
7. Vidal, O.; De Andrade, V.; Lewin, E.; Munoz, M.; Parra, T.; Pascarelli, S. P-T-Deformation-Fe^{3+}/Fe^{2+} Mapping at the Thin Section Scale and Comparison with XANES Mapping: Application to a Garnet-Bearing Metapelite from the Sambagawa Metamorphic Belt (Japan). *J. Metamorph. Geol.* **2006**, *24*, 669–683. [CrossRef]
8. Bourdelle, F.; Parra, T.; Chopin, C.; Beyssac, O. A New Chlorite Geothermometer for Diagenetic to Low-Grade Metamorphic Conditions. *Contrib. Mineral. Petrol.* **2013**, *165*, 723–735. [CrossRef]
9. Lanari, P.; Wagner, T.; Vidal, O. A Thermodynamic Model for Di-Trioctahedral Chlorite from Experimental and Natural Data in the System MgO-FeO-Al^2O^3- SiO^2-H^2O: Applications to P-T Sections and Geothermometry. *Contrib. Mineral. Petrol.* **2014**, *167*, 968. [CrossRef]
10. Inoue, A.; Meunier, A.; Patrier-Mas, P.; Rigault, C.; Beaufort, D.; Vieillard, P. Application of Chemical Geothermometry to Low-Temperature Trioctahedral Chlorites. *Clays Clay Miner.* **2009**, *57*, 371–382. [CrossRef]
11. Inoue, A.; Inoue, S.; Utada, M. Application of Chlorite Thermometry to Estimation of Formation Temperature and Redox Conditions. *Clay Miner.* **2018**, *53*, 143–158. [CrossRef]
12. Bourdelle, F. Low-Temperature Chlorite Geothermometry and Related Recent Analytical Advances: A Review. *Minerals* **2021**, *11*, 130. [CrossRef]
13. de Caritat, P.; Hutcheon, I.; Walshe, J. Chlorite Geothermometry—A Review. *Clays Clay Miner.* **1993**, *41*, 219–239. [CrossRef]
14. Essene, E.J.; Peacor, D.R. Clay Mineral Thermometry—A Critical Perspective. *Clays Clay Miner.* **1995**, *43*, 540–553. [CrossRef]
15. Beaufort, D.; Rigault, C.; Billon, S.; Billault, V.; Inoue, A.; Inoue, S.; Patrier, P. Chlorite and Chloritization Processes through Mixed-Layer Mineral Series in Low Temperature Geological Systems A—Review. *Clay Miner.* **2015**, *50*, 497–523. [CrossRef]
16. Jiang, W.; Peacor, D.; Buseck, P. Chlorite Geothermometry—Contamination and Apparent Octahedral Vacancies. *Clays Clay Miner.* **1994**, *42*, 593–605. [CrossRef]
17. Hillier, S.; Wilson, M.J.; Merriman, R.J. Clay Mineralogy of the Old Red Sandstone and Devonian Sedimentary Rocks of Wales, Scotland and England. *Clay Miner.* **2006**, *41*, 433–471. [CrossRef]
18. Ryan, P.C.; Reynolds, R.C. The Chemical Composition of Serpentine/Chlorite in the Tuscaloosa Formation, United States Gulf Coast: EDX vs. XRD Determinations, Implications for Mineralogic Reactions and the Origin of Anatase. *Clays Clay Miner.* **1997**, *45*, 339–352. [CrossRef]
19. Mosser-Ruck, R.; Pignatelli, I.; Bourdelle, F.; Abdelmoula, M.; Barres, O.; Guillaume, D.; Charpentier, D.; Rousset, D.; Cathelineau, M.; Michau, N. Contribution of Long-Term Hydrothermal Experiments for Understanding the Smectite-to-Chlorite Conversion in Geological Environments. *Contrib. Mineral. Petrol.* **2016**, *171*, 97. [CrossRef]
20. Biernacka, J. Pore-Lining Sudoite in Rotliegend Sandstones from the Eastern Part of the Southern Permian Basin. *Clay Miner.* **2014**, *49*, 635–655. [CrossRef]
21. Burton, J.H.; Krinsley, D.H.; Pye, K. Authigenesis of Kaolinite and Chlorite in Texas Gulf Coast Sediments. *Clays Clay Miner.* **1987**, *35*, 291–296. [CrossRef]
22. Masalehdani, M.N.-N.; Mees, F.; Dubois, M.; Coquinot, Y.; Potdevin, J.-L.; Fialin, M.; Blanc-Valleron, M.-M. Condensate Minerals from a Burning Coal-Waste Heap in Avion, Nothern France. *Can. Mineral.* **2009**, *47*, 573–591. [CrossRef]
23. Bourdelle, F.; Parra, T.; Beyssac, O.; Chopin, C.; Moreau, F. Ultrathin Section Preparation of Phyllosilicates by Focused Ion Beam Milling for Quantitative Analysis by TEM-EDX. *Appl. Clay Sci.* **2012**, *59–60*, 121–130. [CrossRef]
24. Raabe, J.; Tzvetkov, G.; Flechsig, U.; Boege, M.; Jaggi, A.; Sarafimov, B.; Vernooij, M.G.C.; Huthwelker, T.; Ade, H.; Kilcoyne, D.; et al. PolLux: A New Facility for Soft x-Ray Spectromicroscopy at the Swiss Light Source. *Rev. Sci. Instrum.* **2008**, *79*, 113704. [CrossRef] [PubMed]
25. Hitchcock, A.P. Soft X-Ray Imaging and Spectromicroscopy. In *Handbook of Nanoscopy*; Van Tendeloo, G., Van Dyck, D., Pennycook, S.J., Eds.; Wiley-VCH Verlag GmbH & Co. KGaA: Weinheim, Germany, 2012; pp. 745–791. ISBN 978-3-527-64186-4.
26. Bourdelle, F.; Benzerara, K.; Beyssac, O.; Cosmidis, J.; Neuville, D.R.; Brown, G.E.; Paineau, E. Quantification of the Ferric/Ferrous Iron Ratio in Silicates by Scanning Transmission X-Ray Microscopy at the Fe L2,3 Edges. *Contrib. Mineral. Petrol.* **2013**, *166*, 423–434. [CrossRef]

27. Roedder, E. *Fluid Inclusions: An Introduction to Studies of All Types of Fluid Inclusions, Gas, Liquid, or Melt, Trapped in Materials from Earth and Space, and Their Application to the Understanding of Geologic Processes*; Reviews in Mineralogy; Mineral Society of America: Washington, DC, USA, 1984; ISBN 978-0-939950-16-4.
28. Goldstein, R.H.; Reynolds, T.J. *Systematics of Fluid Inclusions in Diagenetic Minerals*; SEPM (Society for Sedimentary Geology): Tusla, OK, USA, 1994; ISBN 978-1-56576-008-0.
29. Bodnar, R.J. Revised Equation and Table for Determining the Freezing Point Depression of H^2O-Nacl Solutions. *Geochim. Cosmochim. Acta* **1993**, *57*, 683–684. [CrossRef]
30. Dowling, K.; Morrison, G. Application of Quartz Textures to the Classification of Gold Deposits Using North Queensland Examples. In *The Geology of Gold Deposits: The Perspective in 1988*; Economic Geology Monograph Series; Society of Economic Geologists (SEG): Littleton, CO, USA, 1989. [CrossRef]
31. Dong, G.; Morrison, G.; Jaireth, S. Quartz Textures in Epithermal Veins, Queensland; Classification, Origin and Implication. *Econ. Geol.* **1995**, *90*, 1841–1856. [CrossRef]
32. Beaufort, D.; Cassagnabere, A.; Petit, S.; Lanson, B.; Berger, G.; Lacharpagne, J.C.; Johansen, H. Kaolinite-to-Dickite Reaction in Sandstone Reservoirs. *Clay Miner.* **1998**, *33*, 297–316. [CrossRef]
33. Goldstein, R. Petrographic Analysis of Fluid Inclusions. In *Petrographic Analysis of Fluid Inclusions*; Mineralogical Association of Canada: Quebec City, QC, Canada, 2003; pp. 9–53.
34. Wiewiora, A.; Weiss, Z. Crystallochemical Classifications of Phyllosilicates Based on the Unified System of Projection of Chemical Composition; II, The Chlorite Group. *Clay Miner.* **1990**, *25*, 83–92. [CrossRef]
35. Lanson, B.; Beaufort, D.; Berger, G.; Bauer, A.; Cassagnabère, A.; Meunier, A. Authigenic Kaolin and Illitic Minerals during Burial Diagenesis of Sandstones: A Review. *Clay Miner.* **2002**, *37*, 1–22. [CrossRef]
36. Hrischeva, E.; Gier, S. Clay Mineralogy and Geochemistry of Early Jurassic Sedimentary Rocks from the Moezian Platform, Northern Bulgaria. *Clay Miner.* **2002**, *37*, 413–428. [CrossRef]
37. Huang, W.L.; Bishop, A.M.; Brown, R.W. The Effect of Fluid/Rock Ratio on Feldspar Dissolution and Illite Formation under Reservoir Conditions. *Clay Miner.* **1986**, *21*, 585–601. [CrossRef]
38. James, R.B.; Stephen, G.F. Clay Diagenesis in Wilcox Sandstones of Southwest Texas: Implications of Smectite Diagenesis on Sandstone Cementation. *SEPM J. Sediment. Res.* **1979**, *49*. [CrossRef]
39. Ruiz Cruz, M.D.; Reyes, E. Kaolinite and Dickite Formation during Shale Diagenesis: Isotopic Data. *Appl. Geochem.* **1998**, *13*, 95–104. [CrossRef]
40. Baronnet, A. Silicate Microstructures at the Sub-Atomic Scale-Web of Science Core Collection. *Comptes Rendus Acad. Sci. Ser. II Fasc. Sci. Terre Planetes* **1997**, *324*, 157–172.
41. Yau, Y.-C.; Peacor, D.R.; Beane, R.E.; Essene, E.J.; Mcdowell, S.D. Microstructures, Formation Mechanisms, and Depth-Zoning of Phyllosilicates in Geothermally Altered Shales, Salton Sea, California. *Clays Clay Miner.* **1988**, *36*, 1–10. [CrossRef]
42. Aagaard, P.; Jahren, J.S.; Harstad, A.O.; Nilsen, O.; Ramm, M. Formation of Grain-Coating Chlorite in Sandstones. Laboratory Synthesized vs. Natural Occurrences. *Clay Miner.* **2000**, *35*, 261–269. [CrossRef]
43. Velde, B.; Raoult, J.-F.; Leikine, M. Metamorphosed Berthierine Pellets in Mid-Cretaceous Rocks from North-Eastern Algeria. *J. Sediment. Res.* **1974**, *44*, 1275–1280. [CrossRef]
44. Iljima, A.; Matsumoto, R. Berthierine and Chamosite in Coal Measures of Japan. *Clays Clay Miner.* **1982**, *30*, 264–274. [CrossRef]
45. Meunier, A. *Clays*; Springer: Berlin/Heidelberg, Germany, 2005; ISBN 978-3-540-21667-4.
46. Velde, B. *Clay Minerals: A Physico-Chemical Explanation of Their Occurrence*; Elsevier: Amsterdam, The Netherlands, 1985; ISBN 978-0-444-41238-6.
47. Hillier, S. Origin, Diagenesis, and Mineralogy of Chlorite Minerals in Devonian Lacustrine Mudrocks, Orcadian Basin, Scotland. *Clays Clay Miner.* **1993**, *41*, 240–259. [CrossRef]
48. Banfield, J.F.; Bailey, S.W. Formation of Regularly Interstratified Serpentine-Chlorite Minerals by Tetrahedral Inversion in Long-Period Serpentine Polytypes. *Am. Mineral.* **1996**, *81*, 79–91. [CrossRef]
49. Ryan, P.C.; Reynolds, R.C. The Origin and Diagenesis of Grain-Coating Serpentine-Chlorite in Tuscaloosa Formation Sandstones, U.S. Gulf Coast. *Am. Mineral.* **1996**, *81*, 213–225. [CrossRef]
50. Hornibrook, E.R.C.; Longstaffe, F.J. Berthierine from the Lower Cretaceous Clearwater Formation, Alberta, Canada. *Clays Clay Miner.* **1996**, *44*, 1–21. [CrossRef]
51. Xu, H.; Veblen, D.R. Interstratification and Other Reaction Microstructures in the Chlorite-Berthierine Series. *Contrib. Mineral. Petrol.* **1996**, *124*, 291–301. [CrossRef]
52. Billault, V.; Beaufort, D.; Baronnet, A.; Lacharpagne, J.-C. A Nanopetrographic and Textural Study of Grain-Coating Chlorites in Sandstone Reservoirs. *Clay Miner.* **2003**, *38*, 315–328. [CrossRef]
53. Inoue, S.; Kogure, T. High-Resolution Transmission Electron Microscopy (HRTEM) Study of Stacking Irregularity in Fe-Rich Chlorite from Selected Hydrothermal Ore Deposits. *Clays Clay Miner.* **2016**, *64*, 131–144. [CrossRef]
54. Jahren, J.S.; Aagaard, P. Compositional Variations in Diagenetic Chlorites and Illites, and Relationships with Formation-Water Chemistry. *Clay Miner.* **1989**, *24*, 157–170. [CrossRef]
55. Walker, J.R.; Thompson, G.R. Structural Variations in Chlorite and Illite in a Diagenetic Sequence from the Imperial Valley, California. *Clays Clay Miner.* **1990**, *38*, 315–321. [CrossRef]

minerals

MDPI

Article

Efflorescent Sulphates with M^+ and M^{2+} Cations from Fumarole and Active Geothermal Fields of Mutnovsky Volcano (Kamchatka, Russia)

Elena S. Zhitova [1,*], Dmitry A. Khanin [2], Anton A. Nuzhdaev [1], Maria A. Nazarova [1], Rezeda M. Ismagilova [3], Vladimir V. Shilovskikh [4], Anastasia N. Kupchinenko [1], Ruslan A. Kuznetsov [1] and Pavel S. Zhegunov [1]

1 Institute of Volcanology and Seismology, Russian Academy of Sciences, Bulvar Piypa 9, Petropavlovsk-Kamchatsky 683006, Russia; nuzhdaev@gmail.com (A.A.N.); nazarovamar@mail.ru (M.A.N.); kupchasta@yandex.ru (A.N.K.); ruslanalexeevich@yandex.ru (R.A.K.); pavel.zhegunov@bk.ru (P.S.Z.)
2 Institute of Experimental Mineralogy, Russian Academy of Sciences, Academica Osypyana ul. 4, Chernogolovka 142432, Russia; mamontenok49@yandex.ru
3 Department of Crystallography, St. Petersburg State University, Universitetskaya Nab. 7/9, St. Petersburg 199034, Russia; rezeda.m.ismagilova@gmail.com
4 Department of Colloid Chemistry, St. Petersburg State University, Universitetskaya Nab. 7/9, St. Petersburg 199034, Russia; vova_bel@mail.ru
* Correspondence: zhitova_es@mail.ru

Abstract: In this study, sulphate efflorescent minerals covering the surface of the Donnoe and Dachnoe fields of the Mutnovsky volcano are described. The minerals were precipitated on the argillic facies as the result of water–rock interaction and fumarole emission. A chemical composition of Ca, Ba, $(NH_4)^+$, Na-Fe^{3+}, $(NH_4)^+$-Al, $(NH_4)^+$-Fe^{3+}, Na-Al, K-Al, and K-Fe^{3+} sulphates was reported. Elements such as Sr, Mg, Co, Ni, Ti and P were found as isomorphic impurities. Ammonia species were concentrated around fumaroles. The mineral assemblage described herein is unique in relation to other geological settings and reflects the process of low-temperature mineral formation associated with volcanism. The thermal water contains cations such as H, Na, K, NH_4, Ca, Mg, Fe^{2+}, Fe^{3+}, and Al in different proportions with pH ranging from 2.4 to 6.5 and the dominance of acidic waters. The gas condensate bears such cations as $(NH_4)^+$, Ca, and Mg and has a pH of ~5. Thus, the rest of the main cations are derived from the leaching of the host rocks. Among the identified phases, the alunite-supergroup minerals are more prone to isomorphism. The Ti, Co, and Ni impurities mark the unique geochemistry of thermal water at the Mutnovsky volcano. We postulate that the chemical composition of alunite-supergroup minerals reflects the types of hydrothermal occurrences and contains important information on the geochemistry of the hydrothermal process.

Keywords: jarosite; alunite; alum; amarillite; tschermigite; lonecreekite; volcano; geothermal field; fumarole; thermal water; steam-emitting vent; Mutnovsky; Kamchatka; hot spring; efflorescence

1. Introduction

The mineralogy of active fumaroles and geothermal fields is interesting from several different points of view. These geological settings can be considered integral companions of active volcanism, which is supposed to take place from the moment plate tectonics appeared. Fumaroles and gas-steam vents (common at geothermal fields) mark tectonically weakened zones and are responsible for the flow of deep elements (magmatic and/or from host rocks) to the day surface. High temperatures and the presence of aggressive fluids lead to the transformation of the initial rocks, the redistribution of chemical elements, and the formation of a definite mineral association depending on many thermodynamic and physicochemical characteristics of the environment.

In volcano-hydrothermal systems, acidic waters are produced due to the interaction of volcanic gases SO_2, H_2S, and HCl with water, resulting in H^+ release [1–3]. The circulation of acidic solution produces leaching of host rocks and remobilization of the elements contained in them, as well as fumarole-emitted elements that lead to the formation of efflorescent minerals that have been the subjects of recent investigations [4–8]. Efflorescent minerals often form finely dispersed poorly crystalline material, intergrowths, which makes it difficult to reliably determine such phases. This is the defining issue for the lack of mineralogical data on the efflorescence of most hydrothermal systems, in particular Kamchatka (Far Eastern Russia). Although such minerals play an important role in the life of the hydrothermal system, scientists generally emphasize the importance of the results of studies on efflorescent minerals for the exploration of Mars, since sulphate-rich mineral associations have been found there [4–6,8]. In particular, in this work, we present the chemical compositions of the alunite-group minerals identified in the efflorescence of the thermal fields of the Mutnovsky volcano, which attract the attention of researchers in the context of the study of Mars. For instance, the finding of jarosite on Mars [9–11] actualized the study of a variety of its aspects: formation conditions [12], thermodynamic stability [13], atomic arrangement at low temperature [14], and band assignment at visible and infrared spectra [15–17]. This increased interest in the study of minerals of the alunite group indicates that our data can be used in the study of the mineralogy of other planets and other objects of the solar system. In particular, the mineral association discovered by us is unique and reflect the geochemical conditions of low-temperature active geothermal and fumarole fields. If such an association of minerals is found, say, on Mars, this would indicate the presence of hydrothermal activity on the surface of the planet in the past.

The present work aims to describe the surface sulphate efflorescence with K, Na, Ba, NH_4, and Ca that precipitated from thermal water and fumarolic gas condensate at two different localities: the Dachnoe thermal field and the Donnoe fumarole field (both associated with the Mutnovsky volcano in Kamchatka, Russia). The description of geochemical conditions is based on our field documentation and the previously reported results on the hydrogeochemistry of the Donnoe fumarole field [18]. Knowledge of the mineral parageneses formed at the fumarole and geothermal fields can be applied in various areas of geology. For example, the transport and concentration of chemical elements are relevant for ecology, geochemistry, and ore geology and may provide information about deep magmatic bodies, the heat of which can be potentially explored as a geothermal energy source.

Finally, we note that the mineral association described in this work is close (but not identical) in element composition to the associations formed in environments such as acid mine drainage (AMD) [19,20] and burning coal heaps [21,22]. Thus, the results obtained in this work may be of interest for comparison with other non-volcanic environments.

2. Materials and Methods

2.1. Geological Setting

South Kamchatka is a part of the Kuril-Kamchatka island arc system, articulated in the north with the Aleutian arc, and in the southwest with the Japanese arc. Similar to other island arcs, it is characterized by high tectonic-magmatic activity, which is expressed on the surface by modern volcanism. Seventy kilometers south of Petropavlovsk-Kamchatsky, there is an area where intense and varied volcanic activity is widespread. The Mutnovsky volcano is located in this area (Figure 1a). Mutnovsky is a Middle Pleistocene thyroid basalt volcano representing a complex uneven-aged massif (Figure 1b,c). This is one of the most active volcanoes in South Kamchatka. The area of Mutnovsky is located within the regional sub-deep fault zone, at the junction of the asymmetric step depression of the volcanic zone of South Kamchatka with the uplift of the Eastern horst-anticlinorium [23]. In the Mutnovsky district, several types of modern thermal manifestations can be distinguished:

1. The active funnel of the Mutnovsky volcano, where the measured temperature reached 750 °C in the 1960s [23–25], 570–600 °C in 2000–2005 [26–28], and 620 °C in 2006 [29].

2. Fumarole fields of the northern crater of the Mutnovsky volcano: Verkhnee and Donnoe, where the measured temperatures reached 300 °C and 320 °C in 1974, respectively [23], and 150 °C on Donnoe in the 2000s [28].
3. Thermal fields of the Severo-Mutnovsky volcano-tectonic zone: Severo-Mutnovsky, Dachnoe, and Verkhnezhirovskiy, with temperatures up to 110 °C [23].
4. Sources of overheated waters: Nizhnezhirovskie and Voinovskie, where temperatures have never been reported higher than 100 °C [23].

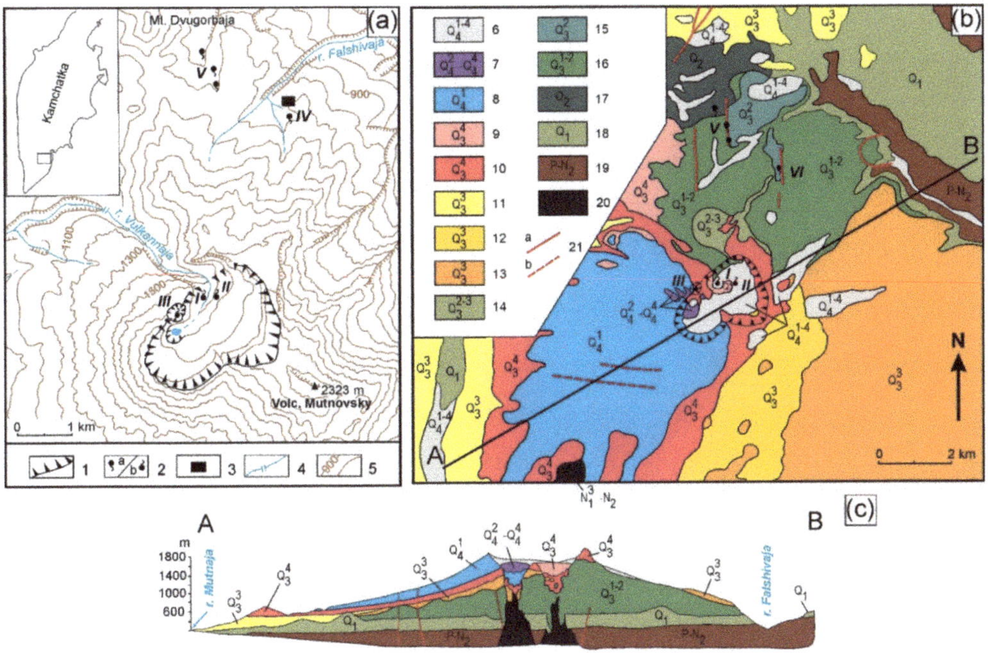

Figure 1. The scheme of the Mutnovsky volcano area (**a**), geological map (**b**) and (**c**) representative A-B (line) cross-section (modified after [23]). Symbols: 1—scarps of craters and calderas of Mutnovsky volcano; 2—the largest thermal manifestations (a—thermal fields; b—fumarole fields); 3—Mutnovsky geothermal power plant; 4—rivers; 5—contours of the relief; 6—modern and undissected Holocene sediments (alluvium, deposits of rockslides, mudflows, lahars, glaciers and glacial sediments); 7—basalts, andesite-basalts, tuffs, and volcanic scoria (8.9 ka BP-present); 8—basalts (10.1–8.9 ka BP); 9—dacite pumice of the caldera-forming eruption of Mutnovsky volcano; 10—basalts; 11—dacite, andesidacite, andesite pumice and ignimbrites from the main stage of the caldera-forming eruptions of Gorely volcano; 12—caldera lake deposits and lava-pyroclastic formations; 13—lava-pyroclastic formation of the basalt-andesite-dacite-rhyolite series; 14—lava flows and residual outcrop of stratovolcanoes basalts, andesites, dacite pumice, and ignimbrites; 15—side breakthroughs and monogenic volcanoes of the precaldera stage; 16—lava-pyroclastic formation of the basalt-andesite-dacite-rhyolite series; 17—basalt-andesite-dacite-rhyolite formation of the pre-caldera structure of Gorely volcano; 18—caldera lake deposits and lava-pyroclastic formations; 19—basement-dislocated strata of green-metamorphosed and silicified volcanic and volcanogenic-sedimentary deposits of the Pribrezhnyj formation and slightly dislocated strata of contrasting volcanites of the basalt-rhyolite series with extrusions and subvolcanic bodies of the Yuzhno-Bystrinskj formation; 20—effusive domes, extrusions, subvolcanic bodies, and lava vents, in section—basalts, andesite-basalts, andesites, dacites, and rhyolites; 21—(a) faults with defined displacement and (b) asymmetric, damped, and buried; I—Donnoe fumarole field; II—Verkhnee fumarole field; III—active funnel; IV—Dachnoe thermal field; V—Severo-Mutnovsky thermal fields. The arrow in (**b**) shows the North direction.

In this work, we consider minerals sampled within the Donnoe fumarole field of the Northern crater of the Mutnovsky volcano (Figure 1a) and within the Southern group of the Dachnoe geothermal field.

The Donnoe fumarole field is located close to the active funnel of the Mutnovsky volcano (the active crater of the volcano). It is located on the smooth surface (Figure 2a) and the temperature at a depth of 20 cm is 20–110 °C. The temperature of the hottest fumaroles are about 120–150 °C. The Donnoe fumarole field is a former boiling crater lake, which was a solution of hydrochloric and sulfuric acids with a pH < 1. The lake existed until 1956 and ceased to exist as a result of a breakthrough of the crater wall.

Figure 2. Sampling schemes of Donnoe fumarole field (**a**), Dachnoe thermal field (**b**), the view to Donnoe fumarole field (**c**), efflorescent minerals from Dachnoe thermal field (**d**), and Donnoe fumarole field (**e**), bubbling pools from Dachnoe thermal field (**f**).

The Dachnoe thermal field (also known as the small geyser field) is located on the slope of the Mutnovsky volcano (Figure 2b), 8 km north from the Donnoe fumarole field. It is located in a rounded basin with steep walls up to 30 m high. The geothermal activity is realized through gas-steam vents, bubbling pools, and steaming ground. The temperature of gas-steam vents does not exceed 99 °C.

In the area around the Dachnoe thermal field, pumice tuffs of dacitic composition come to the surface (Figure 1b). Below the pumice tuffs of the area of Dachnoe lies the lava-pyroclastic complex of the Dvugorby volcano, which is represented by tuffs and tuff breccias from basic to acidic composition (Q_2-Q_3). They are underlain by deposits of the Skalisty volcano: basalts, basaltic andesites, their tuffs, and tuff breccias (Q_2-Q_3). The ignimbrites of the Gorely volcano (Q_2-Q_3) occur below [30]. The surface of the field itself is composed of argillic alteration that is the product of host (primary) rock alteration by volcanic gases and thermal water. The argillic facies surface is covered by efflorescent minerals.

The Donnoe (in some references the name is translated as "Bottom") fumarole field is located in the caldera of the Mutnovsky volcano. The Mutnovsky volcano is composed of rocks in the basalt-andesite-rhyodacite series; modern rocks have a basaltic composition. Basalts, andesite-basalts, andesites, their tuffs, and tuff breccias of the Zhirovsky volcano (Q_1), underlain by volcanic rocks of the Yuzhnobystrinsky andesite-basalt-andesite and Pribrezhny basalt-andesite volcanic complexes (P-N_2) serve as the foundation of the modern volcanic edifice of Mutnovsky volcano [31] (Figure 1b). On the territory of the field, the primary rocks are altered to argillic facies forming the day surface. Efflorescent minerals grow on the argillic facies.

2.2. Materials and Sample Preparation

The mineral samples studied in this work were collected from the surface of the Donnoe fumarole field and Dachnoe thermal field (Mutnovsky volcano) (Figure 2c–f), in particular, from heated ground areas (temperature 40–65 °C), near steam-emitting vents (~70 °C) and fumarole (~120 °C). At both thermal sites, the samples are represented by white, yellow, and ochre yellow efflorescent minerals and crust.

The samples were partly ground in an agate mortar and investigated by powder X-ray diffraction. The other parts of the samples were mounted in epoxy blocks and polished using a set of diamond pastes; these samples were carbon-coated and studied with scanning electron microscopy (SEM) and energy-dispersive X-ray spectroscopy (EDS).

Water samples were collected from bubbling pools and filtered through a membrane with a porosity of 0.45 μm into sterile plastic bottles with a volume of 1 L. The condensates of the steam-emitting vents were sampled by forced pumping with cooling; the sample volume was 1 L. The main physicochemical parameters were measured with portable Hanna pH meters (pH measurement accuracy ±0.1). The temperature was measured with a k-type thermocouple (±0.1 °C).

2.3. Methods

The phase analyses of samples were carried out with the powder X-ray diffractometer Shimadzu XRD-7000 using the following conditions: Cu$K\alpha$ radiation, 30 kV/30 mA, scan speed 1°/min, step width 0.1°, and 2-theta range 5–65° and the powder X-ray diffractometer Rigaku Miniflex II with Cu$K\alpha$ radiation at 30 kV/15 mA, scan speed 1°/min, step width 0.1°, and 2-theta range 4–60°. The data were processed using ICDD database. The samples available in small quantities were subjected to X-ray diffraction using a Rigaku R-Axis Rapid II diffractometer equipped with a curved image plate detector and a rotating anode X-ray source with Co$K\alpha$ radiation, scan speed 1°/min, step width 0.02°, and 2-theta range 5–70°. The data were integrated using the software package Osc2Tab/SQRay [32]. All X-ray diffraction powder analyses were carried out at room temperature. In total, 35 samples were subjected to powder X-ray diffraction.

The chemical composition was analyzed using SEM Camscan MV2300, nanoampermeter, and INCA Energy 450 EDS (Oxford Instruments Ltd., Great Britain) and SEM Hitachi S3400N equipped with an Oxford X-Max 20 EDS (Oxford Instruments Ltd., Great Britain). The operation conditions (for Camscan/Hitachi) were an accelerating voltage of 15/20 kV and a beam current of 30/1.5 nA and the spectra were obtained at spot mode for 70/30 s each. The following standards were used (Camscan/Hitachi): Na—chkalovite/albite, Mg—diopside/MgO, Al—tremolite/albite, Si—chkalovite/albite, P—synthetic $KTiPO_5$/InP, S—$SrSO_4$/FeS_2, K—potassium feldspar/KCl, Ca—diopside/$CaSO_4$, Ti—synthetic $FeTiO_3$/Ti, Fe—hornblende/FeS_2, Co—synthetic CoAsS/Co, Ni—synthetic NiO/Ni, Sr—$SrSO_4$/SrF_2, Ba—synthetic $BaSO_4$/BaF_2, and N—BN. The uncertainty of a single measurement did not exceed 0.1/0.3 wt. %. The advantages of energy-dispersive (ED) spectrometer application are the possibility of analyzing small-size grains of distinct minerals found in tight association and completion of chemical analyses in a short time. Both short-time and low-probe current significantly contribute to non-distraction of the studied material (see, e.g., [33]).

In total, about 1000 chemical analyses were obtained from polished sections, of which about 220 analyses contained considerable content of S, Na, K, Ca, and/or Ba, i.e., corresponded to sulphates of alkali and alkali earth elements. These analyses were grouped by the content of chemical elements. The analyses of each group were averaged and used to calculate the ratios between the elements.

When studying the composition of hydrothermal waters, the following analytical methods (complete chemical analysis) were used: potentiometric (pH, HCO, and F); photometric (SO_4, NH_4, Al, SiO_2 and Fe); titrimetric (Cl); atomic absorption spectrophotometric (Ca and Mg); flame emission spectrophotometry (Na, K, and Li).

3. Results

3.1. Powder X-ray Diffraction

The bulk composition of efflorescent minerals from the Donnoe fumarole field is represented by halotrichite, alunogen, gypsum, native sulphur, melanterite, mascagnite, letovicite, and alunite-type minerals. The efflorescent from the Dachnoe thermal field is composed of halotrichite, alunogen, gypsum, alum- and alunite-type minerals, copiapite group mineral, and melanterite. In both cases, alunite-type minerals are closely associated with alunogen. The detailed information showing phase analyses, *d*-values, line intensity and *hkl* for six representative samples is given in Table S1. Figure 3 shows a schematic visual representation of the results from the Table S1. Table 1 lists identified sulphates relative to the frequency of their occurrence in samples.

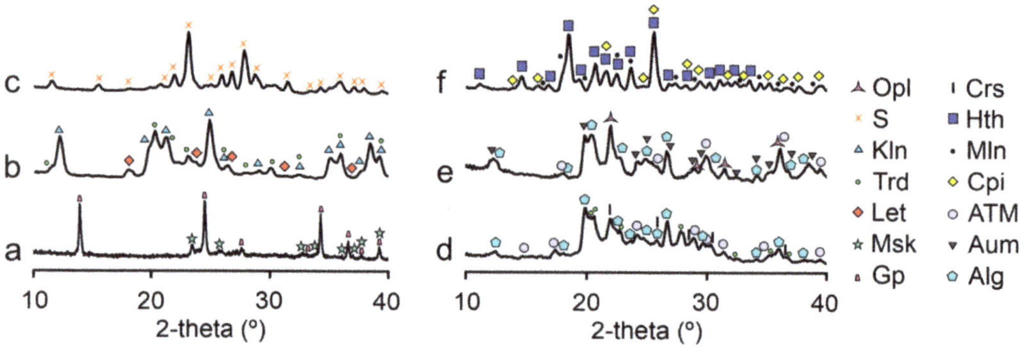

Figure 3. Powder X-ray diffraction patterns of efflorescent minerals from Donnoe fumarole field (**a**–**c**) and Dachnoe thermal field (**d**–**f**). Hth—halotrichite, Mln—melanterite, Cpi—copiapite, ATM—alunite-type compounds, Aum—alum, Alg—alunogen, Opl—opal, S—native Sulphur, Kln—kaolinite, Trd—tridymite, Crs—crystoballite, Let—letovicite, Msk—mascagnite, Gp—gypsum.

Table 1. Sulphate efflorescence minerals from Mutnovsky volcano detected by powder X-ray diffraction analyses.

Mineral, Ideal Chemical Formula	Donnoe Fumarole Field	Dachnoe Thermal Field
Halotrichite, $Fe^{2+}Al_2(SO_4)_4{\cdot}22H_2O$	****	****
Alunogen, $Al_2(SO_4)_3{\cdot}17H_2O$	***	***
Gypsum, $CaSO_4{\cdot}2H_2O$	**	***
Alum-type minerals, $A^+M^{3+}(SO_4)_2{\cdot}12H_2O$	-	**
Alunite-type minerals, $DG_3(TX_4)_2X'_6$	*	*
Melanterite, $Fe^{2+}(H_2O)_6SO_4{\cdot}H_2O$	*	*
Copiapite-group mineral, $M^{2+}Fe^{3+}{}_4(SO_4)_6(OH)_2{\cdot}20H_2O$	-	*
Letovicite, $(NH_4)_3H(SO_4)_2$	*	-
Mascagnite, $(NH_4)_2SO_4$	*	-

The number of asterisks indicates the frequency of the mineral occurrence in the analyzed samples: ****—very widespread; ***—common; **—regular; and *—rare.

According to phase analyses by powder X-ray diffraction, kaolinite is found in tight association with efflorescent minerals since it is the substrate on which and from which these minerals grow.

It is tempting to divide identified minerals into separate associations and correlate them with the processes of either acidic alteration or fumarole emission. However, such a division is impossible due to the complexity of the natural process. The mineralogy of the efflorescent is heterogeneous and changes on a centimeter scale. However, we report some tendencies: ammonium minerals concentrate close to gas-emitting fumaroles or vents, letovicite is found in association with kaolinite and tridymite, and mascagnite is associated with gypsum. The following main associations were found: (i) halotrichite and alunogen; (ii) alunogen and kaolinite; (iii) halotrichite, melanterite, and copiapite-group mineral, (iv) tschermigite/lonecreekite (alum-type minerals), alone or with alunogen; (v) gypsum; (vi) native sulphur; (vii) alunogen, gypsum, and alunite-type minerals; (viii) halotrichite and alunite-type minerals. All sulphate minerals can be found in association with underlying clay minerals and/or rather abundant SiO_2 or $SiO_2 \times nH_2O$ polymorphs: opal, tridymite, and cristobalite, all derived as the result of an alteration of silicates.

3.2. Energy-Dispersive Spectroscopy

The energy-dispersive spectroscopy data agree with powder X-ray diffraction analyses: Fe-Al, Fe and Al sulphates are abundant in the studied samples. Apart from sulphates, silicates corresponding to the remnants of the initial magmatic rocks converted to argillite as a result of hydrothermal activity are found in association with sulphates. In addition, the samples contain pyrite crystals and SiO_2 or $SiO_2 \times nH_2O$ polymorphs. The chemical composition of Na-, K-, Ca-, Ba-, and (NH_4)-sulphates are given in Tables 2 and 3.

Table 2. Chemical composition of baryte, gypsum, and alunite supergroup minerals: jarosite, alunite, natrojarosite and natroalunite.

Wt.%	Baryte	Gypsum		Jarosite	Alunite	Na-Rich Jarosite	K-Rich Natrojarosite	Natro-Alunite	Natrojarosite	
Location	Dachnoe	Donnoe	Dachnoe			Donnoe			Donnoe	Dachnoe
n	6	41	16	13	17	9	23	17	12	16
Na_2O	n.d.	n.d.	n.d.	1.25	1.28	3.15	3.08	5.04	3.96	4.74
K_2O	n.d.	n.d.	n.d.	6.55	6.77	4.91	4.55	1.87	1.81	0.77
CaO	n.d.	32.28	32.67	1.65	0.52	0.25	0.43	0.17	0.04	0.08
SrO	1.19	n.d.	n.d.	0.29	1.72	n.d.	n.d.	0.05	0.09	n.d.
BaO	61.55	n.d.	n.d.	0.05	0.49	n.d.	n.d.	n.d.	n.d.	n.d.
MgO	0.20	n.d.	0.07	0.16	0.02	0.07	0.12	0.07	0.20	0.42
FeO (1)	n.d.	0.49	0.37	n.d.	n.d.	n.d.	n.d.	n.d.	n.d.	n.d.
CoO	n.d.	n.d.	n.d.	0.22	0.07	0.23	0.26	0.11	0.31	n.d.
NiO	n.d.	n.d.	n.d.	0.10	0.04	0.11	n.d.	0.04	n.d.	n.d.
TiO_2	n.d.	n.d.	n.d.	n.d.	n.d.	0.12	0.25	0.04	0.22	0.44
Al_2O_3	n.d.	n.d.	n.d.	6.92	28.18	6.53	5.31	25.15	4.77	5.55
Fe_2O_3 (1)	n.d.	n.d.	n.d.	36.67	10.29	38.72	39.36	15.91	41.11	39.19
SO_3	36.00	47.81	47.16	32.92	34.72	33.34	32.14	37.20	33.25	33.97
P_2O_5	n.d.	n.d.	n.d.	0.68	2.77	0.26	0.37	0.21	0.57	0.93
H_2O (2)	n.d.	21.05	21.12	10.86	12.30	11.15	11.05	12.54	11.16	11.08
$(NH_4)_2O$ (3)	n.d.	n.d.	n.d.	n.d.	n.d.	n.d.	n.d.	0.69	0.81	0.9
Total	**98.95**	**101.63**	**101.39**	**98.32**	**99.16**	**98.82**	**97.53**	**99.04**	**98.76**	**98.07**
					apfu calculated on the basis of					
	Σcat + S = 2	Ca + Fe = 1				Fe + Al + Mg + Co + Ni = 3				
Na	n.d.	n.d.	n.d.	0.20	0.18	0.49	0.49	0.70	0.62	0.75
K	n.d.	n.d.	n.d.	0.69	0.63	0.50	0.47	0.17	0.19	0.09
Ca	n.d.	0.99	0.99	0.15	0.04	0.01	0.04	0.01	0.04	0.01
Sr	0.03	n.d.	n.d.	0.01	0.07	n.d.	n.d.	0.00	0.00	n.d.
Ba	0.93	n.d.	n.d.	0.00	0.01	n.d.	n.d.	n.d.	n.d.	n.d.

Table 2. *Cont.*

Wt.%	Baryte	Gypsum		Jarosite	Alunite	Na-Rich Jarosite	K-Rich Natrojarosite	Natro-Alunite	Natrojarosite	
Location	Dachnoe	Donnoe	Dachnoe			Donnoe			Donnoe	Dachnoe
NH_4	n.d.	n.d.	n.d.	n.d.	n.d.	n.d.	n.d.	0.11	0.15	0.17
Mg	0.01	n.d.	0.00	0.02	0.00	0.01	0.02	0.01	0.02	0.05
Co	n.d.	n.d.	n.d.	0.01	0.00	0.01	0.02	0.01	0.02	n.d.
Ni	n.d.	n.d.	n.d.	0.01	0.00	0.01	0.02	0.00	n.d.	n.d.
Ti	n.d.	n.d.	n.d.	n.d.	n.d.	0.01	n.d.	n.d.	0.01	0.03
Fe^{2+}	n.d.	0.01	0.01	n.d.	n.d.	n.d.	n.d.	n.d.	n.d.	n.d.
Al	n.d.	n.d.	n.d.	0.67	2.43	0.62	0.51	2.13	0.45	0.53
Fe^{3+}	n.d.	n.d.	n.d.	2.28	0.57	2.35	2.44	0.86	2.49	2.39
S	1.04	1.03	1.00	2.11	1.90	2.08	2.06	2.00	2.01	2.02
P	n.d.	n.d.	n.d.	0.05	0.17	0.02	0.03	0.01	0.04	0.06
H_2O	n.d.	2.00	2.00	n.d.	n.d.	n.d.	n.d.	n.d.	n.d.	n.d.
OH	n.d.	n.d.	n.d.	6.00	6.00	6.00	6.00	6.00	6.00	6.00

Note: the oxidation state of iron corresponds to that in the ideal formula of the identified mineral. n—Number of analyses, n.d.—not detected. (1) the oxidation state of Fe is taken in agreement with ideal chemical formula; (2) H_2O/OH is taken as in ideal chemical formula; (3) the content of NH4 is calculated based on the charge balance.

Table 3. Chemical composition of amarillite and alum-group minerals: tschermigite, lonecreekite and alum-(Na) from Dachnoe thermal field normalized to 100%.

Wt.%	Tschermigite	Lonecreekite	Alum-(Na)	Amarillite
n	**13**	**9**	**11**	**16**
Na_2O	n.d.	n.d.	4.18	8.90
MgO	n.d.	n.d.	0.26	0.72
Al_2O_3	10.48	0.37	9.80	2.07
Fe_2O_3 (1)	0.83	16.19	1.07	17.12
SO_3	36.51	33.12	32.84	41.70
P_2O_5	n.d.	n.d.	0.17	0.23
H_2O (2)	46.60	45.35	50.00	29.50
$(NH_4)_2O$ (3)	5.58	4.99	1.70	n.d.
Total	**100.00**	**100.00**	**100.00**	**100.00**
	apfu calculated on the basis of Fe + Al + Mg = 1			
NH_4	0.99	0.91	0.31	n.d.
Na	n.d.	n.d.	0.64	1.05
Mg	n.d.	n.d.	0.03	0.07
Al	0.95	0.03	0.91	0.14
Fe^{3+}	0.05	0.97	0.06	0.79
S	2.11	1.97	1.94	1.91
P	n.d.	n.d.	0.01	0.03
H_2O	12.00	12.00	12.00	6.00

n—Number of obtained analyses; n.d.—not detected. (1) the oxidation state of Fe is taken in agreement with ideal chemical formula; (2) H_2O/OH is taken as in ideal chemical formula; (3) the content of NH_4 is calculated based on the charge balance.

Gypsum is abundant in both fields (Donnoe and Dachnoe), its chemical composition (Tables 2 and 3, Figure 4) almost corresponds to the ideal chemical formula, $CaSO_4 \times 2H_2O$, and negligible Fe content up to 0.01 *apfu* may be present.

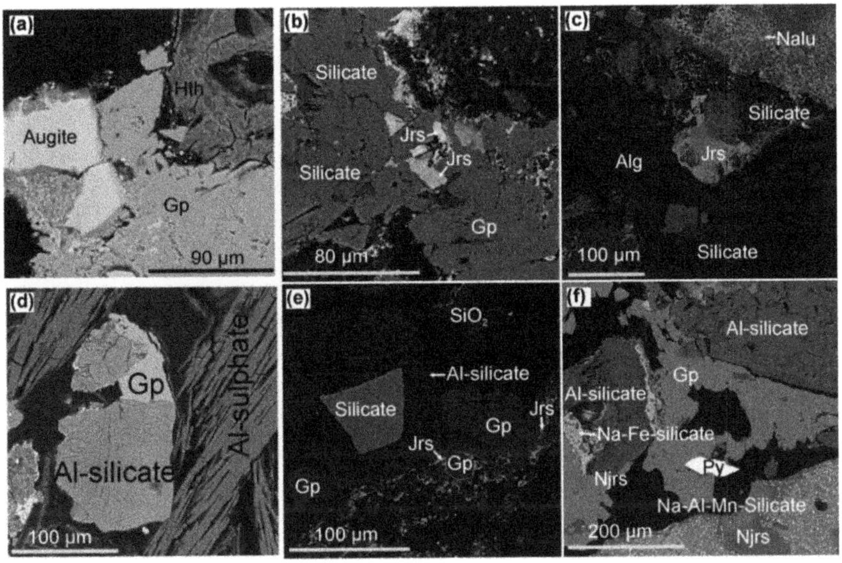

Figure 4. BSE images of efflorescent samples from Donnoe fumarole field: (**a**) halotrichite with gypsum, (**b**) jarosite with gypsum, (**c**) jarosite, natroalunite and alunogen, (**d**) gypsum with Al-silicate and Al-sulphate, (**e**) jarosite with gypsum and (**f**) natrojarosite with gypsum. Jrs—jarosite; Njrs—natrojarosite; Py—pyrite; Alg—alunogen; Gp—gypsum; Nalu—natroalunite, Hth—halotrichite.

In samples from the Dachnoe thermal field, we found three baryte grains with sizes of 30–40 × 20 µm. Trace amounts of Sr and Mg were found in the composition of baryte as 0.03 and 0.01 *apfu*, respectively (Table 2). Baryte was found in association with Al-sulphates identified as alunogen.

Within the alunite-supergroup, four minerals were found: alunite, jarosite, natroalunite, natrojarosite (Figure 4), and intermediate chemical varieties with Na/K ~1:1 and corresponding to Na-rich jarosite and K-rich natrojarosite (Table 2). All four detected alunite supergroup minerals were found at the Donnoe fumarole field, whereas at the Dachnoe thermal field only natrojarosite (Figure 5) hase been identified.

Jarosite and alunite have very similar Na/K ratios, being ~0.2/0.7 and 0.2/0.6, respectively. Jarosite is enriched by Ca (0.15 *apfu*), while alunite has Sr, Ca, and Ba impurities (0.07, 0.04 and 0.01 *apfu*, respectively). The potential mechanism of extra charge (produced by divalent cation in the *D* site) compensation is deprotonation of some OH groups and/or phosphorous (P) incorporation to sulphate tetrahedra. Natrojarosite and natroalunite also have similar Na/K ratios that are ~0.7/0.2 and 0.6/0.2, respectively. At the same time, divalent cations (Ca, Ba, and Sr) at the *D* site are nearly absent (only 0.04 *apfu* of Ca occurs in natroalunite) and the sum of *D* site cations is 0.83–0.89 *apfu*; this results in the deficit of positive charge (Table 2). Thus, we suppose that 0.11 (for natrojarosite) and 0.15/0.17 (for natroalunite) *apfu* of $(NH_4)^+$ cation occurs in the chemical composition of these minerals. Two varieties have the same content of K and Na that correspond to Na-rich jarosite with Na/K/Ca (*apfu*) = 0.49/0.50/0.01 and K-rich natrojarosite with Na/K/Ca = 0.49/0.47/0.04.

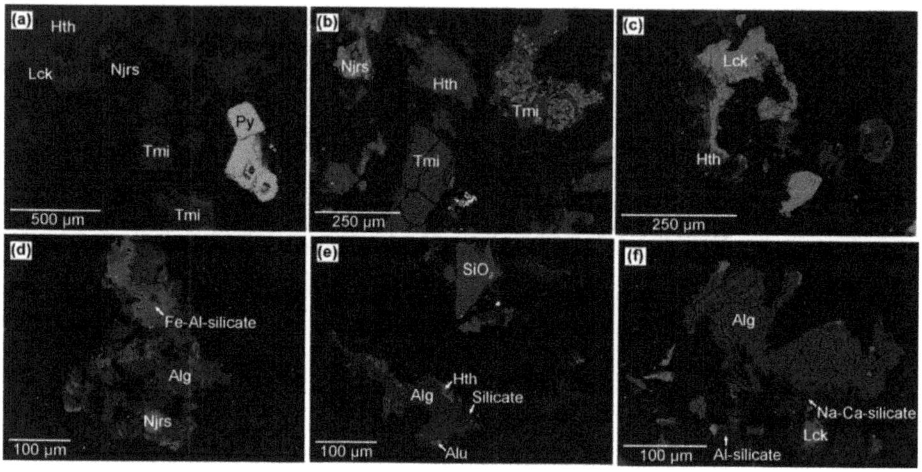

Figure 5. BSE images of efflorescent samples from Dachnoe thermal field: (**a**) alum-group minerals (tschermigite and lonecreekite), natrojarosite, halotrichite; (**b**) tschermigite, natrojarosite, halotrichite; (**c**) lonecreekite, halotrichite; (**d**) natrojarosite and alunogen; (**e**) alunite, alugonen and halotrichite; (**f**) lonecreekite with alunogen. Hth—halotrichite, Njrs—natrojarosite, Lck—lonecreekite, Py—pyrite, Tmi—tschermigite, Alg—alunogen, SiO_2—silica forms.

Even though both the Fe^{3+}- and Al-dominant alunite supergroup minerals were detected, the Fe^{3+}/Al ratios do not show wide variation in each of the minerals. For Fe^{3+}-dominant minerals, the highest Fe^{3+}/Al ratio is ~5:1, and the lowest is 3.4:1. For Al^{3+}-dominant minerals, the Al/Fe^{3+} ratios are 4.3:1 and 2.5:1. All samples bear minor Mg impurity from 0.01 to 0.02 *apfu* for the Donnoe fumarole field and 0.05 *apfu* for the Dachnoe thermal field. Remarkably, alunite-group minerals from the Donnoe fumarole field bear small contents of Co and Ni in the amount of 0.01–0.02 *apfu* for each of the elements. Minor impurity of Ti up to 0.01 *apfu* occur in samples from the Donnoe fumarole field, while Ti in the sample from the Dachnoe thermal field is 0.03 *apfu*. The P impurity is present in all alunite supergroup minerals and its content varies from 0.01 to 0.17 *apfu* in samples from the Donnoe fumarole field and it is 0.06 *apfu* for samples from the Dachnoe thermal field.

At the Dachnoe thermal field, two Na-sulphates were found: amarillite and alum-Na (Figure 6). Since two ammonium minerals of the alum group: tschermigite and lonecreekite are found in association with the alum-(Na), their chemical composition is also given (Table 3).

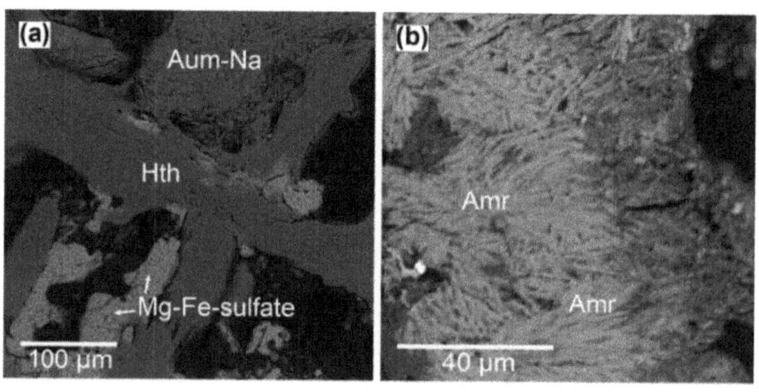

Figure 6. BSE images of (**a**) alum-(Na), (aum-Na), and (**b**) amarillite (Amr) from Dachnoe thermal field. Hth—halotrichite.

Amarillite is found in association with halotrichite and gypsum. The mineral has a rather well-defined stoichiometry Na/(Fe^{3+} + Al + Mg)/(S + P) = 1.05/1.00/1.94 (Table 3). The mineral is P bearing (0.03 *apfu*) and contains Al and Mg (in Fe^{3+}-site) as 0.15 and 0.07 *apfu*, respectively.

Alum-(Na) is characterized by the presence of Na and NH_4 univalent cations in the ratio~2:1 (Table 3). The content of Al (0.91 *apfu*) strongly prevails over Fe (0.06 *apfu*) and Mg (0.03 *apfu*), the mineral contains 0.01 *apfu* of P. Tschermigite has a similar feature, with the strong predominance of Al (0.95 *apfu*) over Fe (0.05 *apfu*). However, solely NH_4 occurs as a monovalent cation. In lonecreekite, Fe^{3+} (0.97 *apfu*) strongly prevails over Al (0.03 *apfu*). Only NH_4 occurs as a univalent cation and no substitution is observed for SO_4 tetrahedra, as for tschermigite.

3.3. Chemical Composition of Thermal Water

The thermal waters from the Dachnoe thermal field have been divided into 4 groups: water with pH < 3, pH in the range 3.5–4.7, pH > 6, and gas condensate of gas-steam vents (pH ~5). The main cations and anions of thermal water and their relative contents are given in Table S2. The relative content of the main cations represented by H, Na, K, NH_4, Ca, Mg, Fe^{2+}, Fe^{3+} and Al in different proportions (Table S2) is shown as pie charts in Figure 7 with resemblance to the data on the chemical composition of thermal water from the Donnoe fumarole field reported previously [17]. The main anion is sulphate in thermal water and sulphate or bicarbonate in gas condensate (Table S2). The strongly dominant cation in the gas condensate is NH_4 in quantities subordinate to Ca and Mg (Figure 7).

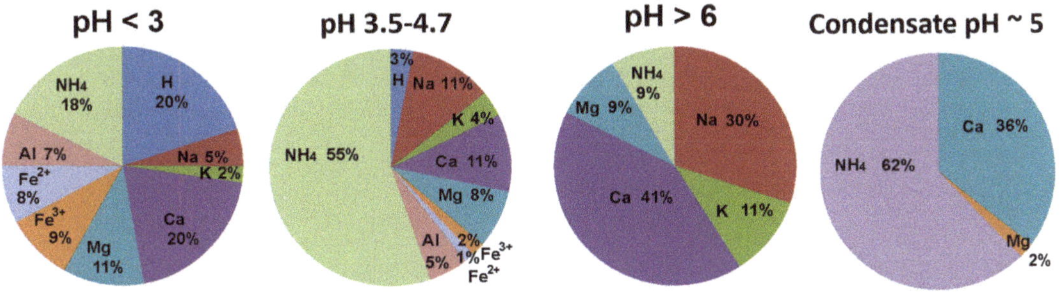

Figure 7. Fractional ratios of basic cations in thermal waters and gas-condensate from Dachnoe thermal field.

4. Discussion

4.1. Chemical Composition of Sulphate Efflorescent Minerals

From a mineralogical point of view, jarosite, alunite, natroalunite and natrojarosite belong to the alunite supergroup [34], which includes about 60 minerals. The general formula of alunite-supergroup minerals is represented as $DG_3(TX_4)_2X'_6$, where D is a tetravalent, trivalent, divalent, or monovalent cation or a partial vacancy; G is a trivalent cation and minor divalent cations; T is a hexavalent, pentavalent cation and minor Si^{4+}; and X/X' is O, (OH), minor F and possibly H_2O. The common constituents are D = Na, K, Pb, Rb, NH_4, Ag, H_3O or H_2O; G = Al, Fe^{3+}, V^{3+} or Ga^{3+}; minor Cu, Zn, Ge; and T = S, As, P [34,35]. The majority of alunite-supergroup minerals is isotypic and crystallize in $R{-}3m$ or $R3m$ space groups of the so-called alunite structure type. It has been shown by numerous previous studies, summarized in [36], that the alunite structure type shows outstanding flexibility in terms of incorporation of different size and charge cations that has even been referred to as a "garbage-can" structure.

The alunite-supergroup minerals show wide variation in Na/K ratio, resulting in the formation of both Na-dominant (natroalunite and natrojarosite) and K-dominant (jarosite and alunite) varieties. Interestingly, two chemical varieties of alunite supergroup minerals with Na/K ~1:1 were found. A wide variation of Na/K ratios are observed for samples from the Donnoe fumarole field, whereas at the Dachnoe thermal field only (K-depleted) natrojarosite was found. Jarosite and alunite are enriched by divalent *A*-site cations, Ba and Sr, while natrojarosite and natroalunite are enriched by the $(NH_4)^+$ cation (Table 3). There is clear compositional segregation between Al- and Fe^{3+}-minerals, so the Fe^{3+} content in the Al-dominated *D* site does not exceed 1/4, while the highest Fe^{3+} content in the Al site is 1/6. In the experimental study of volcanic rocks from Chile and Argentina [37], it was shown that factors responsible for Na/K variation in alunite-group minerals are temperature (a significant difference occurs above 100 °C, which was not the subject of the studied localities) and Na concentrations in the source fluid. The latter depends on the composition of host rocks: higher Na concentrations are observed when andesite is leached, while K-varieties are confined to rocks with a composition from dacitic to granitic.

The thermal waters and pore solutions from the Donnoe fumarole field have varying Na/K ratios; in the same type of water, both K and Na can dominate at different sampling points [18]. In all types of water collected from the Dachnoe thermal field (Figure 7) the Na concentration is 2–3 times higher than that of K. The higher content of K at the Donnoe fumarole field may be explained by the presence of rhyodacite series among the host rocks. The wide variation of the Na/K ratio in the samples from the Donnoe fumarole field reflects the complexity of the crystallization process, which is associated with the constantly changing (locally) composition of pore solutions and physicochemical parameters, which, in particular, are sensitive to weather conditions such as the level and frequency of precipitation, temperature, and humidity. Based on data from the literature [38,39], we also suppose that the formation of alunite-group minerals at the Mutnovsky volcano is determined by very local pH parameters of the pore solution: Fe^{3+}-members (jarosite and natrojarosite) crystallize at lower pH in comparison to Al^{3+}-species (alunite and natroalunite). Alunite-supergroup minerals are commonly found in intimate association with primary silicates, and they replace them (Figure 4) [38,40]. The residue obtained from the alteration of silicates minerals leads to the formation of silicate dioxide minerals: opal, cristobalite, and tridymite. The jarosite rim around shaped gypsum crystal (Figure 4) reflects jarosite formation after gypsum.

The general formula of alum-group minerals can be expressed as $AM^{3+}(SO_4)_2 \times 12H_2O$, where *A*—monovalent cation; *M*—trivalent cation. The *A*-site is represented by Na and $(NH_4)^+$ in alum-(Na) with Na/NH_4 = 2:1 and solely by $(NH_4)^+$ cation in tschermigite and lonecreekite. There is clear compositional segregation between Al and Fe^{3+} cations: tschermigite and alum-(Na) are Al-dominant with Fe content about 10%, while lonecreekite is Fe-dominant with Al content < 6%. Ammonium minerals differ from others in that they do not contain either Na or P. It is of interest to note that the chemical formula of tschermigite from the Nizhny-Koshelevsky thermal field (associated with Koshelevsky volcano, Kamchatka, Russia) almost corresponds to the ideal; the mineral has a minor content of K (0.02 *apfu*) and Na (0.02 *apfu*) in the $(NH_4)^+$-dominant *A* site and the M^{3+} cation is represented exclusively by Al [41]. Alum-group minerals show very limited Fe^{3+} content in Al-dominant members and *vice versa*; the anion is represented almost exclusively by SO_4 and the isomorphism of monovalent cation is observed only for Na-dominant members. The crystallization of ammonium minerals occurs either near ammonia-emitting low-temperature gas-steam vents (<100 °C) where tschermigite is found in or near fumaroles (100–150 °C) around which mascagnite and letovicite are observed. The source of nitrogen has not yet been determined.

Taking into account the chemical composition (Tables 2 and 3) of Fe^{3+}- and Al-sulphates and the existence of alunite with jarosite and tschermigite with lonecreekite at the microscale (Figure 5a), we suggest that the presence of Al and Fe^{3+} at the fumarole and thermal fields of the Mutnovsky volcano leads to the formation of different minerals either Fe^{3+}- or Al-dominant rather than a complete series of solid solutions. Similar suggestions on the compositional segregation of Fe^{3+}- and Al in alunite supergroup members have been observed previously for samples from Broken Hill (New South Wales, Australia) [42], while a large compositional gap between alunite-natroalunite and jarosite-natrojarosite in terms of Na/K ratio has been shown in samples from Goldfield, Nevada (USA) and explained as features of the crystallization process [40].

Amarillite has the following chemical formula: $Na_{1.05}(Fe^{3+}{}_{0.79}Al_{0.14}Mg_{0.07})_{\Sigma 1.00}[(S_{0.955}P_{0.015})O_4]_2 \times 6H_2O$. Amarillite does not show the substitution of Na by other monovalent cations; however, Al and Mg are found in addition to Fe^{3+} and some P substitutes for S. Amarillite is a rare mineral that, in the majority of cases, occurs as a sulphide-oxidation product. However, findings of this mineral, where its formation is associated with volcanic activity, are also known. The mineral has been described at the solfataric field at Campi Flegrei, Italy [43] and in fumaroles of the Tolbachik volcano (Kamchatka, Russia) [44].

Only one Ca sulphate has been identified that is abundant at both the Donnoe fumarole field and Dachnoe thermal field—gypsum. Gypsum has a very persistent chemical composition that does not reflect a difference in the conditions of its formation, which is consistent with existing observations [45]. In general, gypsum is a common mineral of low-temperature volcanic environments since the main dehydration of gypsum occurs above 95 °C [46]. The presence of gypsum in a medium-temperature fumarole in association with mascagnite (Figure 3) indicates the possibility that the sample originally was partially dehydrated. It has been recently shown that the morphology of terrestrial gypsum from hot springs (such as the example of Uzon Kaldera, Kamchatka, Russia) is, to some extent, shaped by the microbial community that is widespread at thermal fields [47]. Thus, in accord with [47], SEM images of gypsum from volcanic environments may serve as a reference for the detection of microbial life on Mars. It is worth noting that gypsum can form in slightly different conditions at thermal fields: as surface efflorescent, well-formed crystals in argillic strata, and massive deposits [48]. In the case of efflorescent, gypsum is found as shapeless masses or fine-grained precipitate.

Baryte occurs as a very minor phase at the Dachnoe thermal field and it plays the role of a Sr concentrator. Regarding volcanic settings, baryte is described at various low-temperature environments (thermal fields) and in medium- and high-temperature fumaroles [49,50]. The enrichment of thermal water by Ba and Sr simultaneously may be due to seawater entering the deep geothermal reservoir, which agrees with [51]; however, this requires more detailed study.

4.2. Mineral-Forming Chemical Elements

The reported suite of surface efflorescent minerals herein is produced by fluid/rock interaction and appears to be the last stage of mineral formation. These phases are in constant dynamics and undergo numerous circles of recrystallization due to high variability in the location of thermal manifestations and the impact of atmospheric precipitation. The (volcanic) steam bears such main (mineral-forming) chemical elements as N (NH_4), Ca, minor Mg, S, O, and H (Table S2) and has a pH of ~5. Oxidation of the discharging fluid occurs on the day surface, so the thermal water from the fields is more acidic with a pH below 3 in most cases (Table S2), and it is enriched by elements such as Na, K, Fe, and Al that, as in the literature [52–54], are supposed to come from the leaching of the host rocks. It is interesting that the main elements in the chemical composition of minerals fully correspond to the main elements of thermal water (Tables 1–3, Figure 7). The high concentration of NH_4 is characteristic of pore solution near fumaroles. Ammonium sulphates of the Dachnoe thermal field are highly hydrated and represented by alums (tschermigite and lonecreekite), while at the Donnoe fumarole field, they are represented by anhydrous species letovicite and mascagnite, indicating the higher temperature formation (deposited around fumarole with temperature ~120–150 °C) of the latter paragenesis. The composition of the thermal water at each sampling point is unique, as it is determined by the proportion of the volcanic component, meteor water, pH, temperature, precipitation, surrounding rocks, and other parameters. Variations in these parameters lead to the formation of different mineral associations: ATM-alunogen-SiO_2, alum-gypsum-alunogen/halotrichite-SiO_2, mascagnite-gypsum, letovicite-kaolinite-tridymite, and alum-voltaite-alunogen [55] (where ATM—alunite-type minerals; alum—alum-group minerals; and SiO_2—silica dioxide minerals as opal, cristobalite, and tridymite). Taking into account the variability of the environment and the conditions of crystallization, it is practically impossible to trace the relationship between the mineral association and certain parameters. To solve such a problem, it is necessary to carry out a large number of model experiments on the influence of various parameters on phase formation.

Of particular interest is the analysis of impurity elements in minerals of the alunite group. The presence of Ca, Ba, Sr, and NH_4 in Na and K sites; Mg, Co, Ni, and Ti in Al and Fe^{3+} sites; and P in the S site agrees with the crystal chemistry of alunite-supergroup minerals [36]. At the same time, chemical elements such as Co, Ni, and Ti look the most exotic. Previously, it was shown [18] that the thermal waters of the Mutnovsky volcano are unique due to the high content of trace elements such as Ti, V, Cr, Co, and Ni, and their concentration is several times higher than at other volcanos such as Etna (Sicily, Italy), Vulcano (Aeolian Islands, Italy), El Chichon (Chiapas, Mexico), and Ebeko (Kuril islands, Russia). It has been suggested [18] that these chemical elements come from deep and uncommon sources and that they should not concentrate in the surface minerals. Our study shows the opposite; that Ti, Co, and Ni concentrate as impurities (as tenths of wt.%) in the alunite-group minerals found at the field's surface. It is noteworthy that the high concentration of Ppresents in all types of waters from the Mutnovsky volcano [18], and the highest concentration of P was found at the Donnoe fumarole field [18] that is in a good correlation with chemical analyses obtained for sulphates.

4.3. *Implication of Efflorescent Sulphates for Volcanology and Planetary Science*

Considering the main elements, the geochemical specificity lies in the presence of ammonia brought by steam and the precipitation of ammonium minerals. Of course, the discovery of ammonium minerals on Mars or other planets would be a big step, but such phases have not yet been reliably recorded. The rest of the cations, Na, K, Ca, and Ba, are likely delivered from the host rocks via leaching. This process of hydrothermal alteration by acidic solution is common for volcanic settings and artificially produced acid-mine drainage (AMD) systems [56]. The Ca sulphate is represented by gypsum, which is a very common mineral and does not show the more or less significant substitution of the Ca cation. Baryte is found in various environments; regarding volcanic systems, it tends to be found at a low-temperature. The Sr impurity in baryte is not a rarity and is more characteristic of environments where seawater has evaporated and produced baryte [57]. The K-Na alunite-type minerals (alunite, jarosite, natroalunite, and natrojarosite) are common in volcanic environments and AMD systems. The samples from the Donnoe fumarole field show that samples with different K/Na and/or Fe/Al ratios can be found in close association. This indicates that structurally similar phases with different ratios of K/Na and/or Fe/Al (referring to different mineral species) can precipitate from the same solution, probably as a result of the complex crystallization process when one mineral leads to a change in the composition of the solution and its physicochemical parameters triggering subsequent crystallization of other phases. Thus, in certain conditions, instead of obtaining one mineral of the alunite group, which could subsequently be used as a reference to specific environments, we obtain 2–4 similar minerals with different K/Na and Fe/Al ratios and even intermediate forms with K/Na ~1:1. It is unlikely that we can fully use the composition of such a complex system for a direct comparison with other terrestrial and extraterrestrial environments.

However, if we consider the specifics of the Mutnovsky volcano thermal waters in terms of trace elements, we will find that it is enriched by Co, Ni, Ti, and P [18]. The impurity of these elements is found in the surface efflorescent, in particular, in alunite-group minerals that have structural flexibility for uptake of different sizes and charge cations to the crystal structure [36]. Thus, the trace-element composition of alunite-group minerals can be considered as typomorphic to volcanic thermal waters of Mutnovsky volcano. The ability to capture and concentrate elements by alunite-type minerals from aqueous solutions is well known and even served as the basis for proposals for the use of jarosite and natroalunite for removal of heavy metals from aqueous solutions [40,58] and divalent alunite-group minerals as a host for radionuclides [58]. These phases have also been considered as matrices for the uptake of heavy metals in AMD systems, which can limit the environmental impact [58]. Although these phases are known and widely described in altered volcanic rocks in the process of hydrothermal or fumarolic activity [58–62], features of their chemical composition relative to thermal water or specification of typomorphic features have not previously been carried out, to the best of our knowledge. Thus, minerals of the alunite group formed under these conditions can carry information about the chemical composition of the contacting medium. In this case, the chemical analysis of alunite-group minerals can be used as a tool to reconstruct the conditions of mineral formation in the volcanic and paleovolcanic (fossil) settings of Earth, Mars, and other objects in the Solar system. The potential use of alunite-supergroup minerals is enhanced by their extraordinary stability over a wide range of pH values (from 1 to 10) and at temperatures at least from −140 (characteristic for Mars) up to 400 °C (characteristic for volcanic settings) [36,40]. As the detailed chemical examination of Martian rocks (either returned or in situ) shortly becomes possible, the terrestrial hot spring efflorescent sulphates, which carry information on specific geological environments that are influenced by hydrothermal activity, can be used as references for mineral assemblages, the chemical composition of minerals in terms of main and trace elements for the reconstruction of Martian geology, and possible volcanic activity. At the same time, paleohydrothermal and paleofumarole systems are frequent in areas of active volcanism. They may have varying degrees of alteration, but they contain clay minerals, minerals of the alunite group. Often, when studying such systems, it becomes necessary to obtain information about circulation in the past hydrothermal solution and its similarity with modern solutions. Our study shows that minerals of the alunite group are potentially informative phases for establishing the chemical features of such systems.

5. Conclusions

We have described modern efflorescent mineralogy of active low-temperature geothermal fields, including ammonia-emitting fumarole, which have an unusual assemblage of sulphate minerals. The most abundant minerals are sulphates of Fe-Al, Al, and Ca: halotrichite, alunogen, and gypsum. In addition, sulphates of alkaline and alkaline earth cations, Na, K, Ba, and NH_4, are widespread but form crystals and precipitates of a smaller size. These are alunite-jarosite quadrilateral defined by alunite, natroalunite, jarosite, and natrojarosite: alum-group minerals, such as tschermigite, lonecrekite, and alum-Na; amarillite; baryte; letovicite and mascagnite. The discovered mineral associations are unique and characteristic of low-temperature geothermal fields and fumaroles, where mineral formation is a low-temperature process in an acidic environment. The broadest ability for isomorphism among the identified minerals is observed for alunite-supergroup minerals that contain Ca, Ba, Sr, and NH_4 in addition to Na and K as well as Mg, Co, Ni, Ti in addition to Al and Fe, while P substitutes for S. The presence of Ni, Co, Ti, P, and NH_4 impurities in the composition of alunite-supergroup minerals reflects the unique composition of the thermal water of the Mutnovsky volcano. Taking into account the crystal-chemical variability and good chemical resistance of alunite-group minerals, they can be considered as indicators of the geochemistry of the medium leading to the alteration of volcanic rocks. In this case, the detailed chemical analyses of alunite-supergroup minerals may help to reconstruct geochemical conditions in ancient hydrothermal systems.

Supplementary Materials: The following are available online at https://www.mdpi.com/article/10.3390/min12050600/s1, Table S1: The powder X-ray diffraction data for samples shown in Figure 3 and Table S2: The chemical composition of the thermal water and gas-condensate from the Dachnoe thermal field.

Author Contributions: Conceptualization, E.S.Z. and D.A.K.; methodology, E.S.Z., D.A.K., M.A.N., A.A.N. and R.M.I.; software, E.S.Z., D.A.K., M.A.N. and R.M.I.; validation, E.S.Z. and A.A.N.; formal analysis, E.S.Z., D.A.K., M.A.N., R.M.I., V.V.S. and P.S.Z.; investigation, E.S.Z., D.A.K., A.A.N., M.A.N., R.M.I., V.V.S., A.N.K., R.A.K. and P.S.Z.; data curation, E.S.Z., D.A.K., A.A.N., M.A.N., R.M.I. and V.V.S.; writing—original draft preparation, E.S.Z., D.A.K., A.A.N., M.A.N., R.M.I. and R.A.K.; writing—review and editing, E.S.Z., D.A.K., A.A.N., M.A.N., R.M.I., V.V.S., A.N.K., R.A.K. and P.S.Z.; visualization, E.S.Z., A.A.N. and R.M.I.; project administration, E.S.Z.; funding acquisition, E.S.Z. All authors have read and agreed to the published version of the manuscript.

Funding: This research was funded by the Russian Fund for Basic Research (field work, experiments), grant number 20-35-70008; the President of Russian Federation Grants MK-451.2022.1.5 (for the study of complex Fe sulphates); and Nsh-1462.2022.1.5 (for the access to SPbU equipment in order to characterize phases available in trace amounts).

Acknowledgments: This research has been carried out using the facilities of the XRD and Geo-Environmental and Modelling Research Centers of Saint Petersburg State University (SPbU). We would like to thank our colleagues for taking part in field work and for the photos that they provided. We would like to thank the five reviewers and the academic editor for their critical comments.

Conflicts of Interest: The authors declare no conflict of interest.

References

1. Varekamp, J.C.; Ouimette, A.P.; Herman, S.W.; Flynn, K.S.; Bermudez, A.; Delpino, D. Naturally acid waters from Copahue volcano, Argentina. *Appl. Geochem.* **2009**, *24*, 208–220. [CrossRef]
2. Henley, R.W. Hyperacidic volcanic lakes, metal sinks and magmatic gas expansion in arc volcanoes. In *Volcanic Lake*; Rouwet, D., Christenson, B.W., Tassi, F., Vandemeulebrouck, J., Eds.; Springer: Berlin/Heidelberg, Germany, 2015; pp. 155–178. [CrossRef]
3. Inguaggiato, C.; García, M.Á.P.; Maldonado, L.F.M.; Peiffer, L.; Pappaterra, S.; Brusca, L. Precipitation of secondary minerals in acid sulphate-chloride waters traced by major, minor and rare earth elements in waters: The case of Puracé volcano (Colombia). *J. Volcanol. Geotherm. Res.* **2020**, *407*, 107106. [CrossRef]
4. Adams, P.M.; Lynch, D.K.; Buckland, K.N.; Johnson, P.D.; Tratt, D.M. Sulfate mineralogy of fumaroles in the Salton Sea Geothermal Field, Imperial County, California. *J. Volcanol. Geotherm. Res.* **2017**, *347*, 15–43. [CrossRef]
5. McHenry, L.J.; Carson, G.L.; Dixon, D.T.; Vickery, C.L. Secondary minerals associated with Lassen fumaroles and hot springs: Implications for martian hydrothermal deposits. *Am. Mineral.* **2017**, *102*, 1418–1434. [CrossRef]
6. McCollom, T.M.; Hynek, B.M.; Rogers, K.; Moskowitz, B.; Berquó, T.S. Chemical and mineralogical trends during acid-sulfate alteration of pyroclastic basalt at Cerro Negro volcano and implications for early Mars. *J. Geophys. Res. Planets* **2013**, *118*, 1719–1751. [CrossRef]
7. Ciesielczuk, J.; Żaba, J.; Bzowska, G.; Gaidzik, K.; Głogowska, M. Sulphate efflorescences at the geyser near Pinchollo, southern Peru. *J. S. Am. Earth Sci.* **2013**, *42*, 186–193. [CrossRef]
8. Rodríguez, A.; Van Bergen, M.J. Volcanic hydrothermal systems as potential analogues of Martian sulphate-rich terrains. *Neth. J. Geosci.* **2016**, *99*, 153–169. [CrossRef]
9. Klingelhöfer, G.; Morris, R.V.; Bernhardt, B.; Schröder, C.; Rodionov, D.S.; De Souza, P.A.; Yen, A.; Gellert, R.; Evlanov, E.N.; Zubkov, B.; et al. Jarosite and hematite at Meridiani Planum from Opportunity's Mössbauer spectrometer. *Science* **2004**, *306*, 1740–1745. [CrossRef]
10. Morris, R.V.; Squyres, S.; Arvidson, R.E.; Bell, J.F., III; Christensen, P.C.; Gorevan, S.; Herkenhoff, K.; Klingelhöfer, G.; Rieder, R.; Farrand, W.; et al. A first look at the mineralogy and geochemistry of the MER-B landing site in Meridiani Planum. In Proceedings of the 35th Lunar and Planetary Science Conference, League City, TX, USA, 15–19 March 2004. Abstract No. 2179.
11. Farrand, W.H.; Glotch, T.D.; Rice, J.W., Jr.; Hurowitz, J.A.; Swayze, G.A. Discovery of jarosite within the Mawrth Vallis region of Mars: Implications for the geologic history of the region. *Icarus* **2009**, *204*, 478–488. [CrossRef]
12. Madden, M.E.E.; Bodnar, R.J.; Rimstidt, J.D. Jarosite as an indicator of water-limited chemical weathering on Mars. *Nature* **2004**, *431*, 821–823. [CrossRef]
13. Navrotsky, A.; Forray, F.L.; Drouet, C. Jarosite stability on Mars. *Icarus* **2005**, *176*, 250–253. [CrossRef]
14. Mills, S.J.; Nestola, F.; Kahlenberg, V.; Christy, A.G.; Hejny, C.; Redhammer, G.J. Looking for jarosite on Mars: The low-temperature crystal structure of jarosite. *Am. Mineral.* **2013**, *98*, 1966–1971. [CrossRef]
15. Bishop, J.L.; Murad, E. The visible and infrared spectral properties of jarosite and alunite. *Am. Mineral.* **2005**, *90*, 1100–1107. [CrossRef]
16. Sergeeva, A.V. Infrared Spectra of Alunite-Group Minerals Formed on Thermal Fields. *J. Appl. Spectrosc.* **2019**, *86*, 371–378. [CrossRef]
17. McCollom, T.M.; Ehlmann, B.L.; Wang, A.; Hynek, B.M.; Moskowitz, B.; Berquó, T.S. Detection of iron substitution in natroalunite-natrojarosite solid solutions and potential implications for Mars. *Am. Mineral.* **2014**, *99*, 948–964. [CrossRef]

18. Bortnikova, S.B.; Gavrilenko, G.M.; Bessonova, E.P.; Lapuchov, A.S. The hydrogeochemistry of thermal springs on Mutnovskii Volcano, southern Kamchatka. *J. Volcanol. Seismol.* **2009**, *3*, 388–404. [CrossRef]
19. Košek, F.; Culka, A.; Fornasini, L.; Vandenabeele, P.; Rousaki, A.; Mirao, J.; Bersani, D.; Candeias, A.; Jehlička, J. Application of a handheld Raman spectrometer for the screening of colored secondary sulfates in abandoned mining areas—The case of the São Domingos Mine (Iberian Pyrite Belt). *J. Raman Spectrosc.* **2020**, *51*, 1186–1199. [CrossRef]
20. Rull, F.; Guerrero, J.; Venegas, G.; Gázquel, F.; Medina, J. Spectroscopic Raman study of sulphate precipitation sequence in Rio Tinto mining district (SW Spain). *Environ. Sci. Pollut. Res. Int.* **2014**, *21*, 6783–6792. [CrossRef]
21. Žáček, V.; Škoda, R.; Laufek, F.; Ek, F.K.; Jehlička, J. Complementing knowledge about rare sulphates lonecreekite, $NH_4Fe^{3+}(SO_4)_2{\cdot}12H_2O$ and sabieite, $NH_4Fe^{3+}(SO_4)_2$: Chemical composition, XRD and RAMAN spectroscopy (Libušín near Kladno, the Czech Republic). *J. Geosci.* **2019**, *64*, 149–159. [CrossRef]
22. Košek, F.; Edwards, H.G.M.; Jehlička, J. Raman spectroscopic vibrational analysis of the complex iron sulfates clairite, metavoltine, and voltaite from the burning coal dump Anna I, Alsdorf, Germany. *J. Raman Spectrosc.* **2019**, *51*, 1454–1461. [CrossRef]
23. Vakin, E.A.; Pilipenko, G.F. Mutnovsky geothermal area. In *Kamchatka in Exploration and Use of Geothermal Resources in Volcanic Areas*, 1st ed.; Aladiyev, I.T., Sugrobov, V.M., Eds.; Nauka: Moscow, Russia, 1979; pp. 36–45. (In Russian)
24. Vakin, E.A.; Kirsanov, I.T.; Pronin, A.A. Active funnel of Mutnovsky volcano. In *Bulletin of the Kamchatka Volcanological Stations*, 1st ed.; Gushchenko, I.I., Ed.; Nauka: Moscow, Russia, 1966; Volume 40, pp. 25–35. (In Russian)
25. Serafimova, E.K. Features of the chemical composition of fumarole gases of Mutnovsky volcano. In *Bulletin of the Kamchatka Volcanological Stations*, 1st ed.; Gushchenko, I.I., Ed.; Nauka: Moscow, Russia, 1966; Volume 42, pp. 56–65. (In Russian)
26. Zelensky, M.E. Transport of Elements and Minersl Forming Conditions at High-Temperature Gas Dischargees from Mutnovsky Volcano (Kamchatka). Ph.D. Thesis, Institute of Volcanology, Petropavlovsk-Kamchatsky, Russia, 2003. (In Russian).
27. Kiryukhin, A.V.; Tranbenkova, A.G.; Bortnikova, S.B.; Fazlullin, S.M. Gas and Chemical Monitoring of the Mutnovsky (Dachny) Geothermal Field Exploration (Kamchatka, Russia). In Proceedings of the World Geothermal Congress, Antalya, Turkey, 24–29 April 2005.
28. Selyangin, O.B. Structure, substance and subsurface magma chambers of the volcanoes Mutnovsky and Gorely (Mutnovsky geothermal area, Kamchatka). I. geological position of the volcanoes). *GIAB №11, «Kamchatka-3»* **2016**, *31*, 365–400. (In Russian)
29. Bortnikova, S.B.; Sharapov, V.N.; Bessonova, E.P. Hydrothermal Composition of Springs at the Donnoe Fumarole Field, Mutnovsky Volcano (Southern Kamchatka) and Problems of Their Relation with Supercritical Magmatic Fluids. *Dokl. Earth Sci.* **2007**, *3*, 410–414. [CrossRef]
30. Leonov, V.L. *Structural Conditions of High-Temperature Fluids' Localization*, 1st ed.; Nauka: Moscow, Russia, 1989; p. 104. (In Russian)
31. Slyadnev, B.I.; Shapovalenko, V.N.; Krikun, N.F. *State Geological Map of the Russian Federation, Scale 1:1,000,000 (Third Generation). Series Koryaksko-Kurilskaya. Sheet N-57-Petropavlovsk-Kamchatsky*; VSEGEI: St. Petersburg, FL, USA, 2006.
32. Britvin, S.N.; Dolivo-Dobrovolsky, D.V.; Krzhizhanovskaya, M.G. Software for processing of X-ray powder diraction data obtained from the curved image plate detector of Rigaku RAXIS Rapid II diractometer. *Zap. Ross. Mineral. Obs.* **2017**, *146*, 104–107, (In Russian with English Abs.).
33. Balić-Žunić, T.; Garavelli, A.; Jakobsson, S.P.; Jonasson, K.; Katerinopoulos, A.; Kyriakopoulos, K.; Acquafredda, P. Fumarolic Minerals: An Overview of Active European Volcanoes. In *Updates in Volcanology, from Volcano Modelling to Volcano Geology*; Nemeth, K., Ed.; InTech: Rijeka, Croatia, 2016; pp. 267–322. [CrossRef]
34. Bayliss, P.; Kolitsch, U.; Nickel, E.H.; Pring, A. Alunite supergroup: Recommended nomenclature. *Mineral. Mag.* **2010**, *74*, 919–927. [CrossRef]
35. Stoffregen, R.E.; Alpers, C.N.; Jambor, J.L. Alunite-jarosite crystallography, thermodynamics, and geochronology. *Rev. Mineral. Geochem.* **2000**, *40*, 453–479. [CrossRef]
36. Kolitsch, U.; Pring, A. Crystal chemistry of the crandallite, beudantite and alunite groups: A review and evaluation of the suitability as storage materials for toxic metals. *J. Mineral. Petrol. Sci.* **2001**, *96*, 67–78. [CrossRef]
37. Deyell, C.L.; Dipple, G.M. Equilibrium mineral-fluid calculations and their application to the solid solution between alunite and natroalunite in the El Indio–Pascua belt of Chile and Argentina. *Chem. Geol.* **2005**, *215*, 219–234. [CrossRef]
38. Hladky, G.; Slansky, E. Stability of alunite minerals in aqueous solutions at normal temperature and pressure. *Bull. Mineral.* **1981**, *104*, 468–477. [CrossRef]
39. Sergeeva, A.V.; Zhitova, E.S.; Nuzhdaev, A.A.; Nazarova, M.A. Modeling the Mineral Formation Process on Thermoanomalies with Ammonium-Sulphate Thermal Waters: The Role of Acidity (pH). *J. Volcanol. Seismol.* **2022**, *1*, 29–53.
40. Papike, J.J.; Karner, J.M.; Spilde, M.N.; Shearer, C.K. Terrestrial analogs of martian sulfates: Major and minor element systematics of alunite-jarosite from Goldfield, Nevada. *Am. Mineral.* **2006**, *91*, 1197–1200. [CrossRef]
41. Zhitova, E.S.; Sergeeva, A.V.; Nuzhdaev, A.A.; Krzhizhanovskaya, M.G.; Chubarov, V.M. Tschermigite from thermal fields of Southern Kamchatka: High-temperature transformation and peculiarities of IR-spectrum. *Zap. Ross. Mineral. Obs.* **2019**, *148*, 100–116. (In Russian)
42. Rattray, K.J.; Taylor, M.R.; Bevan, D.J.M.; Pring, A. Compositional segregation and solid solution in the lead-dominant alunite-type minerals from Broken Hill, NSW. *Mineral. Mag.* **1996**, *60*, 779–785. [CrossRef]
43. Piochi, M.; Mormone, A.; Balassone, G.; Strauss, H.; Troise, C.; De Natale, G. Native sulfur, sulfates and sulfides from the active Campi Flegrei volcano (southern Italy): Genetic environments and degassing dynamics revealed by mineralogy and isotope geochemistry. *J. Volcanol. Geotherm. Res.* **2015**, *304*, 180–193. [CrossRef]

44. Pekov, I.V.; Koshlyakova, N.N.; Agakhanov, A.A.; Zubkova, N.V.; Belakovskiy, D.I.; Vigasina, M.F.; Turchkova, A.G.; Sidorov, E.G.; Pushcharovsky, D.Y. New arsenate minerals from the Arsenatnaya fumarole, Tolbachik volcano, Kamchatka, Russia. XV. Calciojohillerite, $NaCaMgMg_2(AsO_4)_3$, a member of the alluaudite group. *Mineral. Mag.* **2021**, *85*, 215–223. [CrossRef]
45. Africano, F.; Bernard, A. Acid alteration in the fumarolic environment of Usu volcano, Hokkaido, Japan. *J. Volcanol. Geotherm. Res.* **2000**, *97*, 475–495. [CrossRef]
46. Deer, W.A.; Howie, R.A.; Zussman, J. An Introduction to the Rock-Forming Minerals. *Mineral. Mag.* **1992**, *56*, 617–619. [CrossRef]
47. Tang, M.; Ehreiser, A.; Li, Y.L. Gypsum in modern Kamchatka volcanic hot springs and the Lower Cambrian black shale: Applied to the microbial-mediated precipitation of sulfates on Mars. *Am. Mineral.* **2014**, *99*, 2126–2137. [CrossRef]
48. Van Hinsberg, V.J.; Berlo, K.; Pinti, D.L.; Ghaleb, B. Gypsum precipitating from volcanic effluent as an archive of volcanic activity. *Front. Earth Sci.* **2021**, *9*, 764087. [CrossRef]
49. Inostroza, M.; Aguilera, F.; Menzies, A.; Layana, S.; González, C.; Ureta, G.; Sepulveda, J.; Sheller, S.; Böehm, S.; Barraza, M.; et al. Deposition of metals and metalloids in the fumarolic fields of Guallatiri and Lastarria volcanoes, northern Chile. *J. Volcanol. Geotherm. Res.* **2020**, *393*, 106803. [CrossRef]
50. Pekov, I.V.; Kovrugin, V.M.; Siidra, O.I.; Chukanov, N.V.; Belakovskiy, D.I.; Koshlyakova, N.N.; Yapaskurt, V.O.; Turchkova, A.G.; Möhn, G. Antofagastaite, $Na_2Ca(SO_4)_2 \cdot 1.5H2O$, a new mineral related to syngenite. *Mineral. Mag.* **2019**, *83*, 781–790. [CrossRef]
51. Zelenski, M.; Taran, Y. Geochemistry of volcanic and hydrothermal gases of Mutnovsky volcano, Kamchatka: Evidence for mantle, slab and atmosphere contributions to fluids of a typical arc volcano. *Bull. Volcanol.* **2011**, *73*, 373–394. [CrossRef]
52. Inioue, A.; Utada, M. Hydrothermal Alteration in the Kamikita Kuroko Mineralization Area. *Min. Geol.* **1991**, *41*, 203–218. [CrossRef]
53. Martin, R.; Rodgers, K.A.; Browne, P.R.L. The nature and significance of sulphate-rich, aluminous efflorescences from the Te Kopia geothermal field, Taupo Volcanic Zone, New Zealand. *Mineral. Mag.* **1999**, *63*, 413–419. [CrossRef]
54. Zhitova, E.S.; Siidra, O.I.; Belakovsky, D.I.; Shilovskikh, V.V.; Nuzhdaev, A.A.; Ismagilova, R.M. Ammoniovoltaite, $(NH_4)_2Fe^{2+}{}_5Fe^{3+}{}_3Al(SO_4)_{12}(H_2O)_{18}$, a new mineral from the Severo-Kambalny geothermal field, Kamchatka, Russia. *Mineral. Mag.* **2018**, *82*, 1057–1077. [CrossRef]
55. Carbone, C.; Dinelli, E.; Marescotti, P.; Gasparotto, G.; Lucchetti, G. The role of AMD secondary minerals in controlling environmental pollution: Indications from bulk leaching tests. *J. Geochem. Explor.* **2013**, *132*, 188–200. [CrossRef]
56. Hein, J.R.; Zierenberg, R.A.; Maynard, J.B.; Hannington, M.D. Barite-forming environments along a rifted continental margin, Southern California Borderland. *Deep-Sea Res. Part II* **2007**, *54*, 1327–1349. [CrossRef]
57. Luo, Z.; Mu, W.; Zhou, X.; Chen, Z. Removal and immobilization of arsenic from wastewater via arsenonatroalunite formation. *Environ. Technol.* **2021**, *42*, 1–10. [CrossRef]
58. Hudson-Edwards, K.A. Uptake and release of arsenic and antimony in alunite-jarosite and beudantite group minerals. *Am. Mineral.* **2019**, *104*, 633–640. [CrossRef]
59. Owen, N.D.; Cook, N.J.; Rollog, M.; Ehrig, K.J.; Schmandt, D.S.; Ram, R.; Brugger, J.; Ciobanu, C.L.; Wade, B.; Guagliardo, P. REE-, Sr-, Ca-aluminum-phosphate-sulfate minerals of the alunite supergroup and their role as hosts for radionuclides. *Am. Mineral.* **2019**, *104*, 1806–1819. [CrossRef]
60. Graham, I.J.; Robinson, B.W. Natroalunite on Ruapehu volcano, New Zealand. *Geochem. J.* **1986**, *20*, 249–253. [CrossRef]
61. Johnston, J.H. Jarosite and akaganéite from White Island volcano, New Zealand: An X-ray and Mössbauer study. *Geochim. Cosmochim. Acta* **1977**, *41*, 539–544. [CrossRef]
62. Bhattacharya, S.; Mitra, S.; Gupta, S.; Jain, N.; Chauhan, P.; Parthasarathy, G. Jarosite occurrence in the Deccan Volcanic Province of Kachchh, western India: Spectroscopic studies on a Martian analog locality. *J. Geophys. Res. Planets* **2016**, *121*, 402–431. [CrossRef]

minerals

MDPI

Article

Characterization and Origin of Basalt-Derived Carnelian in the Mesozoic Newark Basin, New Jersey, USA

Richard A. Volkert [1,*,†], Matthew L. Gorring [2], William H. Peck [3] and Scott D. Stanford [1]

[1] New Jersey Geological and Water Survey, Trenton, NJ 08625, USA; scott.stanford@dep.nj.gov
[2] Department of Earth and Environmental Studies, Montclair State University, Upper Montclair, NJ 07043, USA; gorringm@montclair.edu
[3] Department of Geology, Colgate University, Hamilton, NY 13346, USA; wpeck@colgate.edu
* Correspondence: rvolkert@comcast.net
† This author has retired.

Abstract: Carnelian occurs locally in New Jersey in the Newark basin as medium- to coarse-size pebbles in fluvial gravel and alluvium and colluvium formed from erosion of Lower Jurassic Preakness Basalt. Vesicles and molds of glauberite are preserved on lower surfaces and botryoidal textures on the upper surfaces of some pieces. The microstructure consists of length-fast chalcedony characterized by parallel fibrous bundles overlain by repetitive, wavy extinction bands. Only peaks of α-quartz and minor moganite are recognized in X-ray diffraction patterns. Carnelian contains 97–98 wt.% SiO_2, ~1.0 wt.% Fe_2O_3, and 1.0–1.4 wt.% LOI; other major elements are <0.1 wt.%. Trace element abundances are low except for Y, Nb, Ta, W, Th, and U. Rare earth element (REE) patterns display heavy REE enrichment and large negative Eu anomalies. Most trace elements were mobilized from Proterozoic sources, whereas Si was likely derived from the alteration of basaltic glass in the Preakness. Carnelian $\delta^{18}O_{VSMOW}$ values are high and range from +18.3 to +31.2‰, comparable to global occurrences of volcanic rock-derived chalcedony. We propose that carnelian precipitated in the first Preakness flow from the mixing of hydrothermal fluid with meteoric water under conditions of low temperature (20–80 °C) and neutral to slightly alkaline pH.

Keywords: carnelian; moganite; Mesozoic Newark basin; Preakness Basalt; oxygen isotopes

1. Introduction

Carnelian is a variety of silica (SiO_2) that is mainly colored in some shade of red or orange. It is classified as chalcedony on the basis of its microcrystallinity and its fibrous texture (https://www.gemdat.org; accessed on 24 October 2022). The absence of banding in carnelian distinguishes it from agate. Because carnelian is a semiprecious gemstone that was prized by early civilizations, the focus of many studies has been mainly from an archaeological perspective, e.g., [1–3]. Some regional geological reports mention carnelians briefly [4,5]; however, studies that focus strictly on the mineralogical and geological characterization of carnelians [6,7] are scarce, especially compared to the numerous studies of other types of microcrystalline silica such as agate, opal, and chert.

A review of the archaeological and geologic literature and the mineral website Mindat (https://www.mindat.org; accessed on 24 October 2022) reveals that globally carnelian occurs most commonly in weakly to unmetamorphosed volcanic rocks, with a large number of occurrences in basalt of varying geologic ages. Carnelian is also present in sedimentary rocks, such as the Miocene conglomerate of the Babaguru Formation in India [4] and the Proterozoic Salmi Formation in Karelia, Russia [8], which is composed of material eroded from the underlying Deccan and Priozersk basalts, respectively, and also in stream gravels derived from erosion of these host rocks.

Carnelian in New Jersey is present only in the central part of the Mesozoic Newark rift basin (Figure 1) at a few closely spaced locations mainly along Green Brook and Stirling

Brook (Figure 2). Information is lacking on the first find of carnelian in the study area, or on the early history of its collection there, beyond a brief mention of its occurrence in mineral collecting guides from the middle 1900s [9,10]. Varieties of quartz reported from the Newark basin in New Jersey include amethyst, smoky quartz, clear quartz crystals, pale-colored chalcedony, and agate in the Jurassic basalt formations [11], and amethyst, milky quartz, clear quartz crystals, pale-colored chalcedony, and opal in Jurassic diabase and references therein [12]. No reference is made to carnelian or other synonymous names such as cornelian, red chalcedony, or sard in any of the New Jersey mineral inventories [13–15] or studies on the occurrence and diversity of zeolites and associated secondary minerals in basalt and diabase in the Newark basin [11,12,16–21].

We are unaware of any prior scientific inquiries undertaken to study carnelians in New Jersey, and this paper presents the results of the first detailed study of samples collected from locations at Green Brook and Stirling Brook (Figure 2). This study was designed to characterize carnelian in the context of the field relationships of host bedrock and surficial lithologies, in combination with results from optical petrographic microscopy, X-ray diffraction (XRD) analysis, major and trace element geochemistry, and oxygen isotope analysis. Collectively, these data provide important constraints on the provenance of the carnelian, as well as the characteristics and source of the mineralizing fluid from which it precipitated. Although carnelian has not been recognized in situ in bedrock in the study area, the results of this study provide adequate information from which reasonable interpretations regarding its origin may be made. We refer to the samples of this study as carnelian because of their predominant red color, although we recognize they could also correctly be termed a red-colored variety of chalcedony.

2. Geologic Setting

The Mesozoic Newark basin in northern and central New Jersey is a northeast-trending half-graben formed during Triassic and Jurassic rifting of the eastern North American margin. It is in fault contact on the west with Mesoproterozoic rocks of the New Jersey Highlands and is unconformably overlain on the east by Cretaceous sediments of the Atlantic Coastal Plain (Figure 1). The Newark basin contains a succession of west-dipping, interbedded terrestrial sedimentary formations, and mafic igneous rocks. These include Upper Triassic to Lower Jurassic conglomerate, sandstone, siltstone, and shale of fluvial and lacustrine origin, three Lower Jurassic continental flood basalt formations (Orange Mountain Basalt, Preakness Basalt, and Hook Mountain Basalt), and several Lower Jurassic diabase bodies, the largest of which is the Palisades sill (Figure 1).

Bedrock formations in the area of the carnelian locations include the sedimentary Feltville Formation and the overlying Preakness Basalt (Figure 2). The 145 m thick Feltville Formation consists of reddish-brown or light-grayish-red, fine- to coarse-grained sandstone, siltstone, and silty shale composed of varied amounts of quartz, white mica, and feldspar that was deposited in a fluvial environment [22,23]. Near the base of the Feltville are two 3 m thick sequences of dark gray to black, carbonaceous shale, light-gray limestone, calcareous siltstone, and gray or olive shale deposited in a shallow- to deep-water lacustrine setting [22–24]. About 0.5 m of the uppermost beds of the formation are thermally altered along the contact with the Preakness Basalt.

The 315 m thick Preakness Basalt has a geochemical composition that corresponds to high-iron, quartz-normative tholeiitic basalt (Table 1). It consists of at least three flows of subaerially erupted dark greenish-gray to black, generally fine-grained basalt that yield a zircon U-Pb crystallization age of 201 $\pm$ 0.03 Ma [25]. It contains (modal%) 45%–51% andesine or labradorite, 38%–44% augite, 4%–6% Fe-Ti oxides, and 3%–6% mesostasis composed of fine-grained intergrown plagioclase, K-feldspar, Fe-Ti oxides, apatite, quartz, and altered clinopyroxene [26]. Zeolites are rare in the Preakness compared to their abundance in the stratigraphically overlying Hook Mountain Basalt and the underlying Orange Mountain Basalt. The base of the 125 to 145 m thick Preakness first flow contains vertical, slender, and prismatic columnar joints up to 0.5 m wide and 6 m long [27] that rest on the contact with the Feltville Formation. At several locations, including the study area, the base of the first flow is pillowed and vesicular [22,28,29], indicating a local subaqueous eruption. Where present, pillows are as much as 0.6–0.9 m in diameter and occur over a 3- to 10- m- thick interval [28,29]. They are in contact with an intercalated layer of gray or brown basaltic glass as much as 2.5 cm thick [30]. The pillowed intervals are overlain by slender columnar jointed basalt that elsewhere defines the base of the first flow.

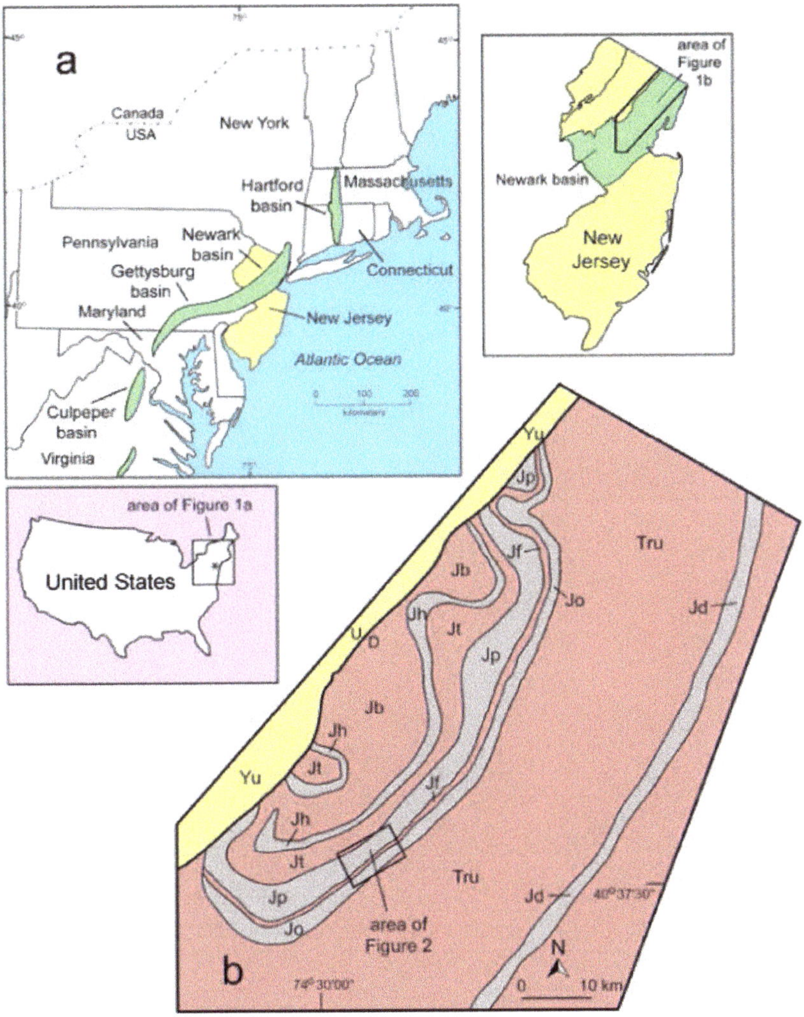

Figure 1. (**a**) Distribution of Mesozoic-age rift basins (green) in eastern North America. Small inset maps locate New Jersey (asterisk) in the United States and the Newark basin in northern New Jersey. (**b**) Simplified bedrock geologic map of igneous (gray) and sedimentary formations (red) in the northern and central parts of the Newark basin modified from [31]. Symbols: Jd, Palisades diabase; Tru, undivided Triassic sedimentary formations; Jo, Orange Mountain Basalt; Jf, Feltville Formation; Jp, Preakness Basalt; Jt, Towaco Formation; Jh, Hook Mountain Basalt; Jb, Boonton Formation; Yu, undivided Mesoproterozoic rocks. U and D show normal movement sense on the Mesozoic border fault separating Mesoproterozoic rocks and Mesozoic rocks in the Newark basin.

Coarse-grained gabbroid (Figure 3a) forms concordant layers that are as much as 80 m thick within the middle and upper parts of the first flow (Figure 2). Gabbroid layers have sharp upper contacts and gradational lower contacts with the host basalt from which it was formed through fractionation [26]. The geochemical composition of gabbroid and intercalated glass-rich gabbroid is given in Table 1. The mineralogy of gabbroid is comparable to that of fine-grained host basalt except for the coarse grain size in which clinopyroxene and plagioclase attain lengths of up to 2 cm. Gabbroid layers 12–36 m beneath the top of the first flow contain dark red or black vitreous glass (Figure 3b). The uppermost part of the first flow consists of blocky and massive basalt overlain at the top by a highly vesicular and amygdaloidal layer ~2.5 m thick (Figure 3c).

Table 1. Geochemical composition of lithologies in the Preakness Basalt.

	Basalt; First Flow Base (n = 4)	Basalt; 20–75 m above Base (n = 17)	Gabbroid; 85–130 m above Base (n = 5)	Glass-Rich Gabbroid (n = 2)	Vesicular Gabbroid 140 m above Base (n = 2)	Basalt; 25 m above Gabbroid (n = 2)	Basalt; Top of First Flow (n = 2)	Basalt; Second Flow Base (n = 1)
SiO_2	51.59	52.43	53.35	53.48	53.93	52.72	50.30	52.75
TiO_2	1.05	0.91	1.38	1.15	1.74	1.05	1.06	1.09
Al_2O_3	14.35	14.42	12.18	10.62	10.58	13.91	12.97	13.95
$Fe_2O_3{}^t$	13.19	11.81	14.63	15.34	16.66	13.59	13.65	13.19
MnO	0.21	0.20	0.22	0.23	0.25	0.21	0.24	0.22
MgO	5.54	6.33	4.75	3.41	3.92	5.48	8.01	6.04
CaO	10.90	9.50	7.05	7.88	5.72	8.52	6.79	10.05
Na_2O	2.98	3.03	3.69	4.51	2.77	3.54	3.97	2.49
K_2O	0.62	0.69	0.46	0.06	1.10	0.79	0.37	0.62
P_2O_5	0.14	0.17	0.20	0.63	0.33	0.13	0.18	0.14
LOI	0.72	0.89	1.77	2.33	2.46	0.85	2.29	n.d.
Rb	18	24	28	n.d.	n.d.	25	n.d.	20
Sr	151	143	114	69	108	152	137	132
Y	27	27	40	n.d.	n.d.	27	27	26
Zr	87	83	122	207	168	87	93	86
Nb	4.5	3.8	7.1	n.d.	n.d.	4.3	3.9	4.7
Hf	2.3	2.1	3.8	n.d.	n.d.	2.3	2.1	2.4
Ta	0.30	0.34	0.58	n.d.	n.d.	0.34	0.33	0.40
Pb	6.0	6.0	5.3	n.d.	n.d.	4.0	7.0	6.0
Th	2.0	2.0	4.0	n.d.	n.d.	2.0	2.0	2.0

Major elements in wt.%; trace elements in ppm; $Fe_2O_3{}^t$, total iron as Fe_2O_3; LOI, loss on ignition; n, number of samples; n.d., not determined. Data sources: [26,32].

The first and second Preakness flows are separated by ~2–3 m of thin-bedded, reddish-brown, fluvial siltstone and shale (Jps1) deposited on the top of the first flow (Figures 2 and 3c). The base of the 70 m thick second flow typically consists of massive basalt containing pipe vesicles overlain by widely spaced, blocky, columnar joints. The rest of the second flow is composed of relatively massive basalt intercalated with a few very thin gabbroid layers.

3. Methods

Three samples of carnelian from Green Brook, two red and one reddish-orange, and two samples from Stirling Brook, one reddish-orange and one yellow, were analyzed at Montclair State University on a Philips X'Pert MPD X-ray diffractometer. Samples were wet ground in a porcelain mortar in acetone and powdered to ≤200 mesh. They were analyzed using CuKα radiation, with runs between 5° and 75° 2θ at 45 kV and 40 mA at a step size of 0.02° and a scan rate of 1.2°/min. The crystallite size of the 101 quartz peak was calculated using the Scherrer equation $K\lambda/(\beta\cos\theta)$, in which K is a dimensionless shape factor = 0.94, λ is the wavelength in Å of CuKα = 1.5418, β is the full width at half maximum (FWHM) °2θ, and θ is $\frac{1}{2}$ of the 101 peak position °2θ. The lattice microstrain was calculated using the equation $\beta/4\tan\theta$, in which β is the FWHM °2θ and θ is the 101 peak position in °2θ.

Because of insufficient sample material from Stirling Brook, three samples of red carnelian from Green Brook were analyzed for major, trace element, and REE abundances at Activation Laboratories in Ontario, Canada. The samples were pulverized to 95% 200 mesh in a mild steel mill. Powders underwent lithium metaborate/tetraborate fusion followed by rapid digestion in a weak nitric acid solution. Whole-rock analysis of major elements, Sc, V, Sr, and Ba, was determined by inductively coupled plasma (ICP) optical emission spectrometry. All other trace elements and REE were analyzed by ICP mass spectrometry. Loss on ignition was determined by gravimetric analysis.

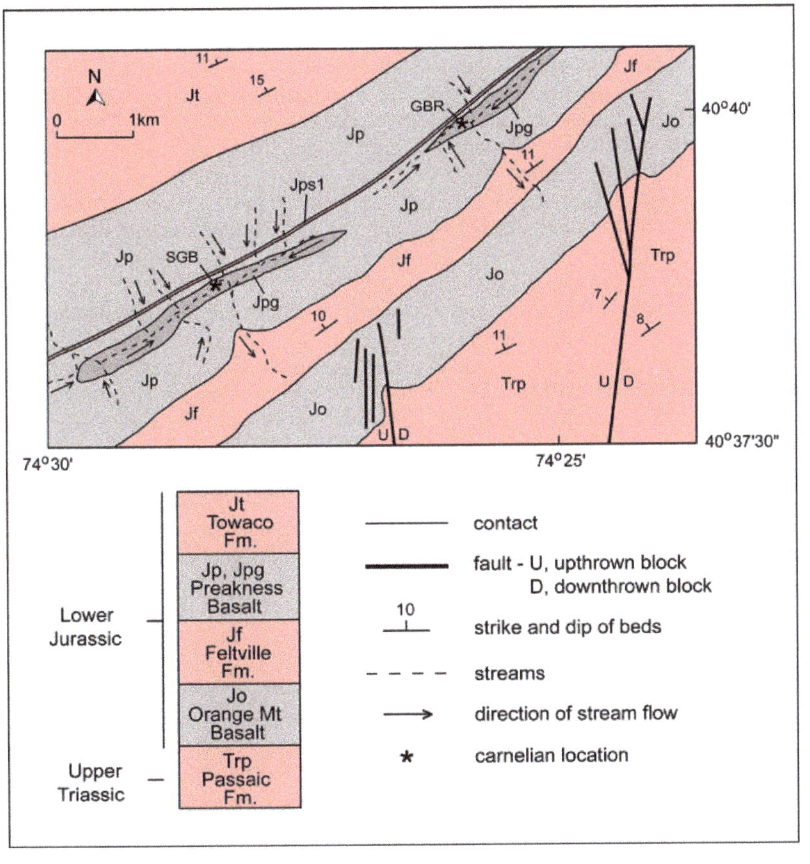

Figure 2. Bedrock geologic relationships in the study area modified from [33]. Carnelian sample locations (shown as asterisks) are labeled GBR (Green Brook) and SGB (Stirling Brook) and are keyed to samples in Tables 2–4. Jpg, gabbroid bodies; Jps1, the sedimentary layer deposited on top of the first Preakness flow. Other symbols as in Figure 1.

Six samples of carnelian from Green Brook, three red, one reddish-orange, and two yellow, were analyzed for oxygen isotope ratios at the Stable Isotope Laboratory at Northern Illinois University. Carnelian separates weighing 1.0–1.5 mg were heated with a PhotonMachines Fusions 10.6μm CO_2 laser in the presence of 10 kPa of BrF_5 to generate O_2 gas [34]. Oxygen was transferred cryogenically by adsorption on a zeolite trap to a Thermo MAT253 isotope ratio mass spectrometer for analysis. $\delta^{18}O$ is reported in conventional per mil values relative to Vienna Standard Mean Ocean Water (VSMOW) and calibrated to IAEA quartz standard NBS-28 ($\delta^{18}O$ = 9.6‰) and an internal quartz standard ($\delta^{18}O$ = 18.2‰).

Five out of the six samples analyzed were duplicated or triplicated. On the day of analysis, NBS-28 had an average precision (1σ) of ±0.11‰ (n = 4), and the internal standard had an average precision of ±0.10‰ (n = 3).

4. Carnelian

4.1. Field Relationships

Carnelian in the study area is recognized from only a few closely spaced locations that are underlain by the Preakness Basalt, and more specifically by coarse-grained gabbroid in the upper part of the first flow (Figure 2). Gabbroid is intermittently exposed at these locations as small outcrops along stream banks or pavement outcrops in stream beds (Figure 3d). Gabbroid attains a calculated thickness of about 82 m at Green Brook and 49 m at Stirling Brook, the former being one of the thickest bodies recorded in the Preakness in the Newark basin. The trend of the carnelian locations falls along a prominent northeast-striking valley that is bordered by ridges to the northwest and southeast. These ridges are underlain by the Preakness and are concordant to the strike of outcropping bedrock formations (Figure 2). The western ridge is underlain by the second flow and the eastern ridge by the first flow and intercalated gabbroid layers. The western part of the valley between the ridge crests marks the eroded top of the first flow that was subaerially exposed at 201 Ma long enough for ~2 m of sediment to have been deposited on the vesicular flow top (Figure 3c) prior to the eruption of the second flow.

Much of the bedrock in the study area is covered by a veneer of weathered basalt and colluvium as much as 18 m thick composed of angular to subangular basalt pebbles and cobbles with a clayey silt to clayey sand matrix [35]. On gentle footslopes and floodplains, colluvium and weathered basalt are overlain by up to 3 m of alluvium composed of silty sand to clayey silt [35]. Carnelian is found principally in fluvial gravel bars deposited on gabbroid in Green Brook (Figure 3d) and Stirling Brook. Gravel bars at the carnelian locations are composed entirely of Preakness Basalt and gabbroid (Figure 3e). Absent in the gravel is reddish-brown sediment eroded from the Feltville Formation or the thin layer of sediment that separates the first and second Preakness flows. At Stirling Brook, carnelian is also present in fine-grained alluvium and slope wash deposited on a gentle terrace adjacent to the stream (Figure 3f).

It is particularly noteworthy that the carnelian locations are present south of the Wisconsinan terminal moraine and occur outside the basins of glacial lakes. The Stirling Brook location (SGB in Figure 2) is south of the limit of an early Pleistocene glaciation, and no glacial erratics have been observed in its drainage basin. The headwaters of the northeastern tributary of Green Brook (northeast of GBR in Figure 2) are within the early Pleistocene glacial limit because there is an occurrence of till of this glaciation there [35]. However, carnelian has not been observed nor reported in this till. The southwestern tributary of Green Brook (on which GBR is located in Figure 2) drains from unglaciated terrain, and no erratic material occurs in its basin. These observations indicate that carnelian was derived from a local source and that it was not glacially eroded and transported from Mesozoic or older bedrock lithologies that crop out to the north or west.

4.2. Provenance of New Jersey Carnelian

Because carnelian has not been observed in situ in bedrock outcrops in the study area, it was necessary to evaluate a number of potential sources. Parts of the Newark basin in New Jersey were inhabited historically by Lenni Lenape Indigenous peoples, so we first considered that carnelian may have been imported to the study area through trade from outside of the region. This seems unlikely for several reasons. If carnelian was introduced as a product of trade, then it should be found over a more widespread area and not confined to the very small area of this study. Furthermore, none of the pieces of carnelian collected show evidence of having been worked, suggesting that smaller pieces are naturally sized and shaped and do not represent debitage. The archaeological literature for New Jersey mentions chert, flint, and jasper but does not indicate that carnelian or red chalcedony were worked as lithic material by the Lenape [36]. We thus conclude that carnelian was not imported from outside of the study area.

Figure 3. (**a**) Outcrop photograph of Preakness gabbroid from the study area representative of bedrock beneath the carnelian locations. The pencil is 14 cm. (**b**) Hand sample of glass-rich gabbroid from the study area. The coin is 19 mm. (**c**) Vesicular and amygdaloidal top of the Preakness first flow overlain above hammer by reddish-brown shale and shaly siltstone. (**d**) Preakness gabbroid along Green Brook in stream bank and stream bed. Angular to subangular gabbroid boulders in the channel were eroded from nearby outcrops. Small gravel bars are deposited on pavement outcrops of gabbroid in stream beds and between the boulders. (**e**) Fluvial gravel bar along Green Brook composed of mainly cobble- to coarse-gravel-size alluvium of Preakness Basalt and gabbroid. Finer gravel has been winnowed out and transported downstream during times of stream flooding. (**f**) Pits dug by collectors into alluvium and colluvium deposited near the base of the slope along Stirling Brook (visible in the top center of the photo). Carnelian is also found in gravel bars in the stream bed.

Carnelian may have formed as a silica-rich layer or part of a clast population within a sedimentary formation in, or near, the study area. However, the available evidence does not support a sedimentary provenance on the basis of the following. Carnelian has not been recognized in nodules or siliceous layers in sedimentary formations of the Newark basin, either in outcrop or in drill-core samples that have penetrated their entire thickness [24,31]. More specifically, it is not recognized in sedimentary formations in the study area [33]. Therefore, the silicification of carbonate-rich beds or the complete replacement of carbonate nodules by silica in a manner described by [37] is not supported by geologic relationships. Furthermore, carnelian or red chalcedony is not a lithologic component of clasts in any of the conglomerates or pebbly sandstones in the Newark basin. Finally, eroded remnants of sedimentary rock that occur stratigraphically above, beneath, or within the Preakness Basalt are absent in the stream gravels in which carnelian is found.

The Orange Mountain Basalt (Figure 2) contains abundant zeolite and secondary silica mineralization in vesicles, vugs, and interstices between pillows. The first flow has undergone locally extensive hydrothermal alteration [38], and glauberite casts are recognized in pillowed intervals in the upper part of the basalt [11,19,21,22]. It is tempting to invoke the Orange Mountain as the source of carnelian, but there are fundamental problems in doing so. Carnelian is not recognized in Orange Mountain Basalt despite excellent exposure of the entire thickness of the basalt formation in a quarry <2 km southeast of the study area. Additionally, the development of the deep valley between Orange Mountain and Preakness Basalt that is underlain by the Feltville Formation preceded the formation of the valleys in which carnelian is present because they are drained by headwater tributaries of the streams in the intermontaine valleys (Figure 2). This would have prevented any river flow or material eroded from Orange Mountain and Feltville from entering the valleys of the study area. This is evidenced by the fact that the basalt colluvium mixed with carnelian is representative of the Preakness but not Orange Mountain or Feltville Formation.

We conclude that Preakness Basalt is the source of the carnelian based on the available evidence. Streams in the valley in which carnelian is found have historically received bedrock eroded from outcrops of the Preakness first and second flows (Figure 2). The location of carnelian occurrences in fluvial gravel deposited on pavement outcrops of Preakness gabbroid and also at the foot of hillslopes underlain by the Preakness in unglaciated terrain strongly suggest that carnelian originated from this basalt formation. Neogene fluvial gravels, containing rounded pebbles of quartz, quartzite, and chert, in upland settings in central New Jersey south of the study area (and south of the Lower Jurassic basalt formations), indicate that a regional river drained southward across northeastern New Jersey in the middle to late Miocene [39]. This river plain was at, or above, the elevation of the highest summits of the basalt ridges and >50 m above the elevation of the carnelian locations. Any gravel carried by this river would have been completely eroded during the formation of the modern valleys during the Pliocene and Pleistocene, and no Neogene fluvial gravel clasts have been observed in the upland areas of the basalt formations. Furthermore, the absence of significant abrasion and rounding of the carnelian pieces, in combination with the preservation of features. such as botryoidal textures and glauberite molds, do not support the long-distance fluvial transport of carnelian and are more compatible with its derivation from a local source.

Two different scenarios are advanced to explain the apparent absence of carnelian in Preakness Basalt or gabbroid outcrops at the present erosion level. (1) Over time, carnelian was largely removed from the bedrock source through erosion, and what remains is preserved in gravel bars and fluvial sediments; or (2) carnelian continues to be eroded from the bedrock source that is now covered in stream beds by a veneer of alluvium and is replenished periodically during times of erosive stream flooding. The second scenario seems more likely given that carnelian continues to be found despite decades of robust collecting.

4.3. Physical Characteristics

Samples of carnelian are massive and compact, translucent, have a waxy luster, and are microcrystalline. They are mainly colored red (Figure 4a) and much less commonly reddish-orange or yellow. Banding is absent in all pieces collected, and none contain macroscopic quartz crystals. Carnelian pieces are subangular to subrounded (Figure 4a), and they vary in diameter from 1.0 to 5.5 cm, corresponding to a medium to very coarse pebble gravel. Various features are recognized in the carnelian samples, but these are not present in all pieces. A vesicular texture that is convex upward into the carnelian (Figure 4b) appears to be an attachment surface to the bedrock, suggesting that it likely precipitated over an amygdaloidal layer. Some pieces display a botryoidal texture (Figure 4c) presumed to have formed on upper surfaces. Molds of dissolved glauberite crystals as much as 4 mm wide and 1.5 cm long (Figure 4d) around which the carnelian formed are preserved on some surfaces. The identification of glauberite was made through comparison with images of similar glauberite crystals from the Newark basin featured in previous publications [11,40].

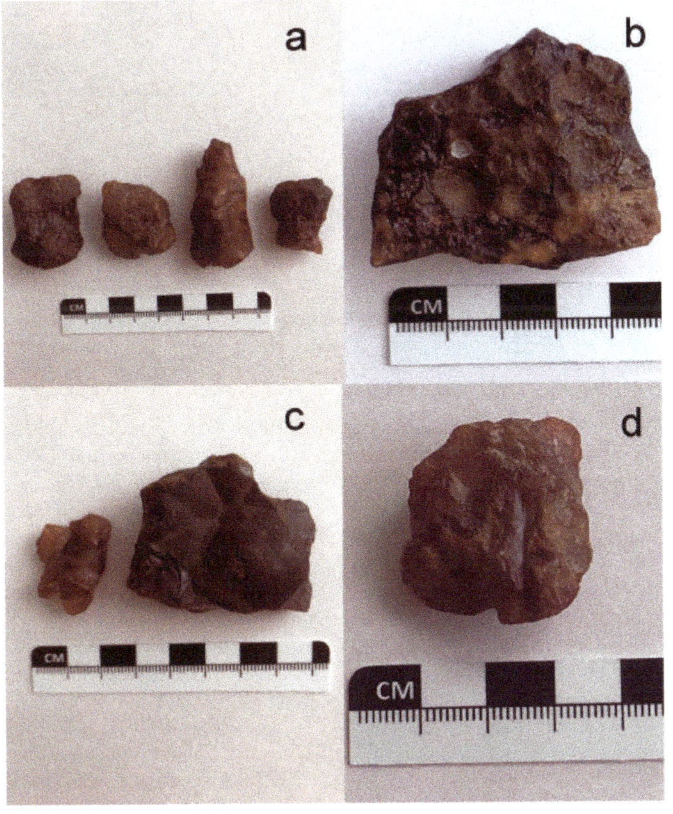

Figure 4. Representative features preserved on surfaces of carnelian from Green Brook. (**a**) Characteristic pieces of reddish-brown to red, subangular to subrounded, pebble gravel. (**b**) The lower surface displays a spheroidal, convex-upward texture presumed to have formed along an attachment surface on amygdaloidal basalt. (**c**) Botryoidal texture preserved on upper surfaces. (**d**) Lower surface displaying molds of glauberite.

4.4. Optical Petrographic Microscopy and X-ray Diffractometry

A representative sample of red carnelian from Green Brook was examined by optical petrographic microscopy, and the micrographs are presented in Figure 5. The sample is composed of microcrystalline quartz in which individual parallel fibrous bundles nucleate from an attachment surface and display elongated growth that intersects adjacent bundles (Figure 5a). These fiber bundles are optically length fast. They are in contact along the direction of elongation with repetitive extinction bands that have a zig-zag appearance due to twisting around the fiber axis (Figure 5b). The thickness, or twist period, remains fairly uniform along the entire fiber length.

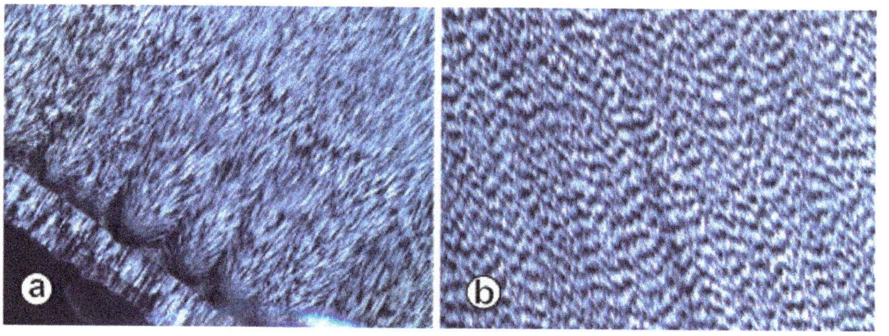

Figure 5. Optical photomicrographs in cross-polarized light showing the microstructure of carnelian from Green Brook (sample GBR 1). (**a**) Length-fast parallel fibrous bundles nucleating from a thin band of microcrystalline quartz presumed to be an attachment surface to basalt. The fiber elongation direction is toward the upper right. (**b**) Parallel aggregates of twisted fibers showing rhythmic extinction bands (wrinkle bands). The direction of elongation is toward the bottom. The horizontal field of view is 2 mm in both photomicrographs.

XRD patterns of all carnelian sampled from the Green Brook and Stirling Brook locations are similar and are dominated by major peaks of α-quartz at 3.34–3.32 Å, 4.25–4.23 Å, and 1.81 Å (Figure 6). Also observed are minor peaks at 4.45 Å, 4.38 Å, 3.09 Å, and 2.88 Å that correspond to the length-slow silica polymorph moganite [41,42]. Between about 20° and 60°, 2θ all of the XRD patterns display additional small peaks that also correspond to moganite (Figure 6). Especially significant is the fact that none of the forms of opal (C, CT, or A), cristobalite, or calcite were identified in the XRD patterns.

Quantification of the moganite content in the carnelian sampled was not performed but must be at least 5% because at contents less than that moganite peaks are unable to be resolved from quartz peaks [43]. The recognition of moganite in association with the sampled carnelian is noteworthy because to the best of our knowledge moganite has not been reported previously in New Jersey. However, a carnelian/moganite association is not unique and has been identified in carnelian analyzed from Libya that contains between 3 and 20% moganite [6]. Moganite is also reported as a common component in global occurrences of agate [43].

The crystallite size of carnelian was calculated from the 101 quartz peak using the Scherrer equation (see the Methods section) that yielded values of 434–542 Å for samples from Stirling Brook and 228–333 Å from Green Brook (Table 2). Calculation of the crystal microstrain (lattice dislocations and point defects; see the Methods section) reveals a similar variation between the two locations, with lower values of 0.078–0.098 in samples from Stirling Brook compared to values of 0.126–0.184 in samples from Green Brook (Table 2). Variation in the crystallite size between the Stirling Brook and Green Brook samples is not due to differences in the temperature or pH of the mineralizing fluid or the age of the samples because the geochemical data indicate that precipitation occurred under comparable conditions, and the field relationships imply that their formation was roughly contemporaneous. We interpret variation in the crystallite size of the samples as a result of the moganite content and amount of lattice microstrain. Moganite is metastable and transforms into a more stable α-quartz through dissolution and loss of internal water as it recrystallizes [43]. Its transformation is also enhanced by decreases in the amount of silica in the fluid and in the amount of lattice defects present [44]. All carnelian samples have similar values of SiO_2, so this is probably not a factor, but samples from Green Brook have a higher percent of lattice microstrain (Table 2). A correlation between the broadening of the

100 quartz peak at ~20° 2θ and the increased amount of moganite was found by [45]. Peak broadening in all varieties of microfibrous quartz is also attributed to small crystallite size and increased lattice microstrain [46]. Carnelian from Green Brook displays a broadened peak width (Figure 6) that is consistent with higher amounts of moganite and lattice microstrain compared to the samples from Stirling Brook.

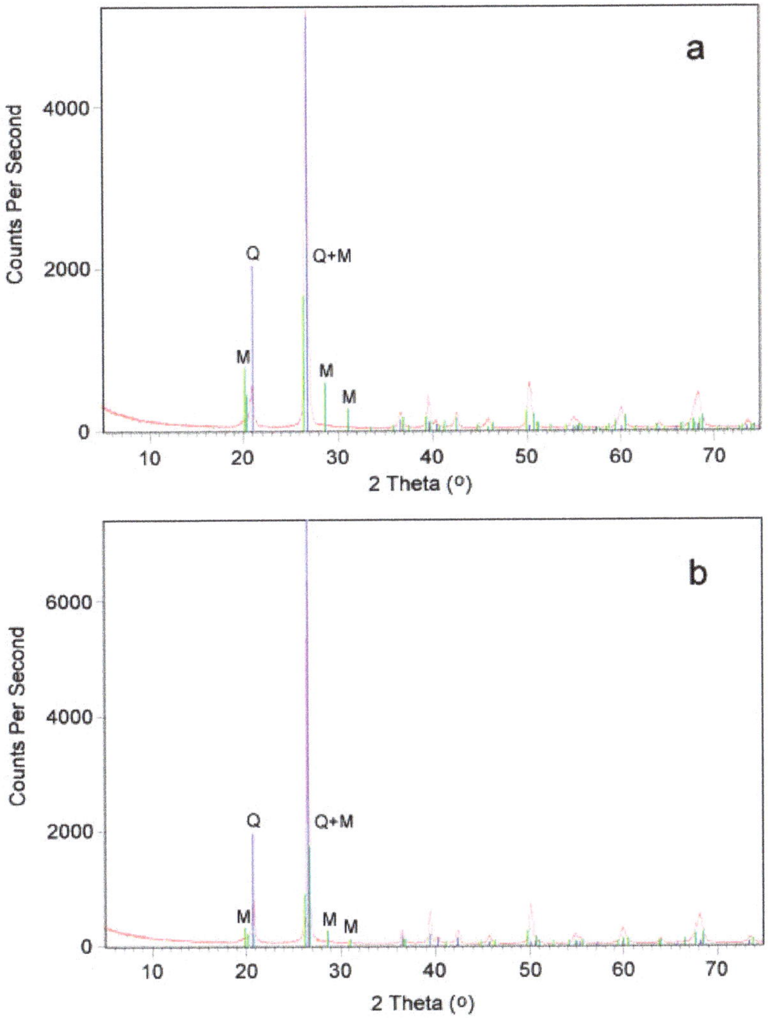

Figure 6. X-ray diffraction patterns of carnelian from New Jersey. (**a**) Sample GBR 1 from Green Brook. (**b**) Sample SGB 1 from Stirling Brook. The principal index peaks of quartz (Q) are shown in blue and moganite (M) is shown in green.

Table 2. Characterization of crystallite size and microstrain from the 101 peaks of the carnelian samples from Green Brook (GBR) and Stirling Brook (SGB).

Sample	Position °2θ	FWHM °2θ	$C_{(s)}$ Å	Lattice Strain %
GBR-1	26.638	0.3346	255	0.165
GBR-2	26.805	0.2558	333	0.126
GBR-3	26.644	0.3739	228	0.184
SGB-1	26.635	0.1968	434	0.098
SGB-2	26.633	0.1574	542	0.078

FWHM, full width at half maximum of the peak; $C_{(s)}$, crystallite size.

4.5. Geochemistry

The geochemical analyses of the carnelian samples are provided in Table 3. Major elements are dominated by SiO_2 and Fe_2O_3, which average 97.51 ± 0.31 and 1.09 ± 0.07 wt.%, respectively. Loss on ignition (LOI) is also high and averages 1.27 ± 0.21 wt.%. Elements such as Al_2O_3 and TiO_2 that readily substitute for Si in the quartz structure are quite low, averaging <0.2 and <0.01 wt.%, respectively, as are the rest of the major elements that are present in amounts <0.07 wt.% (Figure 7). Abundances of transition elements all fall below the analytical detection limits. Values of most other trace elements are low except for Y (5–94 ppm), Nb (68–260 ppm), Ta (14–51 ppm), W (5–19 ppm), Th (8–48 ppm), and U (14–77 ppm). The REE are present in amounts above detection limits, and total REE abundances (ΣREE) range from 15 to 85 ppm. Chondrite-normalized REE patterns are parallel and display a progressive increase from La to Lu that differs markedly from average Preakness Basalt and gabbroid (Figure 8a). All of the samples are significantly enriched in heavy REE (HREE) relative to light REE (LREE) (Figure 8a) and have $(La/Yb)_N$ of 0.01–0.02, $(La/Sm)_N$ of 0.05–0.12, and a $(Gd/Yb)_N$ of 0.11–0.60, normalized to chondrite. The samples have large negative Eu anomalies of 0.020–0.026 and nominal Ce anomalies of 0.87–1.01 (Table 3). Yttrium is fractionated from Ho in all samples and displays negative anomalies in two samples (Y/Ho = 3.9–7.9) and a small positive anomaly in the third (Y/Ho = 29.4) compared to chondrite and MORB that have Y/Ho of 27.7 [47]. Abundances of incompatible elements normalized to the upper continental crust display similar patterns that are characterized by negative anomalies of Rb, Ba, La, Sr, and Ti and positive anomalies of Th, U, Nb, and Ta (Figure 8b). The very low values of CaO and Sr in the carnelian are consistent with the absence of associated carbonate minerals.

Numerous investigations have shown that the red color of chalcedony, and by extension carnelian, is attributed to inclusions of Fe-oxide/hydroxide minerals [48,49], and more specifically to hematite [44,50]. Studies involving spectroscopic analysis of agate reveal a correlation between increased hematite concentrations in layers colored red or reddish-orange and increased goethite concentrations in those colored yellow [44]. We propose that the presence of hematite and/or goethite are responsible for the predominant red and less common yellow color, respectively, in the New Jersey carnelian. The presence of rutile is dismissed because of the extremely low values of TiO_2 (Table 3).

Table 3. Major and trace element concentrations of the carnelian samples from Green Brook.

Sample	GBR-02-1	GBR-02-2	GBR-02-3	Sample	GBR-02-1	GBR-02-2	GBR-02-3
SiO_2 (wt.%)	97.86	97.37	97.29	La	0.76	0.35	0.09
TiO_2	0.012	0.006	0.005	Ce	3.65	2.02	0.58
Al_2O_3	0.18	0.15	0.10	Pr	0.82	0.57	0.14
Fe_2O_3	1.06	1.18	1.04	Nd	4.69	4.46	0.71
MnO	0.013	0.014	0.012	Sm	3.90	4.64	0.67
MgO	0.01	0.01	0.01	Eu	0.033	0.04	0.005
CaO	0.07	0.04	0.03	Gd	3.82	7.92	0.68
Na_2O	0.05	0.07	0.07	Tb	1.69	2.06	0.28
K_2O	0.02	0.03	0.03	Dy	15.60	15.30	2.64
P_2O_5	0.01	0.01	0.01	Ho	3.74	3.19	0.59
LOI	1.03	1.36	1.41	Er	14.10	9.99	2.47
Total	100.31	100.24	100.01	Tm	3.21	1.55	0.556
Ga (ppm)	2.0	1.0	1.0	Yb	25	10.50	5.09
Ge	0.9	4.6	2.9	Lu	3.98	1.44	0.768
Rb	1.0	1.0	1.0	Hf	1.5	0.2	0.3
Sr	4.0	3.0	3.0	Ta	51.1	41.4	14.4
Y	14.6	93.9	4.7	W	18.7	5.1	5.8
Zr	19	4.0	6.0	Pb	15	32	<5.0
Nb	260	158	68.3	Th	48.2	19.9	7.94
Sn	5.0	1.0	1.0	U	76.6	30.9	13.7
Sb	<0.2	<0.2	0.2	ΣREE	84.99	64.03	15.27
Ba	19	12	11	Ce/Ce*	1.00	0.87	1.01
Sr	4.0	3.0	3.0	Eu/Eu*	0.026	0.020	0.023

Fe_2O_3, total iron as Fe_2O_3; LOI, loss on ignition. $Ce/Ce^* = 2Ce_N/(La_N + Pr_N)$; $Eu/Eu^* = Eu_N/[(Sm_N + Gd_N)/2]$, normalized to chondrite values of [51].

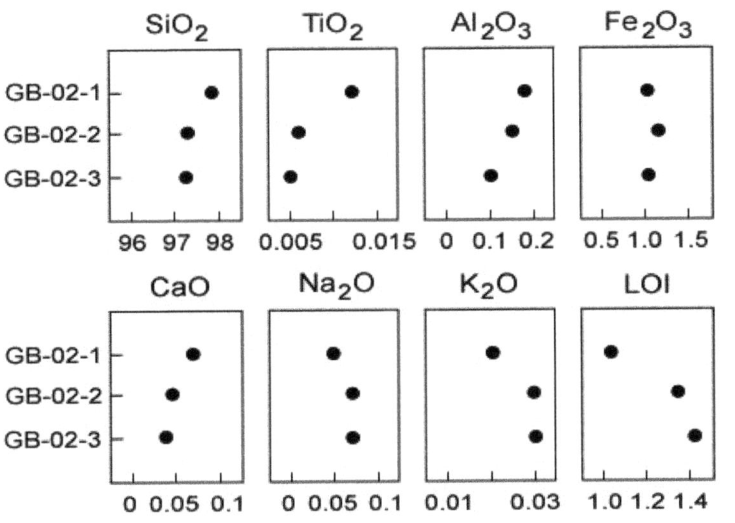

Figure 7. Plots of major element abundances (wt.%) in carnelian from Green Brook. Sample numbers are keyed to Table 3.

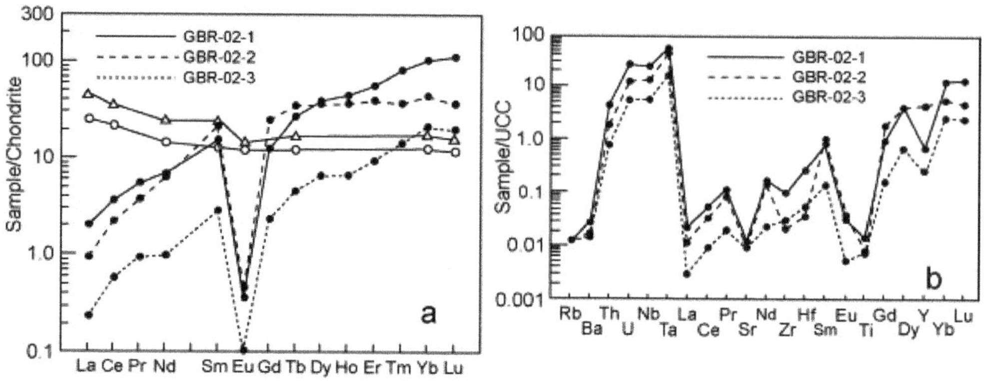

Figure 8. (**a**) Chondrite-normalized REE patterns of carnelian from Green Brook (solid circles) compared to Preakness Basalt (open circles) and Preakness gabbroid (open triangles). REE normalization values are from [51]. Basalt and gabbroid data are from [32]. (**b**) Upper continental crust-normalized multielement diagram of carnelian from Green Brook. Normalization values are from [52].

4.6. Oxygen Isotopes

Oxygen isotope analyses of carnelian samples yield high $\delta^{18}O_{VSMOW}$ values that range from +18.3 to +31.2‰ (Table 4), with all but one sample having an $\delta^{18}O$ of +27.9 to +31.2‰. The analytical precision (1σ) is ±0.06‰ in two samples and ±0.21 to 0.64‰ in the other four (Table 4) compared to ±0.1‰ for the standards, indicating isotopic heterogeneity, particularly in GBR-3 and GBR-4. The reasons for this heterogeneity between the samples are discussed later in this paper (see Section 5.3). It is noteworthy that no correlation was observed between the $\delta^{18}O$ of the samples and their geographic location or their color.

Our data fall within the range of oxygen isotope values of chalcedony hosted by volcanic rocks globally (Figure 9). Because of the paucity of published mineralogical studies of carnelian, it was necessary to compare our $\delta^{18}O$ data to global occurrences of basalt-hosted chalcedony and/or agate, examples of which include the following. $\delta^{18}O$ values of +22.1 to +28.8‰ are reported from agate in Jurassic quartz latite in Namibia [53]. Chalcedony from the Jurassic Karoo basalt in South Africa yielded $\delta^{18}O$ values of +20.1 to +21.2‰ [54]. Agate hosted by Eocene to Miocene volcanic rocks in Iran provided $\delta^{18}O$ values of +23.8 to +26.6‰ [55]. $\delta^{18}O$ values of +24.5 to +30.3‰ are reported from agate in the Cretaceous Paraná basalt in Brazil [56], similar to $\delta^{18}O$ of +21.2 to +31.1‰ from amethyst geodes in the same basalt in Uruguay [57], with early formed chalcedony typically occupying the upper end of this range. $\delta^{18}O$ values of +16.4 to +33.4‰ were obtained from agate occurrences globally in a variety of felsic and mafic host rocks [47,58]. The variation of ~10‰ in the $\delta^{18}O$ values of our study is, therefore, not unusual compared to the similar range of values of other studies.

Table 4. Oxygen isotope results of carnelian from Green Brook.

Sample (# of Analyses)	Color	$\delta^{18}O$ (‰ VSMOW)	Std. Dev. (1σ)	Meteoric Water (°C)	Seawater (°C)	Magmatic Fluid (°C)
GBR-1 (n = 1)	Red	+28.9‰	--	28	50	100
GBR-2 (n = 2)	Red	+31.2‰	0.21	19	39	82
GBR-3 (n = 2)	Red-orange	+29.9‰	0.56	24	45	92
GBR-4 (n = 2)	Red-orange	+27.8‰	0.64	32	56	108
GBR-5 (n = 3)	Yellow	+18.3‰	0.06	82	121	223
GBR-6 (n = 3)	Yellow	+30.5‰	0.06	21	42	88

Calculated formation temperatures (°C) based on $\delta^{18}O$ of meteoric water (−5‰), seawater (0‰), and magmatic fluid (+8‰). See text for details.

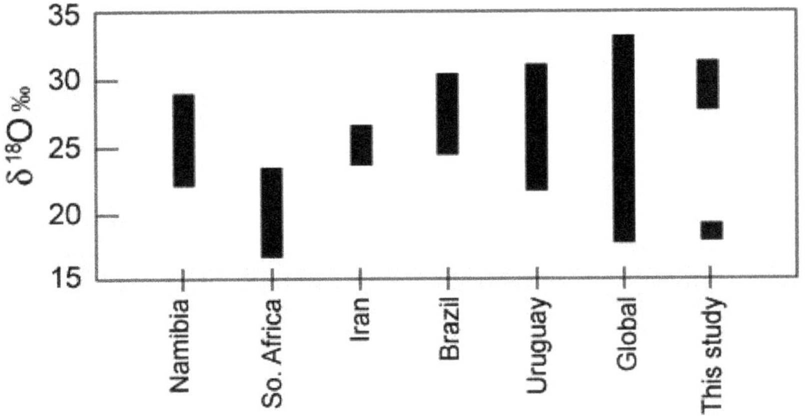

Figure 9. A plot of the $\delta^{18}O$ values of New Jersey carnelian compared to global occurrences of basalt-hosted chalcedony, agate [48,53,55–57], and carnelian [54].

5. Discussion

5.1. Source of Elements in Carnelian

Most trace elements and REE, with the exception of Ga, Ge, and B, are unable to be structurally incorporated into the Si lattice and are incorporated in micro-inclusions in quartz or are adsorbed onto its surface [59]. Therefore, concentrations of trace elements and REE in the carnelian of this study provide important insights into the source of these elements. Nearly all trace elements and REE abundances differ from those in Preakness Basalt or gabbroid (Figure 8). In particular, Preakness Basalt and gabbroid contain much lower abundances of Th (2.0 ppm), Nb (4.2 ppm), and Ta (0.35 ppm) [33] compared to the values in carnelian (Table 3). Also of note are the high values of W and U in the carnelian (5–19 ppm and 14–76 ppm, respectively). Although W and U are not reported for mafic igneous rocks of the Newark basin, these values are typically low in basalt, averaging 0.01–0.09 ppm W and 0.05–0.18 ppm U in mid-ocean-ridge basalt and 0.56 ppm W and 1.02 ppm U in ocean island basalt [47].

In New Jersey, trace element values that are comparable to those in the carnelian are found only in granulite to amphibolite facies Mesoproterozoic rocks and/or greenschist facies Neoproterozoic rocks of the exposed New Jersey Highlands and their buried segments beneath the Newark basin. Widespread and abundant Mesoproterozoic felsic orthogneiss and paragneiss have high REE abundances and as much as 214 ppm U, 650 ppm Th, 100 ppm Nb, 7.0 ppm Ta, and 4.0 ppm W, [60,61], and regionally abundant Mesoproterozoic low-Ti magnetite deposits locally contain elevated abundances of U, Th, and REEs [62–64]. Additionally, Neoproterozoic hematitic ironstone deposits contain as much as 100 ppm U, 45 ppm Th, and 300 ppm W [65]. Elevated Fe abundances in the carnelian were likely sourced from various Proterozoic lithologies and/or ore deposits that oxidized Fe^{2+} to Fe^{3+} under near-surface conditions. The leaching of Fe-rich Mesoproterozoic lithologies at depth by hydrothermal fluids is shown to be the source of Fe and other metals in the Neoproterozoic ironstone deposits in New Jersey [65]. Although high field strength elements are often considered to be immobile, they are shown to have increased solubility in acidic fluids or fluids containing Cl and F [66–68]. Halogens are particularly abundant in Mesoproterozoic orthogneiss that contains Cl in amphibole and widespread occurrences of scapolite, whereas F is present in fluorapatite, which is commonly found in the numerous regional magnetite deposits.

For these reasons, we interpret Proterozoic rocks to be the likely source for the elevated abundance of most elements in the carnelian. This means that during the Jurassic, certain physicochemical conditions existed, such as extensional brittle faults that would have permitted ascending fluids to egress to the Preakness, as well as deeply circulating heated fluids capable of enhancing mineral solubility and mobilizing elements from basement lithologies. The first condition is met by the presence of intrabasinal extensional faults that were active during the Triassic and Jurassic and references therein [69]. The second condition is supported by the downward flow of meteoric water during the Triassic and Jurassic along the border fault separating the Proterozoic rocks from the Newark basin (Figure 1), followed by convective circulation and ascent as hydrothermal fluid at temperatures of 150–280 °C [70] and 137–232 °C [71] from fluid inclusion studies of anhydrite and calcite, respectively, and 100–250 °C from zircon and apatite fission-track analysis [72].

The source of silica in the carnelian is less straightforward, and sources external or internal to the Preakness were considered. Silica is a major component of siliceous sinters formed in a terrestrial hot spring/geyser environment. However, these deposits are mainly composed of some combination of opal-A, opal-CT, opal-C, and macroscopic quartz that typically lack an association with chalcedony [73,74] or do not directly precipitate chalcedony [75]. Opal is absent in all samples of New Jersey carnelian and features consistent with a siliceous sinter, such as those observed in Orange Mountain Basalt [38], are not recognized in the Preakness.

Silica may have been derived through the partial dissolution of detrital quartz grains in the underlying Feltville Formation by alkaline fluids. These reddish-brown clastic rocks are rich in quartz and contain abundant hematite that coats detrital grains and occurs in the rock matrix. A similar interpretation was made for silicified sand layers at the tops of flows in the Paraná continental flood basalt formed from the fluidization of sand in the underlying Botucatu Formation that was transported in ~150 °C hydrothermal fluids [76,77]. The injection of fluidized sand into the Preakness as dikes, sills, or veins would have left remnant patches of non-bedded sediment or sediment-filled fractures in the basalt, which are not observed. Alternatively, the dissolution of detrital quartz grains may have released silica that reacted to form silicic acid, precipitating carnelian in the Preakness as the fluid became silica saturated and leaving no sediment residue. This would have taken place under conditions of higher temperature than cool surface water and at $pH > 4$ necessary for the solubility of silica [37,78].

We favor the devitrification of abundant glass in the gabbroid of the Preakness first flow as the source of silica in the carnelian for the following reasons. Silica-saturated fluids may readily form through the hydrolysis of highly reactive basaltic glass by acidic fluids [37,79]. The breakdown of interstitial glass in the basalt of the Paraná volcanic province was posited as the source of silica that filled large amethyst geodes there [80,81], and the alteration of basalt by a heated fluid is interpreted to be the source of silica in agate in Cuba [82]. Numerous studies of basaltic glass alteration document initial losses in Al_2O_3, MgO, CaO, Na_2O, and K_2O [83–86]. Glass-rich gabbroid in the Preakness displays comparable depletions of up to 4.0 wt.% Al_2O_3, 3.0 wt.% MgO, 3.0 wt.% CaO, and 0.7 wt.% K_2O compared to host Preakness Basalt (Table 1). The fact that carnelian is not found in, or associated with, the Preakness elsewhere in the Newark basin demands that unique conditions must have existed in the study area. We propose that these conditions are a result of the occurrence of the thickest gabbroid bodies recorded in the Preakness that contain a considerable volume of basaltic glass available to supply the silica. Although gabbroid bodies are recognized in the Preakness at other locations, they are much thinner, and the same physicochemical conditions that were present in the study area were not duplicated elsewhere in the Newark basin.

5.2. Movement and Timing of Mineralizing Fluid

We evaluated pathways in the Preakness that were most likely to have controlled the movement of fluid from which carnelian formed. Typical fine-grained basalt in the first flow is sparsely vesicular, with vesicles comprising <0.2 volume% of the rock [26]. The lowest part of this flow is characterized by slender and vertical columnar joints that would have permitted fluid to ascend along a subvertical path during the cooling of the flow. Fluid infiltrating the Preakness may have also utilized intrabasinal extensional faults that were active throughout the evolution of the Newark basin [69] and that are common in the study area as steeply dipping (70–90°), mainly open, and unmineralized brittle faults as wide as 0.5 m [33]. Although it is possible that silica-saturated fluid utilized these pathways, the preservation of vesicle molds on the lower surfaces of some carnelian is not compatible with its precipitation along columnar joints or faults, which lack amygdaloidal features.

Degassing of the volatile phases during the cooling of the Preakness must have been fairly rapid because gas bubbles had insufficient time to form vesicles or coalesce into larger vugs or openings in the lower or middle parts of the first flow. The reduced water–rock interaction also likely explains why this part of the first flow is relatively unaltered and has low values of LOI (Table 1). By contrast, thick gabbroid bodies in the upper part of the first flow contain 3 m thick intervals in which vesicles with diameters of 2–10 mm comprise as much as 19% of the volume of the rock [26], indicating that gabbroid formed a barrier and trapped rising gas bubbles. Gabbroid also preserves petrographic and geochemical features compatible with hydrothermal alteration, which suggests that it had ample opportunity to interact with heated fluid. These include gains in Na_2O due to the albitization of plagioclase, a loss of CaO and Sr that reflect plagioclase alteration, and high values of LOI that range from 1.8 to 2.5 wt.% (Table 1), indicative of precipitation of secondary hydrous minerals. The evidence for hydrothermal fluid movement in the upper part of the first flow is more consistent with the ascent of a heated, syn-eruptive fluid than with infiltration of the basalt by a low-temperature post-eruptive fluid. Thus, we propose that it was in the altered, volatile-rich upper part of the first flow that carnelian most probably precipitated.

The absence of any dateable material in the carnelian renders the timing of its formation to be a challenge. Therefore, we relied on indirect evidence, such as the presence of glauberite molds in some pieces. Studies of the paragenesis of secondary mineralization in basalt formations of the Newark basin indicate that crystallization of the basalt and precipitation of contemporaneous saline minerals were closely followed in time by the albitization of plagioclase and the precipitation of abundant silica [17,21]. The precipitation of carnelian in the Preakness first flow containing glauberite (Figure 4c) is consistent with this paragenetic sequence, suggesting that carnelian may have formed contemporaneously with the crystallization of the Preakness at 201 Ma.

5.3. Characteristics of the Mineralizing Fluid

Multiple lines of evidence provide insights into the temperature, composition, and pH of the fluid from which carnelian formed. One constraint on the syn- to post-depositional thermal history of the Newark basin is provided by geochemical analysis of thermally mature hydrocarbons in the Feltville Formation from beneath the Preakness Basalt that indicate a burial temperature of 90–130 °C [87]. Using this ~100 °C estimate from burial constraints, carnelian samples with $\delta^{18}O$ of +27.9 to +31.1‰ would be in isotopic equilibrium with water having $\delta^{18}O$ of +7 to +10‰ when calculated using the fractionation factor for α-quartz of [88]. If correct, this calculation would indicate that pure magmatic water ($\delta^{18}O$ of ~+8‰) precipitated these carnelian samples. Although Preakness gabbroid preserves evidence compatible with hydrothermal alteration, we do not think the oxygen isotope ratios reflect the formation of the carnelian strictly from a high-temperature (>100 °C) magmatic fluid because calculations at higher temperatures yield unrealistic water $\delta^{18}O$ values, and also because higher temperatures are not supported by the extremely low values of Ti in our samples. Furthermore, in a cooling volcanic system, magmatic fluid commonly mixes with surface water, producing lower fluid $\delta^{18}O$ values.

To explore possible fluid sources, we calculated formation temperatures for the carnelian from end-member fluid compositions (Table 4). Meteoric water for the ~20° paleolatitude of the Newark basin at about 200 Ma [89] is estimated to have been ca. −5‰ using modern latitude versus $\delta^{18}O$ relationships [90]. If water with $\delta^{18}O$ of −5‰ precipitated the higher-$\delta^{18}O$ carnelian samples, then the equilibrium formation temperatures would have been 19–32 °C. The calculated temperatures using a fluid of seawater composition ($\delta^{18}O \approx 0$‰) are 40–55 °C. Seawater may be a viable proxy for a minor component of a brine-like fluid that precipitated glauberite in the Preakness first flow. The upper range of temperatures calculated using only magmatic fluid with $\delta^{18}O$ of +8‰ is 82–108 °C. If the $\delta^{18}O$ of +18.3‰ in the carnelian sample GBR-5 is included, the maximum temperatures are higher and range from 82–223 °C. The lower $\delta^{18}O$ value of GBR-5 more likely reflects a combination of differences in temperature and different fluid or fluid mixtures compared to the other samples. We interpret the formation temperatures for the carnelian in this study to fall mainly in the range of 20–80 °C based on calculations using meteoric water and magmatic (hydrothermal) fluid. A more precise estimate requires knowledge of the proportion of the different fluids involved in mixing, but even then, the maximum temperature would probably not have been much above ~100 °C. Regardless, the formation temperature would not have remained high for very long because chalcedony precipitates from silica-saturated fluids that cool fairly rapidly, with slow cooling favoring the crystallization of macrocrystalline quartz [37].

In terms of the fluid composition, variations in the $\delta^{18}O$ values and geochemical characteristics of the carnelian are most reasonably explained by the mixing of a deeper, higher-temperature hydrothermal fluid and shallower, lower-temperature meteoric water, with variations in these parameters due mainly to differences in the ratio of the fluids. This interpretation is in agreement with [71], who proposed the mixing of heated basement-derived fluid and cooler formational water to be responsible for the precipitation of calcite in veins in Triassic sedimentary rocks in the Newark basin. Our results expand the effects of fluid mixing to include the Jurassic part of the Newark basin and also account for the precipitation of silica mineralization.

We interpret the deep circulating hydrothermal fluid as having been acidic and responsible for mobilizing trace elements from the Proterozoic basement. Total REE abundances are higher in acidic fluids than alkaline fluids [78], and the mobilization of REE also increases when complexed by F- or Cl-rich fluids [66,68]. Such fluids are more strongly enriched in HREE ($La/Yb_N << 1$), comparable to the values of our study (La/Yb_N = 0.01–0.02). In contrast, the shallower meteoric fluid was more likely to have been alkaline. Changing conditions of pH from acidic to alkaline are evidenced in the carnelian geochemistry of samples GBR-02-1 through GBR-02-3. These samples display a progressive decrease in ΣREE (85 to 15 ppm), HREE (50 to 9.5 ppm), Nd + Yb (30 to 6 ppm), and Nb, Ta, and W (Table 3). Studies show that decreases in these elements are attributed to an increase in pH [78,91–93]. The enrichment in HREEs in all of the carnelian samples may result from the increased solubility of HREE over LREEs in hydrothermal fluids under conditions of a neutral to slightly alkaline pH of 6.3–9 [93,94]. Additionally, the adsorption of HREE onto the large surface area of Fe oxide/hydroxide is greater at pH < 8 [93] and is especially significant at a pH of 6.5–8 [78]. Uranium is mobile as U^{VI} at a neutral to alkaline pH of 7–9 and readily adsorbs onto quartz surfaces or in structural sites in quartz [95,96]. Uranium and Th also readily adsorb onto Fe oxides/hydroxides at a pH of 6–9 [78,92,93,97]. The presence of moganite in our carnelian samples is also noteworthy because moganite precipitates from alkaline fluids containing ferric iron [42]. These results, in combination with the data of our study, support the precipitation of carnelian under conditions of a neutral to slightly alkaline pH.

6. Conclusions

Carnelian in New Jersey is recognized from only a few closely spaced locations in the Mesozoic Newark basin. Although carnelian has yet to be observed in situ in bedrock outcrops, we present evidence that supports its precipitation in coarse-grained gabbroid layers in the Lower Jurassic Preakness Basalt roughly contemporaneous with the crystallization of the basalt at 201 Ma. Erosion of the host rock continues to concentrate carnelian in proximal deposits of fluvial gravel, colluvium, and alluvium.

Carnelian is composed of length-fast chalcedony characterized by fibrous bundles overlain by wavy extinction bands. XRD patterns exhibit peaks that are characteristic of α-quartz and minor moganite (>5%). Major element abundances are dominated by SiO_2

(97.5 ± 0.31 wt.%) and Fe_2O_3 (1.09 ± 0.07 wt.%); Al_2O_3 and TiO_2 are low, averaging < 0.2 and <0.01 wt.%, respectively, and the rest of the major elements are <0.07 wt.%. Silica was derived through the devitrification of abundant volcanic glass in coarse-grained gabbroid layers in the upper part of the Preakness first flow. Trace element concentrations are low except for Y, Nb, Ta, Pb, W, Th, and U. We show that most trace elements were likely sourced from Proterozoic lithologies. REE display parallel chondrite-normalized patterns that are enriched in HREE relative to LREE and have large negative Eu anomalies (0.02). $\delta^{18}O$ values of carnelian are high and range from +18.3 to +31.2‰, with all but one sample having an $\delta^{18}O$ of 29.9 ± 1.3‰.

The results of this study are consistent with the formation of carnelian mainly from a mixture of higher-temperature hydrothermal fluid and shallower, lower-temperature, meteoric water. Variations in the $\delta^{18}O$ values and geochemical characteristics of carnelian are due to differences in the ratio of these fluids. Carnelian precipitated under conditions of neutral to slightly alkaline pH and relatively low temperatures of 20–80 °C, which is comparable to other global occurrences of volcanic rock-derived chalcedony.

Author Contributions: Conceptualization—R.A.V. and S.D.S.; methodology—M.L.G., W.H.P. and R.A.V.; formal analysis—M.L.G. and W.H.P.; writing—original draft, R.A.V. and S.D.S.; writing—review and revisions, M.L.G. and W.H.P. All authors have read and agreed to the published version of the manuscript.

Funding: This research received no external funding.

Data Availability Statement: The data presented in this study are contained within the article.

Acknowledgments: Justin Dodd and Anna Buczynska are thanked for performing the oxygen isotope analyses at Northern Illinois University. We are grateful to the two anonymous reviewers whose comments improved the quality of the paper.

Conflicts of Interest: The authors declare no conflict of interest.

References

1. Francis, P., Jr. The stone bead industry of southern India. *Beads* **2000**, *12*, 49–62.
2. Brunet, O. Bronze and Iron Age carnelian bead production in the UAE and Armenia: New perspectives. *Proc. Semin. Arab. Stud.* **2009**, *39*, 57–68.
3. Charpentier, V.; Brunet, O.; Méry, S.; Velde, C. Carnelian, agate, and other types of chalcedony: The prehistory of Jebel al-Ma'taradh and its semiprecious stones, Emirate of Ra's al-Khaimah. *Arab. Arch. Epig.* **2017**, *28*, 175–189. [CrossRef]
4. Geological Survey of India. *Geology and Mineral Resources of Gujarat, Daman and Diu*; Miscellaneous Publication 30, Part 14; Popular Printers: Jaipur, India, 2012; 85p.
5. Malunga, G.W.P. *An Analysis of Mineral Resources of Malawi*; Published in Malawi by the author; 2014; 82p.
6. Gliozzo, E.; Mattingly, D.J.; Cole, F.; Artoli, G. In the footsteps of Pliny: Tracing the sources of Garamantian carnelian from Fazzan, south-west Libya. *J. Archaeol. Sci.* **2014**, *52*, 218–241. [CrossRef]
7. Carter, A.K.; Dussubieux, L. Geologic provenience analysis of agate and carnelian beads using laser ablation-inductively coupled plasma-mass spectrometry (LA-ICP-MS): A case study from Iron Age Cambodia and Thailand. *J. Archaeol. Sci. Rep.* **2016**, *6*, 321–331. [CrossRef]
8. Polekhovsky, Y.S.; Punin, Y.O. Agate mineralization in basaltoids of the northeastern Ladoga Region, South Karelia. *Geol. Ore Depos.* **2008**, *50*, 642–646. [CrossRef]
9. Zodac, P. New Jersey Brook a carnelian locality. *Rocks Miner.* **1950**, *25*, 481–483. [CrossRef]
10. Zeitner, J.C. *Appalachian Mineral and Gem Trails*; Arts and Crafts Press: San Diego, CA, USA, 1968; 134p.
11. Mason, B.H. *Trap Rock Minerals of New Jersey*; Bulletin 64; Geological Survey of New Jersey: Trenton, NJ, USA, 1960; 51p.
12. Peters, J.J. The minerals of Bergen Hill, New Jersey. In *Igneous Rocks of the Newark Basin: Petrology, Mineralogy, Ore Deposits and Guide to Field Trip*; Puffer, J.H., Ed.; Geological Association of New Jersey Field Conference: Trenton, NJ, USA, 1984; pp. 96–102.
13. Seymour, E. List of minerals in New Jersey. In *Geology of New Jersey*; Geological Survey of New Jersey: Trenton, NJ, USA, 1868; pp. 743–750.
14. Canfield, F.A. Catalogue of minerals found in New Jersey. In *Final Report of the State Geologist*; Geological Survey of New Jersey: Trenton, NJ, USA, 1889; pp. 1–24.
15. Valiant, W.S. New Jersey mineral localities. *Min. Coll.* **1904**, *11*, 122–125, 137–141, 150–154.
16. Beck, L.C. Notices of some trappen minerals found in New Jersey and New York. *Am. J. Sci.* **1843**, *44*, 54–60.

17. Fenner, C.N. The Watchung Basalt and the paragenesis of its zeolites and other secondary minerals. *Ann. N. Y. Acad. Sci.* **1910**, *20*, 93–187. [CrossRef]
18. Gordon, S.G. A review of the genesis of the zeolite deposits of First Watchung Mountain, N.J.1. *Am. Mineral.* **1916**, *1*, 72–80.
19. Wherry, E.T. The lozenge-shaped cavities in the First Watchung Mountain zeolite deposits. *J. Wash. Acad. Sci.* **1916**, *6*, 181–184.
20. Manchester, J.G. The minerals of the Bergen Archways. *Am. Mineral.* **1919**, *4*, 107–116.
21. Schaller, W.T. *The Crystal Cavities of the New Jersey Zeolite Region*; Bulletin 832; U.S. Geological Survey: Reston, VA, USA, 1932; 90p.
22. Manspeizer, W. Rift tectonics inferred from volcanic and clastic structures. In *Field Studies of New Jersey Geology and Guide to Field Trips*; Manspeizer, W., Ed.; Rutgers University: Newark, NJ, USA, 1980; pp. 314–350.
23. Fedosh, M.S.; Smoot, J.P. A cored stratigraphic section through the northern Newark basin, New Jersey. In *Studies of the Early Mesozoic Basins of the Eastern United States*; Froelich, A.J., Robinson, G.R., Jr., Eds.; Bulletin 1776; U.S. Geological Survey: Reston, VA, USA, 1988; pp. 19–24.
24. Olsen, P.E.; Kent, D.V.; Cornet, B.; Witte, W.K.; Schlische, R.W. High-resolution stratigraphy of the Newark rift basin (early Mesozoic, eastern North America). *Geol. Soc. Am. Bull.* **1996**, *108*, 40–77. [CrossRef]
25. Blackburn, T.J.; Olsen, P.E.; Bowring, S.A.; McLean, N.M.; Kent, D.V.; Puffer, J.; McHone, G.; Rasbury, E.T.; Et-Touhami, M. Zircon U-Pb geochronology links end-Triassic extinction with the Central Atlantic Magmatic Province. *Science* **2013**, *340*, 941–945. [CrossRef]
26. Puffer, J.H.; Volkert, R.A. Pegmatoid and gabbroid layers in Jurassic Preakness and Hook Mountain Basalts, Newark Basin, New Jersey. *J. Geol.* **2001**, *109*, 585–601. [CrossRef]
27. Faust, G.T. *Joint Systems in the Watchung Basalt Flows, New Jersey*; Professional Paper 864-B; U.S. Geological Survey: Reston, VA, USA, 1978; 46p.
28. Kümmel, H.B. The Newark System of New Jersey. In *Annual Report of the State Geologist*; Geological Survey of New Jersey: Trenton, NJ, USA, 1898; pp. 23–159.
29. Lewis, J.V. The pillow lavas of the Watchung Mountains. In *Administrative Report of the State Geologist*; Bulletin 16; Geological Survey of New Jersey: Trenton, NJ, USA, 1915; pp. 51–56.
30. Lewis, J.V. Petrography of the Newark igneous rocks of New Jersey. In *Annual Report of the State Geologist*; Geological Survey of New Jersey: Trenton, NJ, USA, 1908; pp. 97–167.
31. Drake, A.A., Jr.; Volkert, R.A.; Monteverde, D.H.; Herman, G.C.; Houghton, H.F.; Parker, R.A.; Dalton, R.F. *Bedrock Geologic Map of Northern New Jersey*; Misc. Inves. Series Map I-2540-A; U.S. Geological Survey: Reston, VA, USA, 1996.
32. Tollo, R.P.; Hawkins, D.P.; Gottfried, D. *Petrographic and Geochemical Data for Jurassic Basalts from Eight Passaic Tunnel Cores, Newark Basin, New Jersey*; Open-File Report 90-689; U.S. Geological Survey: Reston, VA, USA, 1990; 32p.
33. Monteverde, D.H.; Volkert, R.A. *Bedrock Geologic Map of the Chatham Quadrangle, Morris, Union and Somerset Counties, New Jersey*; GMS 04-02; New Jersey Geological Survey: Ewing Township, NJ, USA, 2005.
34. Sharp, Z.D. A laser-based microanalytical method for the in situ determination of oxygen isotope ratios of silicates and oxides. *Geochim. Cosmochim. Acta* **1990**, *54*, 1353–1357. [CrossRef]
35. Stanford, S.D. *Surficial Geology of the Chatham Quadrangle Morris, Union and Somerset Counties, New Jersey*; OFM 69; New Jersey Geological Survey: Ewing Township, NJ, USA, 2007.
36. Skinner, A.; Schrabisch, M. *A Preliminary Report of the Archaeological Survey of New Jersey. Bulletin 9*; New Jersey Geological Survey: Ewing Township, NJ, USA, 1913; 94p.
37. Fournier, R.O. The behavior of silica in hydrothermal solutions. *Rev. Econ. Geol.* **1985**, *2*, 45–61.
38. Puffer, J.H.; Laskowich, C. Volcanic diapirs in the Orange Mountain flood basalt: New Jersey, USA. *J. Volcanol. Geotherm. Res.* **2012**, *237–238*, 1–9. [CrossRef]
39. Stanford, S.D. Onshore record of Hudson River drainage to the continental shelf from the Late Miocene through the late Wisconsinan deglaciation, USA: Synthesis and revision. *Boreas* **2010**, *39*, 1–17. [CrossRef]
40. Wherry, E.T. Glauberite crystal-cavities in the Triassic rocks of eastern Pennsylvania. *Am. Mineral.* **1916**, *1*, 37–43.
41. Cady, S.L.; Wenk, H.-R.; Sintubin, M. Microfibrous quartz varieties: Characterization by quantitative X-ray texture analysis and transmission electron microscopy. *Contrib. Mineral. Petrol.* **1998**, *130*, 320–335. [CrossRef]
42. Heaney, P.J. Moganite as an indicator for vanished evaporates: A testament reborn? *J. Sediment. Res.* **1995**, *A65*, 633–638.
43. Moxon, T.; Rios, S. Moganite and water content as a function of age in agate: An XRD and thermogravimetric study. *Eur. J. Mineral.* **2004**, *16*, 269–278. [CrossRef]
44. Zhang, X.; Ji, L.; He, X. Geomological characteristics and origin of the Zhanguohong agate Beipo, Liaoning Province, China: A combined microscopic, X-ray diffraction, and Raman spectroscopic study. *Minerals* **2020**, *10*, 401. [CrossRef]
45. Zhang, M.; Moxon, T. Infrared absorption spectroscopy of SiO_2-moganite. *Am. Mineral.* **2014**, *99*, 671–680. [CrossRef]
46. Graetsch, H.A.; Grünberg, J.M. Microstructure of flint and other chert raw materials. *Archaeometry* **2012**, *54*, 18–36. [CrossRef]
47. Sun, S.S.; McDonough, W.F. Chemical and isotopic systematics of oceanic basalts: Implications for mantle composition and processes. In *Magmatism in the Ocean Basins*; Saunders, A.D., Norry, M.J., Eds.; Special Publication 42; Geological Society of London: London, UK, 1989; pp. 313–345.
48. Götze, J.; Tichomirowa, M.; Fuchs, H.; Pilot, J.; Sharp, Z. Geochemistry of agates: A trace element and stable isotope study. *Chem. Geol.* **2001**, *175*, 523–541. [CrossRef]

49. Pršek, J.; Dumańska-Slowick, M.; Powolny, T.; Natkaniec-Nowak, L.; Tobola, T.; Zych, D.; Skrepnicka, D. Agates from Western Atlas (Morocco)—Constraints from mineralogical and microtextural characteristics. *Minerals* **2020**, *10*, 198. [CrossRef]
50. Dumańska-Slowick, M.; Natkaniec-Nowak, L.; Weselucha-Birczyńska, A.; Gawel, A.; Lankosz, M.; Wróbel, P. Agates from Sidi Rahal, in the Atlas Mountains of Morocco: Geochemical characteristics and proposed origin. *Gems Gemol.* **2013**, *49*, 148–159. [CrossRef]
51. Masuda, A.; Nakamura, N.; Tanaka, T. Fine structures of mutually normalized rare-earth patterns of chondrites. *Geochim. Cosmochim. Acta* **1973**, *37*, 239–248. [CrossRef]
52. Rudnick, R.L.; Gao, S. Composition of the continental crust. In *Treatise on Geochemistry*; Holland, H.D., Turekian, K.K., Eds.; Pergamon: Oxford, UK, 2003; pp. 1–64.
53. Harris, C. Oxygen-isotope zonation of agates from Karoo volcanics of the Skeleton Coast, Namibia. *Am. Mineral.* **1989**, *74*, 476–481.
54. Gliozzo, E.; Cairncross, B.; Vennemann, T. A geochemical and micro-textural comparison of basalt-hosted chalcedony from the Jurassic Drakensberg and Neoarchean Ventersdorp Supergroup (Vaal River alluvial gravels), South Africa. *Int. J. Earth Sci.* **2019**, *108*, 1857–1877. [CrossRef]
55. Rezaei-Kahkaei, M.; Ansarifar, O.; Ghasemi, H. Geochemistry and oxygen stable isotopes of Reza Abad agates, SE Shahrood, Central Iran: An approach to temperature and formation process. *J. Econ. Geol.* **2019**, *11*, 525–541. (In Persian)
56. Fallick, A.E.; Jocely, J.; Hamilton, P.J. Oxygen and hydrogen stable isotope systematics in Brazilian agates. In *Geochemistry and Mineral Formation in the Earth Surface*; Rodriguez-Clemente, R., Taedy, Y., Eds.; Centre Nationale de la Recherche Scientifique: Paris, France, 1987; pp. 99–117.
57. Duarte, L.C.; Hartmann, L.A.; Ronchi, L.H.; Berner, Z.; Theye, T.; Massonne, H.J. Stable isotope and mineralogical investigation of the genesis of amethyst geodes in the Los Catalanes gemological district, Uruguay, southernmost Paraná volcanic province. *Miner. Depos.* **2011**, *46*, 239–255. [CrossRef]
58. Götze, J.; Möckel, R.; Pan, Y. Mineralogy, Geochemistry and Genesis of Agate—A review. *Minerals* **2020**, *10*, 1037. [CrossRef]
59. Götze, J.; Pan, Y.; Müller, A. Mineralogy and mineral chemistry of quartz; A review. *Min. Mag.* **2021**, *85*, 639–664. [CrossRef]
60. Volkert, R.A.; Drake, A.A., Jr. *Geochemistry and Stratigraphic Relations of Middle Proterozoic Rocks of the New Jersey Highlands*; Professional Paper 1565-C; U.S. Geological Survey: Reston, VA, USA, 1999; 77p.
61. Volkert, R.A.; Feigenson, M.D.; Patino, L.C.; Delaney, J.S.; Drake, A.A., Jr. Sr and Nd isotopic compositions, age and petrogenesis of A-type granitoids of the Vernon Supersuite, New Jersey Highlands, USA. *Lithos* **2000**, *50*, 325–347. [CrossRef]
62. Klemic, H.; Heyl, A.V., Jr.; Taylor, A.R.; Stone, J. *Radioactive Rare-Earth Deposit at Scrub Oaks Mine, Morris County, New Jersey*; Bulletin 1082-B; U.S. Geological Survey: Reston, VA, USA, 1959; pp. 29–59.
63. Vassiliou, A.H. Uranium and rare earth mineralization at the Bemco mine near Cranberry Lake, New Jersey. In *Field Studies of New Jersey Geology and Guide to Field Trips*; Manspeizer, W., Ed.; New York State Geological Association: Oswego, NY, USA, 1980; pp. 192–199.
64. Baillieul, T.A.; Indelicato, G.J. Uranium in the New Jersey and New York Highlands of the Reading Prong. *Econ. Geol.* **1981**, *76*, 167–171. [CrossRef]
65. Volkert, R.A.; Monteverde, D.H.; Gates, A.E.; Friehauf, K.C.; Dalton, R.F.; Smith, R.C., II. Geochemistry and origin of Neoproterozoic ironstone deposits in the New Jersey Highlands and implications for the Iapetan rifted margin in the north-central Appalachians. In *From Rodinia to Pangea: The Lithotectonic Record of the Appalachian Region*; Tollo, R.P., Bartholomew, M.J., Hibbard, J.P., Karabinos, P.M., Eds.; Memoir 206; The Geological Society of America: Boulder, CO, USA, 2010; pp. 283–306.
66. Jiang, S.-Y.; Wang, R.-C.; Wu, X.-S.; Zhao, K.-D. Mobility of high field strength elements (HFSE) in magmatic-, metamorphic-, and submarine-hydrothermal systems. *Phys. Chem. Earth* **2005**, *30*, 1020–1029. [CrossRef]
67. Bobos, I.; Gomes, C. Mineralogy and geochemistry (HFSE and REE) of the present-day acid-sulfate types alteration from the active hydrothermal system of Furnas Volcano, São Miguel Island, The Azores Archipelago. *Minerals* **2021**, *11*, 335. [CrossRef]
68. Németh, N.; Kristály, F.; Balassa, C. Hydrothermal high field strength element enrichment in the Bükk Mts. (NE Hungary). *J. Geochem. Explor.* **2023**, *246*, 107159. [CrossRef]
69. Withjack, M.O.; Schlische, R.W.; Malinconico, M.L.; Olsen, P.E. Rift-basin development: Lessons from the Triassic-Jurassic Newark Basin of eastern North America. In *Conjugate Divergent Margins*; Mohriak, W.U., Danforth, A., Post, P.J., Brown, D.E., Tari, G.C., Nemčok, M., Sonha, S.T., Eds.; Special Publications 369; The Geological Society: London, UK, 2013; pp. 301–321.
70. El-Tabakh, M.; Riccioni, R.; Schreiber, B.C. Evolution of Late Triassic rift basin evapotites (Passaic Formation): Newark basin, eastern North America. *Sedimentology* **1997**, *44*, 767–790. [CrossRef]
71. Rddad, L.; Kraemer, D.; Walter, B.F.; Darling, R.; Cousens, B. Unraveling the fluid flow evolution and precipitation mechanisms in calcite veins in relation to Pangea rifting—Newark Basin, USA. *Geochemistry* **2022**, *82*, 125918. [CrossRef]
72. Steckler, M.S.; Omar, G.I.; Karner, G.D.; Kohn, B.P. Pattern of hydrothermal circulation within the Newark basin from fission-track analysis. *Geology* **1993**, *21*, 735–738. [CrossRef]
73. Jones, B.; Renaut, R.W. Microstructural changes accompanying the opal-A to opal-CT transition: New evidence from the siliceous sinters of Geysir, Haukadalur, Iceland. *Sedimentology* **2007**, *54*, 921–948. [CrossRef]
74. Lynne, B.Y.; Campbell, K.A.; James, B.J.; Browne, P.R.L.; Moore, J. Tracking crystallinity in siliceous hot-spring deposits. *Am. J. Sci.* **2007**, *307*, 612–641. [CrossRef]
75. Heaney, P.J. A proposed mechanism for the growth of chalcedony. *Contrib. Mineral. Petrol.* **1993**, *115*, 66–74. [CrossRef]

76. Hartmann, L.A.; Antunes, L.M.; Rosenstengel, L.M. Stratigraphy of amethyst geode-bearing lavas and fault-block structures of the Entre Rios mining district, Paraná volcanic province, southern Brazil. *Ann. Braz. Acad. Sci.* **2014**, *86*, 187–198. [CrossRef]
77. Baggio, S.B.; Hartmann, L.A.; Andrade, R.H.P.; Rizzotto, G.J.; Duarte, S.K.; Knijnik, D.B.; Simões-Neto, J.A. Basalt stratigraphy and silica gossans in Campo Grande and Serra de Maracaju, Mato Grosso do Sul, Paraná Volcanic Province. *Ore Geol. Rev.* **2015**, *69*, 73–87. [CrossRef]
78. Li, B.; Kong, Q.; Wang, G.; Liu, F.; Guo, L.; Liu, C.; Liao, F.; Shi, Z. Controls on the behaviors of rare earth elements in acidic and alkaline thermal springs. *Appl. Geochem.* **2022**, *143*, 105379. [CrossRef]
79. Gislason, S.R.; Oelkers, E.H. Mechanism, rates, and consequences of basaltic glass dissolution: II. An experimental study of the dissolution rates of basaltic glass as a function of pH and temperature. *Geochim. Cosmochim. Acta* **2003**, *67*, 3817–3832. [CrossRef]
80. Gilg, H.A.; Morteani, G.; Kostitsyn, Y.; Preinfalk, C.; Gatter, I.; Streider, A.J. Genesis of amethyst geodes in basaltic rocks of the Serra Geral Formation (Ametista do Sul, Rio Grande do Sul, Brazil): A fluid inclusion, REE, oxygen, carbon, and Sr isotope study on basalt, quartz, and calcite. *Miner. Depos.* **2003**, *38*, 1009–1025. [CrossRef]
81. Commin-Fischer, A.; Berger, G.; Polyé, M.; Dubois, M.; Sardini, P.; Beaufort, D.; Formoso, M. Petrography and chemistry of SiO_2 filling phases in the amethyst geodes from the Serra Geral Formation deposit, Rio Grande do Sul, Brazil. *J. South Am. Earth Sci.* **2010**, *29*, 751–760. [CrossRef]
82. Götze, J.; Stanek, K.; Orozco, G.; Liesegang, M.; Mohr-Westheide, T. Occurrence and distribution of moganite and opal-CT in agates from Paleocene/Eocene tuffs, El Picado (Cuba). *Minerals* **2021**, *11*, 531. [CrossRef]
83. Franzson, H.; Zierenberg, R.; Schiffman, P. Chemical transport in geothermal systems in Iceland: Evidence from hydrothermal alteration. *J. Volcanol. Geotherm. Res.* **2008**, *173*, 217–229. [CrossRef]
84. Jeong, G.Y.; Sohn, Y.K. Mineralogy and microtextures of basaltic glass alteration in hyaloclasite, Jeju Island, Korea. *J. Anal. Sci. Technol.* **2011**, *2*, 13–22. [CrossRef]
85. Ducasse, T.; Gourgiotis, A.; Pringle, E.; Moynier, F.; Frugier, P.; Jollivet, P.; Gin, S. Alteration of basaltic glass in silica saturated conditions: Analogy with nuclear glass. *Appl. Geochem.* **2018**, *97*, 19–31. [CrossRef]
86. Prause, S.; Weisenberger, T.B.; Kleine, B.I.; Monien, P.; Rispoli, C.; Stefánsson, A. Alteration of basaltic glass within the Surtsey hydrothermal system, Iceland—Implication to oceanic crust seawater interaction. *J. Volcanol. Geotherm. Res.* **2022**, *429*, 107581. [CrossRef]
87. Pratt, L.M.; Shaw, C.A.; Burruss, R.C. Thermal histories of the Hartford and Newark basins inferred from maturation indices of organic matter. In *Studies of the Early Mesozoic Basins of the Eastern United States*; Froelich, A.J., Robinson, G.R., Jr., Eds.; Bulletin 1776; U.S. Geological Survey: Reston, VA, USA, 1988; pp. 58–63.
88. Clayton, R.N.; O'Neil, J.R.; Mayeda, T.K. Oxygen isotope exchange between quartz and water. *J. Geophys. Res.* **1972**, *77*, 3057–3067. [CrossRef]
89. Kent, D.V.; Tauxe, L. Corrected Late Triassic latitudes for continents adjacent to the North Atlantic. *Science* **2005**, *307*, 240–244. [CrossRef]
90. Fricke, H.C.; O'Neil, J.R. The correlation between $^{18}O/^{16}O$ ratios of meteoric water and surface temperature: Its use in investigating terrestrial climate change over geologic time. *Earth Planet. Sci. Lett.* **1999**, *170*, 181–196. [CrossRef]
91. Michard, A. Rare earth element systematics in hydrothermal fluids. *Geochim. Cosmochim. Acta* **1989**, *53*, 745–750. [CrossRef]
92. Bau, M. Scavenging of dissolved yttrium and rare earths by precipitating iron oxyhydroxide: Experimental evidence for Ce oxidation, Y-Ho fractionation, and lanthanide tetrad effect. *Geochim. Cosmochim. Acta* **1999**, *63*, 67–77. [CrossRef]
93. Liu, H.; Pourret, O.; Guo, H.; Bonhoure, J. Rare earth elements sorption to iron oxyhydroxide: Model development and application to groundwater. *Appl. Geochem.* **2017**, *87*, 158–166. [CrossRef]
94. Munemoto, T.; Ohmori, K.; Iwatsuki, T. Rare earth elements (REE) in deep groundwater from granite and fracture-filling calcite in the Tono area, central Japan: Prediction of REE fractionation in paleo- to present-day groundwater. *Chem. Geol.* **2015**, *417*, 58–67. [CrossRef]
95. Pan, Y.; Li, D.; Feng, R.; Wiens, E.; Chen, N.; Chernikov, R.; Götze, J.; Lin, J. Uranyl binding mechanism in microcrystalline silicas: A potential missing link for uranium mineralization by direct uranyl co-precipitation and environmental implications. *Geochim. Cosmochim. Acta* **2021**, *292*, 518–531. [CrossRef]
96. Smedley, P.L.; Kinniburgh, D.G. Uranium in natural waters and the environment: Distribution, speciation and impact. *Appl. Geochem.* **2023**, *148*, 105534. [CrossRef]
97. Möller, P.; Dulski, P.; De Lucia, M. REY patterns and their natural anomalies in waters and brines: The correlation of Gd and Y anomalies. *Hydrology* **2021**, *8*, 116. [CrossRef]

minerals

MDPI

Article

Sm–Nd Isochron Age Constraints of Au and Sb Mineralization in Southwestern Guizhou Province, China

Zepeng Wang [1,2], Qinping Tan [3,*], Yong Xia [3,*], Jianzhong Liu [2,4], Chengfu Yang [1,2], Songtao Li [1,2], Junhai Li [1,2], Faen Chen [1], Xiaoyong Wang [1], Qiquan Pan [1] and Dafu Wang [1]

1 Geological Party 105, Bureau of Geology and Mineral Exploration and Development Guizhou Province, Guiyang 550018, China; wangzepengyu@126.com (Z.W.); yzpy1616@163.com (C.Y.); lisongtaozgh@163.com (S.L.); lijunhai2020@163.com (J.L.); chenfaen@126.com (F.C.); xiaoyongw1984@163.com (X.W.); panqiquan2020@126.com (Q.P.); wangdafu89@163.com (D.W.)

2 Innovation Center of Ore Resources Exploration Technology in the Region of Bedrock, Ministry of Natural Resources of People's Republic of China, Guiyang 550018, China; liujianzhong2020@126.com

3 State Key Laboratory of Ore Deposit Geochemistry, Institute of Geochemistry, Chinese Academy of Sciences, Guiyang 550002, China

4 Bureau of Geology and Mineral Exploration and Development Guizhou Province, Guiyang 550004, China

* Correspondence: tanqinping@vip.gyig.ac.cn (Q.T.); xiayong@vip.gyig.ac.cn (Y.X.)

Abstract: Southwestern Guizhou, China, is famous for hosting clusters of Carlin-type Au, Sb, and Hg-Tl deposits. These deposits are thought to be the products of a low-temperature hydrothermal metallogenic event. Calcite and fluorite are common and widespread gangue minerals in Au and Sb deposits, respectively. Ore-related calcite commonly coexists with stibnite, realgar, and orpiment at the periphery of high-grade orebodies in Au deposits, while ore-related fluorite is generally intergrown with stibnite in Sb deposits. In this study, ore-related calcite and fluorite samples from representative Au (Zimudang) and Sb (Dachang) deposits, respectively, were separated, and the rare earth element (REE) concentrations, Sm/Nd isotope ratios, and Sm–Nd isochron ages were analyzed. This study aims to determine the formation ages of the calcite and fluorite and to constrain the age of low-temperature metallogenic event in Southwestern Guizhou. The calcite and fluorite samples contain relatively high total concentrations of REEs (8.21–22.5 μg/g for calcite, 21.7–36.6 μg/g for fluorite), exhibit variable Sm/Nd ratios (0.51–1.01 for calcite, 0.35–0.49 for fluorite), and yield Sm–Nd isochron ages of 148.4 ± 4.8 and 141 ± 20 Ma, respectively. These ages are consistent with the age range constrained by the low-temperature thermochronology of zircon (132–160 Ma), crosscutting relationships of stratigraphy or intrusions (96–160 Ma), and previous dating results (135–150 Ma) in Southwestern Guizhou. Collectively, the ages obtained in this study add new evidence to previous geochronology studies, such that the low-temperature hydrothermal mineralization in Southwestern Guizhou can be constrained to 135–150 Ma, corresponding to the Yanshanian orogeny, which was associated with a weak extensional tectonic environment.

Keywords: geochronology; mineralization; Zimudang; Dachang; Southwestern Guizhou

1. Introduction

The Youjiang basin hosts clusters of Carlin-type Au, Sb, and Hg-Tl deposits (Figure 1) [1–8], such that it is the second largest low-temperature Carlin-type Au mineralized area in the world, after Nevada, USA [9–13]. The Au and Sb deposits in the Youjiang Basin are primarily located within Southwestern Guizhou and account for more than 70% of the deposits and endowments.

Accurate geochronology is a crucial tool for understanding the genesis of ore deposits. Direct dating of Au and Sb deposits in Southwestern Guizhou has been problematic owing to a lack of suitable minerals for radiometric dating. Many studies have reported ages for Au and Sb deposits in Southwestern Guizhou, but the true ages remain controversial because the results span a large age interval, i.e., 235–195 [14,15], 135–150 [16–20], and 83–106 Ma [21,22]. This uncertainty hampers our understanding of the genesis of Au and Sb deposits and their relation to the geotectonic evolution of Southwestern Guizhou.

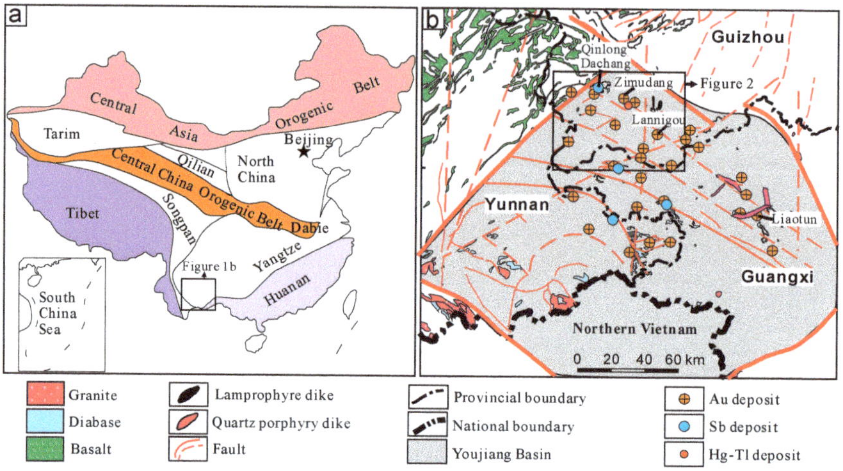

Figure 1. (**a**) Schematic tectonic map of China showing the location of the Youjiang Basin. (**b**) Geologic map of the Youjiang Basin showing the locations of the Carlin-type Au, Sb, and Hg-Tl deposits.

Sm and Nd have similar chemical properties, and the ^{143}Nd daughter, which decays from the parent ^{147}Sm, is easily preserved in a mineral lattice. Sm–Nd isotope systematics are liable to be kept closed and are partially capable of resisting weathering and/or alteration [23]. The Sm–Nd isochron method is an effective tool for dating hydrothermal Ca-bearing minerals and has been used in various hydrothermal deposits for geochronology studies (e.g., calcite [18,19,24–26], scheelite [27–29], fluorite [30–32], and tourmaline [27,28]).

Fluorite and calcite are widespread Ca-bearing minerals in Au and Sb deposits in Southwestern Guizhou, respectively. They usually have higher rare earth element (REE) concentrations and relatively variable Sm/Nd ratios, which are favorable for Sm–Nd isotope dating [18–20,33]. Ore-related calcite is generally intergrown with realgar and orpiment at Au deposits and is thought to be the product of decarbonatization (carbonate dissolution) of host rocks, providing Fe^{2+} to form Au-bearing pyrite [18,19,33]. Calcite deposited from decarbonatization reflects the age of Au mineralization [18,19]. Ore-related stibnite is symbiotic and coeval with fluorite crystallization, such that fluorite Sm–Nd dating can represent the age of Sb mineralization [20].

In this paper, we report the REE patterns, Sm/Nd isotope ratios, and Sm–Nd isochron ages of ore-related calcite and fluorite from representative Au (Zimudang) and Sb (Dangchang) deposits in Southwestern Guizhou, respectively. We aim to explore the temporal relationships of Au and Sb mineralization and constrain the age of low-temperature metallogenic events in Southwestern Guizhou.

2. Geological Setting

The Youjiang Basin is located at the southwestern margin of the Yangtze Craton (Figure 1a). This region is commonly referred to as the "Dian-Qian Gui" Golden Triangle Region (Figure 1b) [14]. The basin overlays a Lower Paleozoic basement, which evolved from a rift basin (Early Devonian to Late Devonian) to a passive continental margin (Early Carboniferous to Early Triassic) and then a foreland basin (Middle Triassic) [34]. The Youjiang Basin was then uplifted by the Indosinian orogeny at the end of the Late Triassic and, subsequently, folded by the Yanshanian orogeny after the Early Jurassic [35,36].

Southwestern Guizhou (Figure 2), situated north of the Youjiang Basin (Figure 1b), contains several super-large or large Au (e.g., Shuiyindong, Lannigou, Zimudang, and Getang) and Sb (e.g., Qinglong and Dachang) deposits. These deposits are thought to have formed during a low-temperature hydrothermal metallogenic event [1,5,17,37], where the Au and Sb orebodies were primarily controlled by short-axis anticlines (domes) and associated fracture zones (Figure 2). The majority of the exposed strata in this region are Devonian to Triassic, with the Triassic system being distributed most widely, which was followed by the Permian, while the Devonian and Carboniferous are sporadically exposed at some cores of anticlines or domes. The magmatic rocks mainly include the Emeishan flood basalts and alkaline ultrabasic pipes and dikes. An unconformity (SBT) between the Middle and Upper Permian, consisting of altered breccia [38], is thought to be the structural conduit that fed ore fluids into the cores of the anticlines (domes) [39]. The Au and Sb mineralization occurs primarily in the SBT and the Upper Permian to Lower Triassic systems.

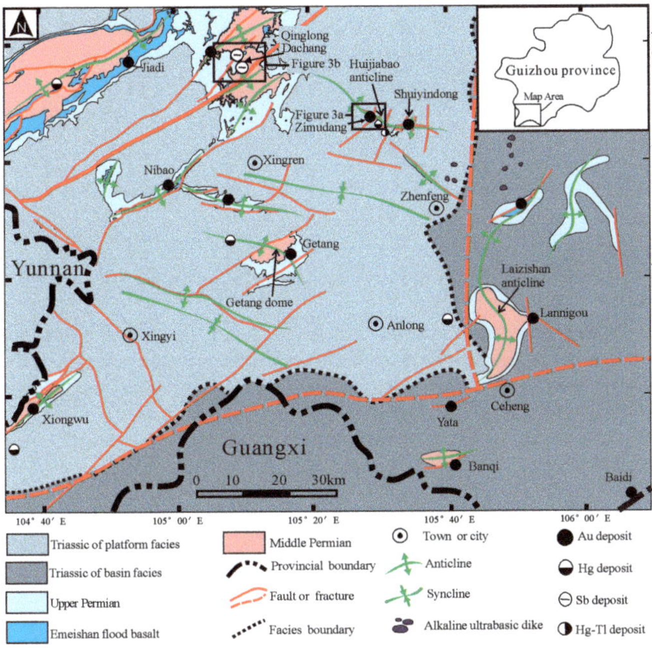

Figure 2. Geological map of Southwestern Guizhou showing the distributions of major Carlin-type Au, Sb, Hg, and Hg-Tl deposits.

The Zimudang large Au deposit, with a total Au reserve of over 72 tons [17], is located west of the Huijiabao anticline. The Au orebodies tend to be found in silicified brecciated argillite and limestone of the SBT and in bioclastic limestone, siltstones, and claystone of the Upper Permian and Lower Triassic (Figure 3a,b). Hydrothermal alterations include silicification, decarbonatization, sulfidation, dolomitization, and argillization. Ore minerals consist primarily of arsenian pyrite and arsenopyrite, with small amounts of orpiment, realgar, stibnite, and cinnabar. Gangue minerals include calcite, quartz, and fluorite. The stibnite, realgar, and orpiment commonly coexist with the ore-related calcite vein (Figure 4a,b).

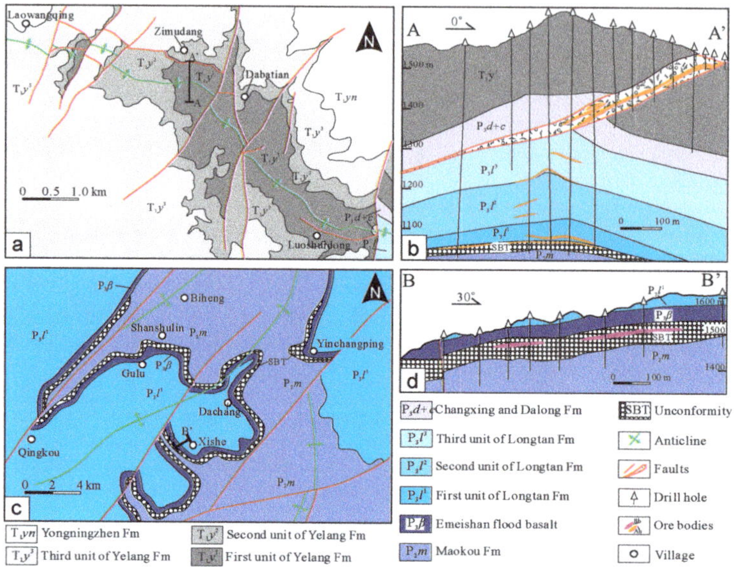

Figure 3. Geologic plan and cross-section of the Zimudang Au deposit (**a**,**b**) and Dachang Sb deposit (**c**,**d**). Abbreviations: Fm = Formation.

The Dachang large Sb deposit contains a proven and inferred Sb resource of approximately 30 ten thousand tons [5]. The Sb orebodies are generally found within silicified brecciated basalt, basaltic volcanic tuff, and limestone of the SBT (Figure 3c,d). Hydrothermal alterations are similar to the Au deposits in the region, including silicification, decarbonatization, sulfidation, argillization, fluoritization, and minor baritization. Stibnite is the dominant ore mineral, and fluorite and quartz are the major gangue minerals. Quartz and fluorite are generally in assemblage with stibnite (Figure 4c,d).

Figure 4. Photographs of calcite (**a**,**b**) and fluorite (**c**,**d**) from the Zimudang Au deposit and Dachang Sb deposit, respectively.

3. Samples and Analytic Methods

A total of five calcite and five fluorite samples were collected from the Zimudang Au and Dachang Sb deposits, respectively. The calcite samples are intergrown with realgar and orpiment in high-grade Au ores (Figure 4a,b), and the fluorite samples coexist with stibnite and quartz (Figure 4c,d) in high-grade Sb ores. The paragenesis suggests that the calcite and fluorite are closely related to Au and Sb mineralization, respectively. The fluorite and calcite samples were crushed into 40 to 60 meshes and then hand-picked under a binocular microscope, yielding a mineral purity of more than 99%. Finally, the pure calcite and fluorite separates were finely crushed to 200 mesh in an agate mortar.

The REE contents for calcite and fluorite were analyzed using an ELAN 6000 inductively coupled plasma quadrupole mass spectrometer at the Institute of Geochemistry, Chinese Academy of Sciences (Guiyang, China). The detailed analytical methods are described by Qi et al. [40], and the analysis accuracy was better than $\pm 5\%$. Sm and Nd concentrations and their isotope ratio measurements were performed using an IsoProbe Thermal Ionization Mass Spectrometer at the Tianjin Institute of Geology and Mineral Resources, Chinese Academy of Geological Sciences (Tianjin, China). Detailed analytical procedures are available in Peng et al. [23], Su et al. [18], and Zhang et al. [41]. The reproducibility of the isotopic ratios was better than 0.005% (2σ). The precision for the Sm and Nd content was less than 0.5% of the quoted values (2σ). Sm–Nd isochron ages were calculated using the Excel add-in ISOPLOT 2.9 (U.S. Geological Survey, Washington, DC, USA) [42], and the decay constant used in the calculation was $\lambda^{147}Sm = 6.54 \times 10^{-12}/year$.

4. Results

The REE concentrations and Sm–Nd isotopic compositions of calcite and fluorite are presented in Tables 1 and 2, respectively. The ΣREE contents of calcite and fluorite range from 8.21–22.5 and 21.7–36.6 μg/g, showing variable Sm/Nd ratios of 0.51–1.01 and 0.35–0.49, respectively. C1 chondrite-normalized [43] REE patterns of calcite (Figure 5a) and fluorite (Figure 5b) exhibit apparent differences, although all samples show middle rare earth element (MREE) enrichment characteristics. The calcite samples from the Zimudang Au deposit are characterized by positive Eu anomalies (δEu = 1.24–1.38), whereas the fluorite samples from the Dachang Sb deposit are characterized by negative Ce anomalies (δCe = 0.67–0.72).

Table 1. Rare earth element (REE) compositions (μg/g) of the calcite and fluorite samples.

Sample No.	La	Ce	Pr	Nd	Sm	Eu	Gd	Tb	Dy	Ho	Er	Tm	Yb	Lu	ΣREE	Y	δEu	δCe
ZMD-03	0.91	2.93	0.55	3.31	1.61	0.76	2.07	0.27	1.33	0.22	0.40	0.04	0.20	0.02	14.6	6.17	1.27	0.99
ZMD-04	0.43	1.60	0.35	2.58	1.52	0.81	2.11	0.29	1.34	0.22	0.42	0.05	0.21	0.03	12.0	7.38	1.38	0.95
ZMD-25	0.51	1.48	0.28	1.78	1.20	0.70	2.22	0.35	1.76	0.28	0.59	0.07	0.34	0.05	11.6	10.6	1.29	0.96
ZMD-26	0.15	0.53	0.14	0.98	1.03	0.59	1.93	0.30	1.45	0.23	0.48	0.06	0.32	0.04	8.21	8.19	1.27	0.82
ZMD-28	0.75	3.05	0.72	4.73	2.53	1.29	3.99	0.59	2.99	0.49	0.86	0.08	0.38	0.04	22.5	15.0	1.24	0.93
DC-02	1.60	2.69	0.65	4.09	1.88	1.14	4.84	1.00	6.35	1.39	3.15	0.34	1.64	0.19	31.0	104	1.10	0.65
DC-06	1.50	2.79	0.61	3.58	1.59	0.76	3.30	0.58	3.41	0.77	1.67	0.18	0.88	0.1	21.7	56.3	1.00	0.72
DC-07	2.13	3.40	0.79	4.64	1.95	0.92	4.29	0.73	4.48	1.01	2.39	0.27	1.33	0.16	28.5	69.6	0.94	0.64
DC-10	3.43	5.99	1.31	7.61	2.61	1.15	4.68	0.76	4.30	0.97	2.17	0.25	1.19	0.16	36.6	80.9	0.99	0.69
DC-15	2.23	3.48	0.77	4.36	1.70	0.79	3.47	0.61	3.61	0.84	2.08	0.22	1.15	0.14	25.5	59.2	0.97	0.65

Table 2. Sm–Nd isotopic compositions of the calcite and fluorite samples.

Sample No.	Deposits/Occurrence	Sm (μg/g)	Nd (μg/g)	Sm/Nd	$^{147}Sm/^{144}Nd$	$^{143}Nd/^{144}Nd(2\delta)$	εNd
ZMD-03	Zimudang Cal + Rlg + Orp	1.7307	3.3831	0.51	0.3093	0.512612 ± 8	−2.5
ZMD-04		1.7147	2.7122	0.63	0.3822	0.512699 ± 9	−2.2
ZMD-25		1.3471	1.9572	0.69	0.4161	0.512734 ± 19	−2.1
ZMD-26		1.1104	1.0996	1.01	0.6105	0.512956 ± 8	−1.3
ZMD-28		2.3163	4.2144	0.55	0.3323	0.512642 ± 7	−2.4
DC-02	Dachang Fl + Stb	1.3930	2.8706	0.49	0.2934	0.512507 ± 8	−4.3
DC-06		1.3047	3.1282	0.42	0.2522	0.512468 ± 10	−4.3
DC-07		1.5371	3.5324	0.44	0.2631	0.512475 ± 3	−4.4
DC-10		2.0384	5.8439	0.35	0.2109	0.512429 ± 8	−4.3
DC-15		1.1420	2.8716	0.40	0.2404	0.512457 ± 15	−4.3

Abbreviations: Cal = Calcite, Rlg = Realgar, Orp = Orpiment, Fl = Fluorite, Stb = Stilbite.

The calcite samples show $^{147}Sm/^{144}Nd$ and $^{143}Nd/^{144}Nd$ values ranging from 0.3093 to 0.6105 and 0.512612 to 0.512956, respectively. The fluorite samples have $^{147}Sm/^{144}Nd$ and $^{143}Nd/^{144}Nd$ values of 0.2109–0.2934 and 0.512429–0.512507, respectively. Calcite and fluorite all display good linear relationships in the $^{147}Sm/^{144}Nd$-$^{143}Nd/^{144}Nd$ diagram (Figure 5c,d). Calcite yields an Sm–Nd isochron age of 148.4 ± 4.8 Ma, with a low mean square of weighted deviates (MSWD) of 0.82 and an initial $^{143}Nd/^{144}Nd$ ratio of 0.512362 ± 0.000013 (initial $\varepsilon_{Nd}(t) = -2.1$). Fluorite yields an Sm–Nd isochron age of 141 ± 20 Ma, with a low MSWD of 0.43 and an initial $^{143}Nd/^{144}Nd$ ratio of 0.512233 ± 0.000034 (initial $\varepsilon_{Nd}(t) = -4.3$).

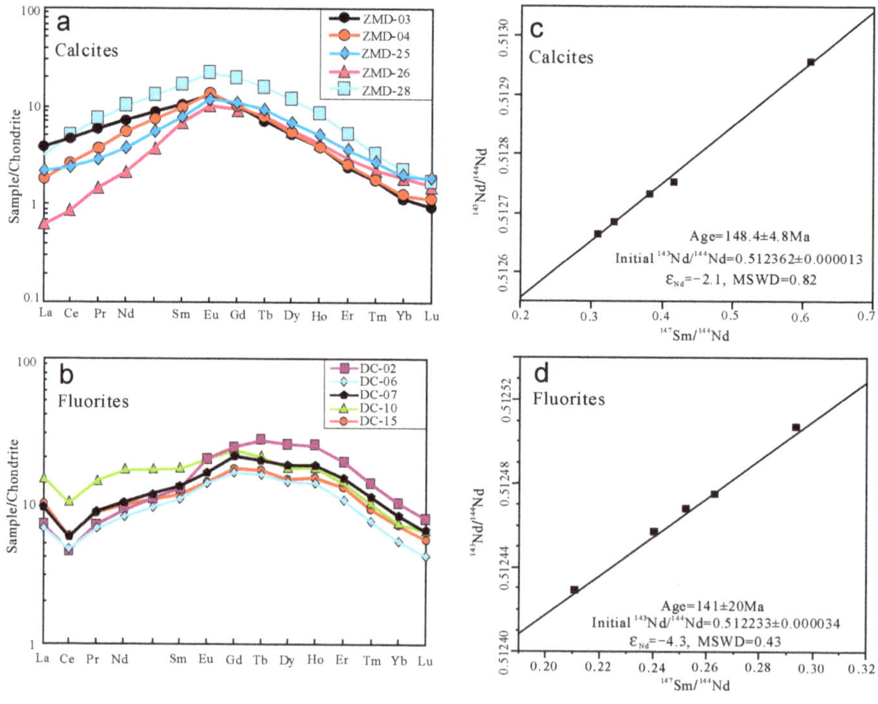

Figure 5. C1 chondrite-normalized [43] rare earth element (REE) patterns (**a**,**b**) and corresponding Sm–Nd isochron ages (**c**,**d**) of the calcite and fluorite.

5. Interpretations and Discussion

5.1. REE Features in Calcite and Fluorite

MREE-enriched patterns of calcite and fluorite are currently not well understood. MREE-enriched calcite and fluorite also occur in other low-temperature hydrothermal deposits in Southwestern Guizhou, such as calcite in the Shuiyindong [18] and Banqi [44] Au deposits, and fluorite in the Qinlong [20] and Banpo [25] Sb deposits. MREE-enriched patterns were also observed in natural terrestrial waters and acidic leachates by Johannesson et al. [45,46]. Johannesson et al. [45] proposed solid-liquid exchange reactions, dissolution of MREE-enriched surface Fe-Mn coatings, particulates, secondary mineral phases, and sulfate complexation as possible mechanisms to develop these MREE-enriched patterns. The ore-forming fluids for the Au and Sb deposits in Southwestern Guizhou may be MREE-enriched, and the calcite and fluorite likely inherited the MREE-enriched pattern of the ore forming fluids from which the calcite and fluorite precipitated [33].

In natural aqueous systems, the occurrence of Eu and Ce anomalies, which are due to redox reactions, may indicate reducing conditions for the former and oxidizing conditions for the latter [47]. Fluid inclusion studies of quartz in the Zimudang Au deposit [48] show that the ore-forming fluids were low temperature (approximately 200–230 °C), CO_2 and CH_4 rich, and reducing. Thermodynamic calculations suggest that, at an elevated temperature (>200 °C) and reducing conditions, Eu^{2+} should be dominant over Eu^{3+} [49] and may preferentially substitute for Ca^{2+} over trivalent REEs in calcite, leading to the positive Eu anomalies observed in this study. Negative Ce anomalies have been extensively investigated in sea and river waters [47]. The Ce anomaly occurs in response to the oxidation of Ce^{3+} to Ce^{4+} and the precipitation of Ce^{4+} from solution as CeO_2 [50]. Fluid inclusion studies for fluorite in the Dachang Sb deposit [48] show that a large volume of meteoric water was involved in the ore-forming fluids at the Sb metallogenic stage, decreasing the fluid temperature (140–160 °C), which may have led to the negative Ce anomalies in fluorite.

The initial $\varepsilon_{Nd}(t)$ values can also be used to obtain evidence on the source of Nd. The relative higher initial $\varepsilon_{Nd}(t)$ value (−2.1) for calcite is similar to calcite from the Shuiyindong Au deposit (−2.0 to −1.1 at 145 Ma; [18,19]), and the lower initial $\varepsilon_{Nd}(t)$ value (−4.3) for fluorite resembles that from the Qinglong Sb deposit (−3.7 to −5.8 at 145 Ma) [20]. The host rocks of the Au and Sb mineralization in Southwestern Guizhou are primarily limestone and basaltic volcanic tuff from the Permian. The initial $\varepsilon_{Nd}(t)$ values of calcite and fluorite in this study are within the range of values for Permian limestone (−6.3) and basaltic volcanic tuff (+1.5) at 145 Ma [20,51]. Thus, the $\varepsilon_{Nd}(t)$ values of the calcite and fluorite may have been sourced from mixtures of basaltic volcanic tuff and bioclastic limestone in the Permian.

5.2. Au and Sb Metallogenic Age in Southwestern Guizhou

The linear relationships shown in Figure 5c,d represent isochrons or mixed lines with two end-members characterized by significantly different $^{143}Nd/^{144}Nd$ and $^{147}Sm/^{144}Nd$ ratios. In the former, the slopes of the straight lines determine the ages of the fluorite and calcite. In the latter, the slopes have no meaning. The calcite and fluorite did not yield linear relationships in the 1/Nd vs. $^{143}Nd/^{144}Nd$ diagram (Figure 6a) and show a consistent Y/Ho ratio in the Y/Ho vs. La/Ho diagram (Figure 6b), respectively. Therefore, the two lines have isochronal significance and can accurately reflect the ages of the calcite (148.4 ± 4.8 Ma) and fluorite (141 ± 20 Ma).

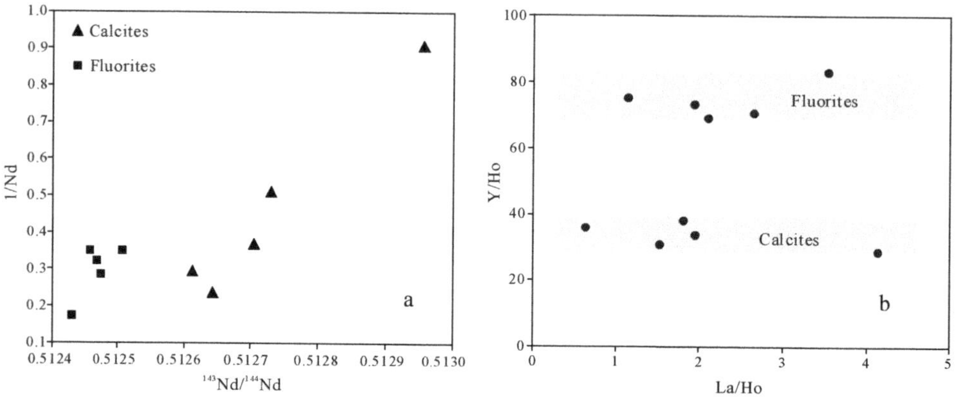

Figure 6. Diagrams of 1/Nd vs. $^{143}Nd/^{144}Nd$ (**a**) and Y/Ho vs. La/Ho (**b**) for the calcite and fluorite.

Hydrothermal deposits are the product of a geological thermal event. Fission tracks and (U–Th)/He ratios are temperature-sensitive radiometric dating techniques. Huang et al. [35] reported fission track and (U–Th)/He ages of detrital zircons collected from the Shuiyindong, Taipingdong, Yata, and Getang Au deposits in Southwestern Guizhou. Two thermal events (i.e., 192–216 and 132–160 Ma) with temperatures higher than the Au ore-formation temperature (~210 °C [9]) were obtained. The Au and Sb metallogenic event in Southwestern Guizhou occurred at 192–216 and/or 132–160 Ma. Tan et al. [19] also reported two instances of hydrothermal fluorite (200.1 $\pm$ 8.6 Ma) and calcite (150.2 $\pm$ 2.2 Ma) minerals at the Shuiyindong Au deposit, corresponding to the Indosinian orogeny and Yanshanian orogeny, respectively.

The crosscutting relationship of stratigraphy or intrusions is the most reliable method for constraining the mineralization epoch. The Au and Sb mineralization in Southwestern Guizhou is primarily located in the Upper Permian to Triassic systems. Wang [52] described a remnant exposure of Jurassic strata near the Zimudang Au deposit that is unconformably parallel to and folded with Late Triassic strata (Figure 7b). We note that many fault-controlled orebodies in the Zimudang Au deposit crosscut the anticline. Therefore, the Au and Sb mineralization is synchronous with or younger than the folding, and the Jurassic is the maximum age limit of the Au and Sb mineralization. The parallel unconformity (ca. 200 Ma) between the Early Jurassic and Late Triassic strata may represent the weak remote response of the Indosinian orogeny that transformed the Youjiang Basin uplift into land. Huang et al. [35] reported that the oldest group of zircon fission tracks and (U–Th)/He ages (192–216 Ma) may record the cooling age of the pre-ore thermal event caused by the Indosinian orogeny. In addition, there is another exposure of Late Cretaceous sedimentary rocks (ca. 100 Ma) north of Guizhou, which is angularly unconformable to Middle Jurassic rocks (ca. 160 Ma) (Figure 7c) [36]. The angular unconformity resulted from the Yanshanian orogeny, which initiated the pervasive folding and faulting of Middle Jurassic and underlying strata and created the present structural appearance of Guizhou Province. Chen et al. [53] described a 96 Ma unaltered quartz porphyry dike that crosscuts the orebodies in the Liaotun Au deposit in Guangxi Province, which represents the minimum age limit of Au mineralization in the Youjiang Basin [4,9,54,55]. In summary, these field surveys indicate that the Au and Sb mineralization in Southwestern Guizhou formed between the Middle Jurassic (ca. 160 Ma) and Late Cretaceous periods (96 Ma).

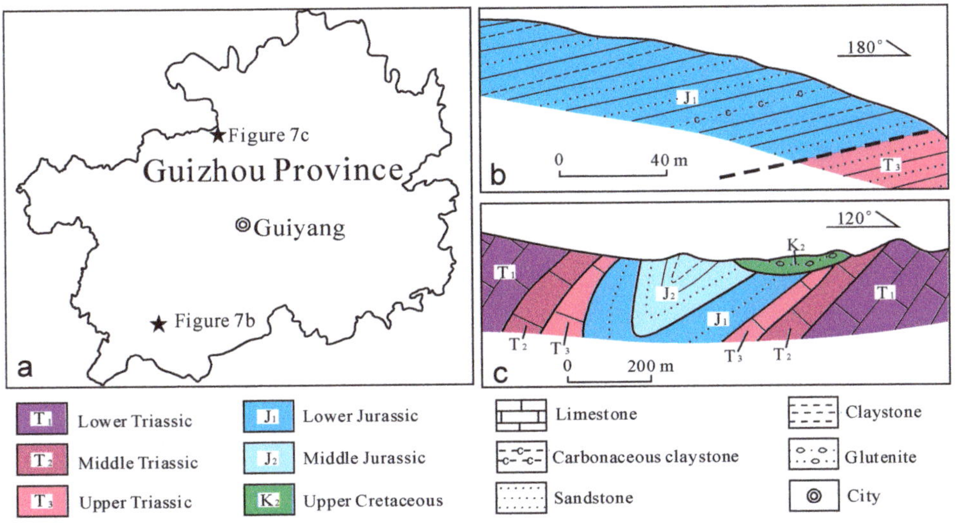

Figure 7. Two cross-sections in Guizhou Province showing the contact relations from the Lower Triassic to the Upper Cretaceous systems. (**a**) Geographical map of Guizhou showing the locations of the geologic cross sections; (**b**) Geologic cross section near the Zimudang Au deposit showing Jurassic strata is unconformably parallel to and folded with Late Triassic strata; (**c**) Geologic cross section north of Guizhous showing Late Cretaceous strata is angularly unconformable to Middle Jurassic rocks.

Over the past few decades, some important progress has been made in dating hydrothermal minerals in Au and Sb deposits in Southwest Guizhou. Chen et al. [15] obtained an ^{40}Ar–^{39}Ar plateau age of 194.6 ± 2 Ma using sericite in quartz veins from the Lannigou Au deposit. Chen et al. [14] reported ages of 204 ± 19 Ma and 235 ± 33 Ma based on the Re–Os isotopes of arsenopyrite from the Lannigou and Shuiyindong deposits, respectively. Sm–Nd isochron ages of 134 ± 3 and 136 ± 3 Ma have been reported for calcite from the Shuiyindong Au deposit [18]. Tan et al. [19] reported another Sm–Nd isochron age of 150.2 ± 2.2 Ma of calcite from the Shuiyindong Au deposit. Peng et al. [20] obtained Sm–Nd isochron ages of 148 ± 8 and 142 ± 16 Ma from fluorite in the Qinglong Sb deposit. Chen et al. [16] obtained a weighted-mean secondary ion mass spectrometry Th–Pb age of 141 ± 3 Ma for apatite from the Nibao Au deposit. Zheng et al. [17] dated fluid inclusions in quartz from the Nibao Au deposit and obtained Rb–Sr isochron ages of 142 ± 3 and 141 ± 2 Ma.

The reported Re–Os isochron ages of the Lannigou and Shuiyindong Au deposits are likely significantly mixed and older than the Au mineralization. Ore-stage arsenopyrite is commonly intergrown with fine-grained diagenetic pyrite and is very difficult to separate [9,16,19,54]. Additionally, the reported Ar–Ar age spectrum of the sericite is highly irregular and has no plateau, indicating that this age is not reliable [9]. Excluding the unreliable ages of arsenopyrite and sericite, the age range obtained by calcite and fluorite Sm–Nd, quartz fluid inclusions Rb–Sr, and apatite Th–Pb methods is 135–150 Ma [9,16,17,19,20], which is consistent with the age interval constrained by low-temperature thermochronology (132–160 Ma) and the crosscutting relationships of stratigraphy or intrusions (96–160 Ma). The ages obtained in this study add new evidence to previous geochronology studies and reveal the regional Au and Sb mineralization age interval of 135–150 Ma corresponding to the Yanshanian orogeny.

Many studies have proposed magmatic origin models for the Au and Sb mineralization in Southwestern Guizhou based on δ^{34}S (ore-related sulfide, approximately −5 to +5‰ [3,4,56–58], Δ^{199}Hg (mineralized rocks, −0.1 to 0.1‰ [59]) and He isotope data (minerals of main ore stage, R/R_A ratios ranging from 0.01 to 0.4 [2]). According to a synthesis of structural, petrological, geochronological, and geochemical studies, South China notably underwent a tectonic transition from contraction to large-scale extension and experienced extensive magmatic activity ca. 135–150 Ma as a result of the westward flat-slab subduction of the Pacific oceanic lithosphere beneath South China [60,61]. Although magmatic outcrops from 135–150 Ma are absent in Southwestern Guizhou, gravity and magnetic geophysical investigations indicate the presence of a pluton ~5 km below the surface of the Huijiabao anticline district [4]. Collectively, the low-temperature hydrothermal Au and Sb mineralization in Southwestern Guizhou is likely related to the Yanshanian magmatic activity associated with a weak, extensional, tectonic environment.

6. Conclusions

Calcite and fluorite samples, collected from the Zimudang Au deposit and the Dangchang Sb deposit, respectively, are symbiotic with sulfides of a metallogenic stage (e.g., realgar, orpiment, and stibnite), and they are thought to be coeval with Au and Sb mineralization. The ore-related calcite and fluorite yielded Sm–Nd isochron ages of 148.4 ± 4.8 and 141 ± 20 Ma, respectively. These results, combined with regional low-temperature thermochronology, crosscutting relationships of stratigraphy or intrusions, and previous dating results, indicate that the Au and Sb deposits in Southwestern Guizhou most likely formed at 135–150 Ma, corresponding to the Yanshanian orogeny associated with a weak, extensional, tectonic environment.

Author Contributions: Q.T. and Y.X. designed the study concept and revised the manuscript. Z.W. contributed to the analysis, data interpretation, and manuscript preparation. J.L. (Jianzhong Liu), C.Y., S.L., J.L. (Jumhai Li), F.C., X.W., Q.P., and D.W. collected the calcite and fluorite samples. Q.T. drew the geologic maps. All authors have read and agreed to the published version of the manuscript.

Funding: This study was financially supported by the National Key R&D Program of Deep penetrating Geochemistry (2016YFC0600607, 2017YFC0601500, and 2020YFC1807700), the National Natural Science Foundation of China (U1812402; 41803046), and the Science & Technology Department of Guizhou Province ([2019]2833, [2019]1424).

Institutional Review Board Statement: Not applicable.

Informed Consent Statement: Not applicable.

Data Availability Statement: Data is contained within the article.

Acknowledgments: The authors thank the anonymous reviewers for their helpful suggestions, which improved the manuscript.

Conflicts of Interest: The authors declare no conflict of interest.

References

1. Li, J.; Hu, R.; Xiao, J.; Zhuo, Y.; Yan, J.; Oyebamiji, A. Genesis of gold and antimony deposits in the Youjiang metallogenic province, SW China: Evidence from in situ oxygen isotopic and trace element compositions of quartz. *Ore Geol. Rev.* **2020**, *116*, 103257. [CrossRef]
2. Jin, X.-Y.; Hofstra, A.H.; Hunt, A.G.; Liu, J.-Z.; Yang, W.; Li, J.-W. Noble gases fingerprint the source and evolution of ore-forming fluids of Carlin-type gold deposits in the golden triangle, south China. *Econ. Geol.* **2020**, *115*, 455–469. [CrossRef]
3. Yan, J.; Hu, R.Z.; Liu, S.; Lin, Y.T.; Zhang, J.C.; Fu, S.L. NanoSIMS element mapping and sulfur isotope analysis of Au-bearing pyrite from Lannigou Carlin-type Au deposit in SW China: New insights into the origin and evolution of Au-bearing fluids. *Ore Geol. Rev.* **2018**, *92*, 29–41. [CrossRef]
4. Xie, Z.J.; Xia, Y.; Cline, J.S.; Pribil, M.J.; Koenig, A.; Tan, Q.P.; Wei, D.T.; Wang, Z.P.; Yan, J. Magmatic origin for sediment-hosted Au deposits, Guizhou Province, China: In situ chemistry and sulfur isotope composition of pyrites, Shuiyindong and Jinfeng deposits. *Econ. Geol.* **2018**, *113*, 1627–1652. [CrossRef]
5. Chen, J.; Yang, R.-D.; Du, L.-J.; Zheng, L.-L.; Gao, J.-B.; Lai, C.-K.; Wei, H.-R.; Yuan, M.-G. Mineralogy, geochemistry and fluid inclusions of the Qinglong Sb-(Au) deposit, Youjiang basin (Guizhou, SW China). *Ore Geol. Rev.* **2018**, *92*, 1–18. [CrossRef]
6. Wei, D.-T.; Xia, Y.; Gregory, D.D.; Steadman, J.A.; Tan, Q.-P.; Xie, Z.-J.; Liu, X.-J. Multistage pyrites in the Nibao disseminated gold deposit, southwestern Guizhou Province, China: Insights into the origin of Au from textures, in situ trace elements, and sulfur isotope analyses. *Ore Geol. Rev.* **2020**, *122*, 103446. [CrossRef]

7. Wei, D.; Xia, Y.; Steadman, J.; Xie, Z.; Liu, X.; Tan, Q.; Bai, L.A. Tennantite-Tetrahedrite-Series minerals and related pyrite in the Nibao Carlin-type gold deposit, Guizhou, SW China. *Minerals* **2020**, *11*, 2. [CrossRef]
8. Wang, J.; Chang, J.; Li, C.; Han, Z.; Wang, T.; Han, H. Significance of calcite trace elements contents and C-O isotopic compositions for ore-Forming fluids and gold prospecting in the Zhesang Carlin-Like gold deposit of Southeastern Yunnan, China. *Minerals* **2020**, *10*, 338. [CrossRef]
9. Su, W.C.; Dong, W.D.; Zhang, X.C.; Shen, N.P.; Hu, R.Z.; Hofstra, A.H.; Cheng, L.Z.; Xia, Y.; Yang, K.Y. Carlin-Type gold deposits in the Dian-Qian-Gui "golden triangle" of Southwest China. In *Diversity of Carlin-Style Gold Deposits, Reviews in Economic Geology*; Muntean, J.L., Ed.; Society of Economic Geologists, Inc.: Littleton, CO, USA, 2018; Volume 20, pp. 157–185.
10. Hu, R.Z.; Fu, S.L.; Huang, Y.; Zhou, M.F.; Fu, S.H.; Zhao, C.H.; Wang, Y.J.; Bi, X.W.; Xiao, J.F. The giant South China Mesozoic low-temperature metallogenic domain: Reviews and a new geodynamic model. *J. Asian Earth Sci.* **2017**, *137*, 9–34. [CrossRef]
11. Muntean, J.L.; Cline, J.S.; Simon, A.C.; Longo, A.A. Magmatic-Hydrothermal origin of Nevada's Carlin-type gold deposits. *Nat. Geosci.* **2011**, *4*, 122–127. [CrossRef]
12. Cline, J.S.; Hofstra, A.H.; Muntean, J.L.; Tosdal, R.M.; Hickey, K.A. Carlin-Type gold deposits in Nevada: Critical geologic characteristics and viable models. In *Economic Geology 100th Anniversary Volume*; Society of Economic Geologists, Inc.: Littleton, CO, USA, 2005; pp. 451–484.
13. Cline, J.S. Nevada's Carlin-type gold deposits: What we've learned during the past 10 to 15 years. In *Diversity of Carlin-Style Gold Deposits, Reviews in Economic Geology*; Muntean, J.L., Ed.; Society of Economic Geologists, Inc.: Littleton, CO, USA, 2018; Volume 20, pp. 7–37.
14. Chen, M.H.; Mao, J.W.; Li, C.; Zhang, Z.Q.; Dang, Y. Re-Os isochron ages for arsenopyrite from Carlin-like gold deposits in the Yunnan–Guizhou–Guangxi "golden triangle", southwestern China. *Ore Geol. Rev.* **2015**, *64*, 316–327. [CrossRef]
15. Chen, M.H.; Huang, Q.W.; Hu, Y.; Chen, Z.Y.; Zhang, W. Genetic type of Phyllosilicate (Micas) and its Ar-Ar dating in Lannigou gold deposit, Guizhou Province, China. *Acta Mineral. Sin.* **2009**, *29*, 353–362, (In Chinese with English abstract).
16. Chen, M.H.; Bagas, L.; Liao, X.; Zhang, Z.Q.; Li, Q.L. Hydrothermal apatite SIMS ThPb dating: Constraints on the timing of low-temperature hydrothermal Au deposits in Nibao, SW China. *Lithos* **2019**, *324–325*, 418–428. [CrossRef]
17. Zheng, L.L.; Yang, R.D.; Gao, J.B.; Chen, J.; Liu, J.Z.; Li, D.P. Quartz Rb-Sr isochron ages of two type orebodies from the Nibao Carlin-Type gold deposit, Guizhou, China. *Minerals* **2019**, *9*, 399. [CrossRef]
18. Su, W.C.; Hu, R.Z.; Xia, B.; Xia, Y.; Liu, Y.P. Calcite Sm-Nd isochron age of the Shuiyindong Carlin-type gold deposit, Guizhou, China. *Chem. Geol.* **2009**, *258*, 269–274. [CrossRef]
19. Tan, Q.; Xia, Y.; Xie, Z.; Wang, Z.; Wei, D.; Zhao, Y.; Yan, J.; Li, S. Two hydrothermal events at the Shuiyindong Carlin-Type gold deposit in Southwestern China: Insight from Sm-Nd dating of fluorite and calcite. *Minerals* **2019**, *9*, 230. [CrossRef]
20. Peng, J.T.; Hu, R.Z.; Jiang, G.H. Samarium-Neodymium isotope system of fluorites from the Qinglong antimony deposit, Guizhou Province: Constrains on the mineralizing age and ore-forming materials' sources. *Acto Petrol. Sin.* **2003**, *19*, 785–791. (In Chinese with English abstract)
21. Zhang, F.; Yang, K.Y. Mineralization time of fission track in disseminated gold deposits in Southwest Guizhou. *Chin. Sci. Bull.* **1992**, *37*, 1593–1595. (In Chinese with English abstract)
22. Su, W.C.; Yang, K.Y.; Hu, R.Z.; Chen, F. Fluid inclusion chronological study of the Carlin-type gold deposits in southwestern China: As exemplified by the Lannigou gold deposit, Guizhou province. *Acta Mineral. Sin.* **1998**, *18*, 359–362. (In Chinese with English abstract)
23. Peng, J.T.; Hu, R.Z.; Burnard, P.G. Samarium-Neodymium isotope systematics of hydrothermal calcites from the Xikuangshan antimony deposit (Hunan, China): The potential of calcite as a geochronometer. *Chem. Geol.* **2003**, *200*, 129–136. [CrossRef]
24. Zou, Z.C.; Hu, R.Z.; Bi, X.W.; Wu, L.Y.; Feng, C.X.; Tang, Y.Y. Absolute and relative dating of Cu and Pb-Zn mineralization in the Baiyangping area, Yunnan Province, SW China: Sm-Nd geochronology of calcite. *Geochem. J.* **2015**, *49*, 103–112. [CrossRef]
25. Wang, J.S.; Wen, H.J.; Fan, H.F.; Zhu, J.J. Sm-Nd geochronology, REE geochemistry and C and O isotope characteristics of calcites and stibnites from the Banian antimony deposit, Guizhou Province, China. *Geochem. J.* **2012**, *46*, 393–407. [CrossRef]
26. Uysal, I.T.; Zhao, J.X.; Golding, S.D.; Lawrence, M.G.; Glikson, M.; Collerson, K.D. Sm-Nd dating and rare-earth element tracing of calcite: Implications for fluid-flow events in the Bowen Basin, Australia. *Chem. Geol.* **2007**, *238*, 63–71. [CrossRef]
27. Anglin, C.D.; Jonasson, I.R.; Franklin, J.M. Sm-Nd dating of scheelite and tourmaline; Implications for the genesis of Archean gold deposits, Val d'Or, Canada. *Econ. Geol.* **1996**, *91*, 1372–1382. [CrossRef]
28. Roberts, S.; Palmer, M.R.; Waller, L. Sm-Nd and REE characteristics of tourmaline and scheelite from the Bjorkdal gold deposit, northern Sweden: Evidence of an intrusion-related gold deposit? *Econ. Geol.* **2006**, *101*, 1415–1425. [CrossRef]
29. Bell, K.; Anglin, C.; Franklin, J. Sm-Nd and Rb-Sr isotope systematics of scheelites: Possible implications for the age and genesis of vein-hosted gold deposits. *Geology* **1989**, *17*, 500–504. [CrossRef]
30. Chesley, J.T.; Halliday, A.N.; Kyser, T.K.; Spry, P.G. Direct dating of Mississippi valley-type mineralization; use of Sm-Nd in fluorite. *Econ. Geol.* **1994**, *89*, 1192–1199. [CrossRef]
31. Munoz, M.; Premo, W.; Courjault-Rade, P. Sm-Nd dating of fluorite from the worldclass Montroc fluorite deposit, southern Massif Central, France. *Miner. Depos.* **2005**, *39*, 970–975. [CrossRef]
32. Xu, W.G.; Fan, H.R.; Hu, F.F.; Santosh, M.; Yang, K.F.; Lan, T.G.; Wen, B.J. Geochronology of the Guilaizhuang gold deposit, Luxi Block, eastern North China Craton: Constraints from zircon U-Pb and fluorite-calcite Sm-Nd dating. *Ore Geol. Rev.* **2015**, *65*, 390–399. [CrossRef]

33. Tan, Q.P.; Xia, Y.; Wang, X.Q.; Xie, Z.J.; Wei, D.T. Carbon-Oxygen isotopes and rare earth elements as an exploration vector for Carlin-type gold deposits: A case study of the Shuiyindong gold deposit, Guizhou Province, SW China. *J. Asian Earth Sci.* **2017**, *148*, 1–12. [CrossRef]
34. Du, Y.; Huang, H.; Yang, J.; Huang, H.; Tao, P.; Huang, Z.; Hu, L.; Xie, X. The basin translation from late Paleozoic to Triassic of the Youjiang basin and its tectonic signification. *Geol. Rev.* **2013**, *59*, 1–11, (In Chinese with English abstract).
35. Huang, Y.; Hu, R.Z.; Bi, X.W.; Fu, S.L.; Peng, K.Q.; Gao, W.; Oyebamiji, A.; Zhaanbaeva, A. Low-Temperature thermochronology of the Carlin-type gold deposits in southwestern Guizhou, China: Implications for mineralization age and geological thermal events. *Ore Geol. Rev.* **2019**, *115*, 103178. [CrossRef]
36. Guizhou Institute of Geological Survey. *The Regional Geology of China, Guizhou Province*; Geological Publishing House: Beijing, China, 2017; (In Chinese with English abstract).
37. Chen, J.; Huang, Z.-L.; Yang, R.-D.; Du, L.-J.; Liao, M.-Y. Gold and antimony metallogenic relations and ore-forming process of Qinglong Sb(Au) deposit in Youjiang basin, SW China: Sulfide trace elements and sulfur isotopes. *Geosci. Front.* **2020**, *12*, 605–623. [CrossRef]
38. Liu, J.; Yang, C.; Xia, Y.; Chen, S.; Chen, F.E.; You, B.; Fu, Z. SBT study and ideas in patform lithofaices area in the Southwest Guizhou. *Guizhou Geol.* **2010**, *27*, 178–184, (In Chinese with English abstract).
39. Tan, Q.P.; Xia, Y.; Xie, Z.J.; Yan, J. Migration paths and precipitation mechanisms of ore-forming fluids at the Shuiyindong Carlin-type gold deposit, Guizhou, China. *Ore Geol. Rev.* **2015**, *69*, 140–156. [CrossRef]
40. Qi, L.; Hu, J.; Gregoire, D.C. Determination of trace elements in granites by inductively coupled plasma mass spectrometry. *Talanta* **2000**, *51*, 507–513.
41. Zhang, J.R.; Wen, H.J.; Qiu, Y.Z.; Zhang, Y.X.; Li, C. Ages of sediment-hosted Himalayan Pb-Zn-Cu-Ag polymetallic deposits in the Lanping basin, China: Re-Os geochronology of molybdenite and Sm-Nd dating of calcite. *J. Asian Earth Sci.* **2013**, *73*, 284–295. [CrossRef]
42. Ludwig, R.K. *ISOPLOT: A Plotting and Regression Program for Radiogenic Isotope Data (Version 2.9)*; U.S. Geological Survey: Washington, DC, USA, 1996; pp. 91–445.
43. Sun, S.S.; McDonough, W.F. Chemical and isotopic systematics of oceanic basalts: Implications for mantle composition and processes. *Geol. Soc. Lond. Spec. Publ.* **1989**, *42*, 313–345. [CrossRef]
44. Wang, J.; Han, Z.; Li, C.; Gao, Z.; Yang, Y.; Zhou, G. REE, Fe and Mn contents of calcites and their prospecting significance for the Banqi Carlin-type gold deposit in Southwestern China. *Geotecton. Metallog.* **2018**, *42*, 494–504, (In Chinese with English abstract).
45. Johannesson, K.H.; Lyons, W.B.; Yelken, M.A.; Gaudette, H.E.; Stetzenbach, K.J. Geochemistry of the rare-earth elements in hypersaline and dilute acidic natural terrestrial waters: Complexation behavior and middle rare-earth element enrichments. *Chem. Geol.* **1996**, *133*, 125–144. [CrossRef]
46. Johannesson, K.H.; Lyons, W.B. Rare-Earth element geochemistry of Colour Lake, an acidic freshwater lake on Axel Heiherg Island, Northwest Territories, Canada. *Chem. Geol.* **1995**, *119*, 209–223. [CrossRef]
47. Bau, M.; Möller, P. Rare earth element fractionation in metamorphogenic hydrothermal calcite, magnesite and siderite. *Mineral. Petrol.* **1992**, *45*, 231–246. [CrossRef]
48. Wang, Z.P.; Xia, Y.; Song, X.Y.; Liu, J.Z.; Yang, C.F.; Yan, B.W. Study on the evolution of ore-formation fluids for Au-Sb ore deposits and the mechanism of Au-Sb paragenesis and differentiation in the southwestern part of Guizhou Province, China. *Chin. J. Geochem.* **2013**, *32*, 56–68. [CrossRef]
49. Sverjensky, D.A. Europium redox equilibria in aqueous solution. *Earth Planet. Sci. Lett.* **1984**, *67*, 70–78. [CrossRef]
50. Rollinson, H.R. *Using Geochemical Data: Evaluation, Presentation, Interpretation*; Longman Scientific and Technical: London, UK, 1993.
51. Chung, S.-L.; Jahn, B.-M. Plume-Lithosphere interaction in generation of the Emeishan flood basalts at the Permian-Triassic boundary. *Geology* **1995**, *23*, 889–892. [CrossRef]
52. Wang, Z. Affirmation of the Jurassic in Longtoushan of Zhenfeng, Guizhou and its geological significance. *Guizhou Geol.* **1997**, *13*, 201–203. (In Chinese with English abstract)
53. Chen, M.H.; Zhang, Y.; Meng, Y.; Lu, G.; Liu, S. Determination of upper limit of metallogenic epoch of Liaotun gold deposit in western Guangxi and its implications for chronology of Carlin-type gold deposits in Yunnan-Guizhou-Gangxi golden triangle area. *Miner. Depos.* **2014**, *33*, 1–13.
54. Xie, Z.J.; Xia, Y.; Cline, J.S.; Koenig, A.; Wei, D.T.; Tan, Q.P.; Wang, Z.P. Are there Carlin-type gold deposits in China? A comparison of the Guizhou, China, deposits with Nevada, USA, deposits. In *Diversity of Carlin-Style Gold Deposits, Reviews in Economic Geology*; Muntean, J.L., Ed.; Society of Economic Geologists, Inc.: Littleton, CO, USA, 2018; Volume 20, pp. 187–233.
55. Zhu, J.J.; Hu, R.Z.; Richards, J.P.; Bi, X.-W.; Stern, R.; Lu, G. No genetic link between Late Cretaceous felsic dikes and Carlin-type Au deposits in the Youjiang basin, Southwest China. *Ore Geol. Rev.* **2017**, *84*, 328–337. [CrossRef]
56. Zhao, J.; Liang, J.; Li, J.; Huang, Y.; Liu, X.; Zhang, J.; Hao, J.; Sun, W.; Li, J.; Xie, J. Gold and sulfur sources of the Taipingdong Carlin-type gold deposit: Constraints from simultaneous determination of sulfur isotopes and trace elements in pyrite using nanoscale secondary ion mass spectroscopy. *Ore Geol. Rev.* **2020**, *117*, 1–11. [CrossRef]
57. Liang, J.; Li, J.; Liu, X.; Zhai, W.; Huang, Y.; Zhao, J.; Sun, W.; Song, M.; Li, J. Multiple element mapping and in-situ S isotopes of Au-carrying pyrite of Shuiyindong gold deposit, southwestern China using NanoSIMS: Constraints on Au sources, ore fluids, and mineralization processes. *Ore Geol. Rev.* **2020**, *123*, 103576. [CrossRef]

58. Tan, Q.P.; Xia, Y.; Xie, Z.J.; Yan, J.; Wei, D.T. S, C, O, H, and Pb isotopic studies for the Shuiyindong Carlin-type gold deposit, Southwest Guizhou, China: Constraints for ore genesis. *Chin. J. Geochem.* **2015**, *34*, 525–539. [CrossRef]
59. Yin, R.; Deng, C.; Lehmann, B.; Sun, G.; Lepak, R.F.; Hurley, J.P.; Zhao, C.; Xu, G.; Tan, Q.; Xie, Z.; et al. Magmatic-Hydrothermal origin of mercury in Carlin-style and epithermal gold deposits in China: Evidence from mercury stable isotopes. *ACS Earth Space Chem.* **2019**, *3*, 1631–1639. [CrossRef]
60. Hu, R.Z.; Zhou, M.F. Multiple Mesozoic mineralization events in South China—An introduction to the thematic issue. *Miner. Depos.* **2012**, *47*, 579–588. [CrossRef]
61. Li, Z.X.; Li, X.H. Formation of the 1300-km-wide intracontinental orogen and postorogenic magmatic province in Mesozoic South China: A flat-slab subduction model. *Geology* **2007**, *35*, 179–182. [CrossRef]

minerals

MDPI

Article

Mineralogy, Fluid Inclusions, and Oxygen Isotope Geochemistry Signature of Wolframite to Scheelite and Fe,Mn Chlorite Veins from the W, (Cu,Mo) Ore Deposit of Borralha, Portugal

Iuliu Bobos [1,*], Carlos Marques de Sá [2] and Fernando Noronha [1]

[1] ICTerra-Porto, Faculdade de Ciências, Universidade do Porto, 4168-007 Porto, Portugal; fmnoronh@fc.up.pt
[2] Grupo de Pesquisa em Recursos Minerais, Departamento de Geologia, Universidade Federal de Sergipe, Aracaju 49060-108, SE, Brazil; dingesster@gmail.com
* Correspondence: ibobos@fc.up.pt

Abstract: Scheelitization of Mn-bearing wolframite, scheelite, quartz, and Fe,Mn-chlorite veins was identified in the W, (Cu,Mo) ore deposits of Borralha, by optical microscopy, electron-microprobe analysis, and stable isotope geochemistry. Fluid inclusions derived scheelite crystallization temperature was compared with the oxygen isotope temperature estimated. Scheelite was formed mainly during stage I from a low salinity aqueous-carbonic fluid dominated by CO_2, where the homogenization temperature (T_h) decreased from 380 °C to 200 °C (average of 284 °C). As temperature decreased further, the aqueous-carbonic fluid became dominated by CH_4 (Stage II; (average T_h = 262 °C)). The final stage III corresponds to lower temperature mineralizing aqueous fluid (average T_h = 218 °C). In addition, salinity gradually decreased from 4.8 wt.% to 1.12 wt.%. The $\delta^{18}O_{Fluid}$ values calculated for quartz-water and wolframite-water fractionation fall within the calculated magmatic water range. The $\Delta_{quartz\text{-}scheelite}$ fractionation occurred at about 350–400 °C. The $\Delta_{chlorite\text{-}water}$ fractionation factor calculated is about +0.05‰ for 330 °C, dropping to −0.68‰ and −1.26‰ at 380 °C and 450 °C, respectively. Estimated crystallizing temperatures based on semi-empirical chlorite geothermometers range from 373 °C to 458 °C and 435 °C to 519 °C. A narrower temperature range of 375 °C to 410 °C was estimated for Fe,Mn-chlorite crystallization.

Keywords: wolframite scheelitization; Fe,Mn-chlorite; H_2O-CO_2 (CH_4) fluid inclusion; stable isotope fractionation; chlorite geothermometry; Borralha; Portugal

1. Introduction

Tungsten mineralization in magmatic-hydrothermal deposits occurs predominantly as wolframite ([Fe,Mn]WO_4) and scheelite ($CaWO_4$). Wolframite is the main dominated tungsten mineral in granite-related systems, whereas in skarn deposits, it is scheelite [1–4]. Given the variability of the geological context in which scheelite occurs in nature- scheelite-quartz vein mineralizations and greisen in granite [5–8], quartz-scheelite veins in metabasite [9], scheelite in orogenic gold [10], and skarn deposits [11], published scheelite fluid inclusions (FIs) are somewhat limited.

Scheelite related to quartz veins and greisen (e.g., Yangjiashan, China) include vapor- and liquid-rich Fis types with detectable CH_4, CO_2, and N_2, where the T_h ranges (mean values) from 230 °C to 250 °C and salinity (mean values) from 9.3 wt.% to 6.6 wt.% NaCl equiv. [12]. In addition, ore-forming fluids show intermediate to low temperatures that belong to the H_2O-NaCl ± CO_2 system in the disseminated scheelite associated with potassic feldspathization and silicification of porphyritic biotite granite from the Wuyi metallogenic belt (China) [13].

In the quartz-scheelite veins hosted in metabasites (e.g., Metaggitsi area, Greece), the most abundant FIs consist of H_2O-CO_2 with highly variable CO_2 contents (20 ± 90 vol.%) and salinity between 0.2 wt.% and 8.3 wt.% NaCl equiv., where the homogenization temperature of aqueous and gas-rich type inclusions has been estimated to about 220 ± 25 °C (mean values) [9]. Furthermore, scheelite has a distinctive trace elements composition from various deposit types and geological environments, being considered an indicator mineral for orogenic gold [14]. Microthermometric and Raman spectroscopic studies of FIs have documented low salinity aqueous-gaseous (H_2O-CO_2-CH_4-NaCl) ore fluids, where gold and scheelite precipitated in orogenic gold deposits from the Dharwar craton (South India) [15].

The FIs of scheelite crystallized during the cooler period of skarn development (e.g., Kara, Tasmania) consist of primary H_2O-CO_2 with moderate salinity (12.0 wt.% and 17.8 wt.% NaCl equiv.) and homogenization temperatures of 350–580 °C [16]. A similar thermal history and scheelite formation were observed for other W skarn deposits (e.g., CanTung Mine, Northwest Territories, Canada) [17,18]. Recently, scheelite-bearing albitite (no FI data are known) was identified in the Zhuxi (China) skarn tungstate deposit as a product of a silicate-poor, H_2O-rich melt that formed by melt-melt-liquid immiscibility in an extremely fractionated residual magma [19].

The scheelite occurrences mentioned above showed a connection with a magmatic-hydrothermal, metamorphic, or hydrothermal-meteoric fluid of low to moderate salinity, where homogenization temperatures ranged from 220 °C to 580 °C. Concerning the source and evolution of ore-forming fluids, the mechanisms and thermodynamic conditions of scheelite deposition have yet to be addressed [20,21]. In addition, in most related cases, scheelite was found associated with quartz veins, greisen, albitite, or secondary K-feldspar. Nevertheless, there are no published data available in the literature regarding the association of scheelite with chlorite.

Scheelite from the W, (Cu,Mo) ore deposit of Borralha is associated with wolframite, Fe,Mn-chlorite, and quartz, where Mn-bearing wolframite has been replaced by scheelite. The relationships "*chlorite* vs. *metal assemblages*" have been revisited and explored, aimed towards using chlorite as a "proxy" for detecting porphyry ore deposits [22] because the large variety of chlorite minerals can serve as a tool to determine the direction towards a mineralized area, to estimate the temperatures under which the ore processes took place [23,24], or to estimate the heat flux away from a magmatic-hydrothermal centre [25].

The stability of the Fe^{2+} end-member wolframite (ferberite) is much more sensitive to the high variation of f_{O2} than the Mn^{2+} end-member wolframite (hübnerite) and the composition of ferberite might also be strongly affected by the oxidation state of the system [26]. The physicochemical parameters (e.g., temperature, Mn^{2+}/Fe^{2+} ratio in the mineralizing solution) during wolframite crystallization influence the hübnerite vs. ferberite proportions of wolframite [27].

Furthermore, redox state exchange between Fe- and Mn-distributions in chlorite associated with wolframite and scheelite from the Borralha deposit showed that its composition is directly influenced by thermal gradients of the evolution of different stages within the magmatic-hydrothermal system [28], where the diversity of chlorite minerals associated with W-mineralization serves as an excellent indicator to identify mineralization areas and to estimate crystallization temperatures. Geothermometers have been developed for the Al-, Fe-, or Mg-rich chlorites and little is known about the specific chlorite systems, like Mn-rich chlorite, where a new question arises on Mn-redox, analogously to Fe-redox processes [29].

The application of empirical geothermometers to ore deposit research depends on complementary data regarding the mineralogical reaction mechanisms, estimation of P-V-T-X parameters of the ore-forming fluids or ones circulated through rocks, and the stable isotope geochemistry of neoformed minerals, helping to understand mineral history. Understanding the principles governing isotopic variations in nature is still limited due to the magnitude of isotopic fractionation that accompanies various geological processes and the diversity of fluids circulation mechanisms in the oceanic or continental crust, where geochemical parameters and the kinetic isotope effects accompanying a conversion of one mineral to another under a variety of conditions.

The main aim of this study is to characterize the mineralogy of scheelite + Fe,Mn-chlorite ± wolframite scheelitized assemblages found in large crystalline Fe,Mn-chlorite masses in a W veins system, the chemistry of the fluids responsible for scheelite crystallization, where FIs data are coupled with Fe,Mn-chlorite geothermometry and the oxygen isotope geochemistry of wolframite, scheelite, quartz, and Fe,Mn-chlorite.

2. Geological Background

The Variscan orogeny is currently explained by an obduction-collision orogenic model [30–32], which distinguishes three main deformation phases (D1, D2, and D3) in northern Portugal [33,34]. The maximum crustal thickening was achieved by the end of the D1/D2 phases, whereas the D3 phase (intra-Westphalian age) corresponds to the final stages of the continental collision process. Most of the granitic intrusions and associated thermal metamorphic peaks are coeval with the D3 phase [30,35,36]. The syn- to late-orogenic magmatic activity generated voluminous granitoid batholiths often with contrasting compositions, where their emplacement was in most cases controlled by major Variscan structures.

The W, (Cu,Mo) deposit of Borralha is located in the Central Iberian Zone in northwestern part of the Iberian Peninsula of the (Figure 1a), where different syn-orogenic Variscan granites (e.g., two mica peraluminous, syn-D3, and biotite rich syn-D3) intruded Palaeozoic metasedimentary units. The ore deposits occur at the contact between metasedimentary formations (Silurian) and the syn-D3 porphyritic Borralha biotite granite (~315 Ma) and a syn-D3 two-mica granite (Figure 1b). The Borralha ore deposit was an important W deposit in Portugal explored for wolframite, scheelite, and argentiferous chalcopyrite during 1903–1985. W mineralization occurs in quartz veins and in the two breccia pipes known as Santa Helena and Venise [37]. The Santa Helena breccia (SHB) occurs in outcrops (Figure 1b), whereas the Venise breccia was found underground during mining works. Rock fragments of variable size, cemented with mineralized coarse-grained quartz aggregates, characterize both breccias. The underground works exploited vertical quartz veins (more than 45° dip, VV) and sub-horizontal veins or "flat veins" (<30° dip, FV). The main vertical veins mineralized with tungstates occur in the most central deposit part, located in the Santa Helena veins (e.g., Santo António, Santa Helena north, and Santa Helena south) and cross several lithologies including porphyroid biotite granite, tonalites, and schists. The dominant micaschist rocks to the east part are cross-cut by the main W veins subdivided into vertical veins sub-concordant with the schistosity, namely: The three north (3N) and three south (3S) veins. However, these lodes contain less mineralizations of tungstates and more sulphides (i.e., chalcopyrite). The ore mineral assemblages consist of tungstate (wolframite and scheelite), various sulphide minerals (chalcopyrite, pyrite, pyrrhotite, molybdenite, sphalerite, galena, bismuthinite, marcasite), native bismuth, and Pb-Bi-Ag sulphosalts. Fluid inclusion studies of quartz related to W mineralization have identified aqueous-carbonic fluids (300 °C < T < 400 °C and 50 MPa < P < 100 MPa) and aqueous fluids (250 °C < T < 300 °C and 30 MPa < P < 50 MPa) in sulphide mineralization [38,39].

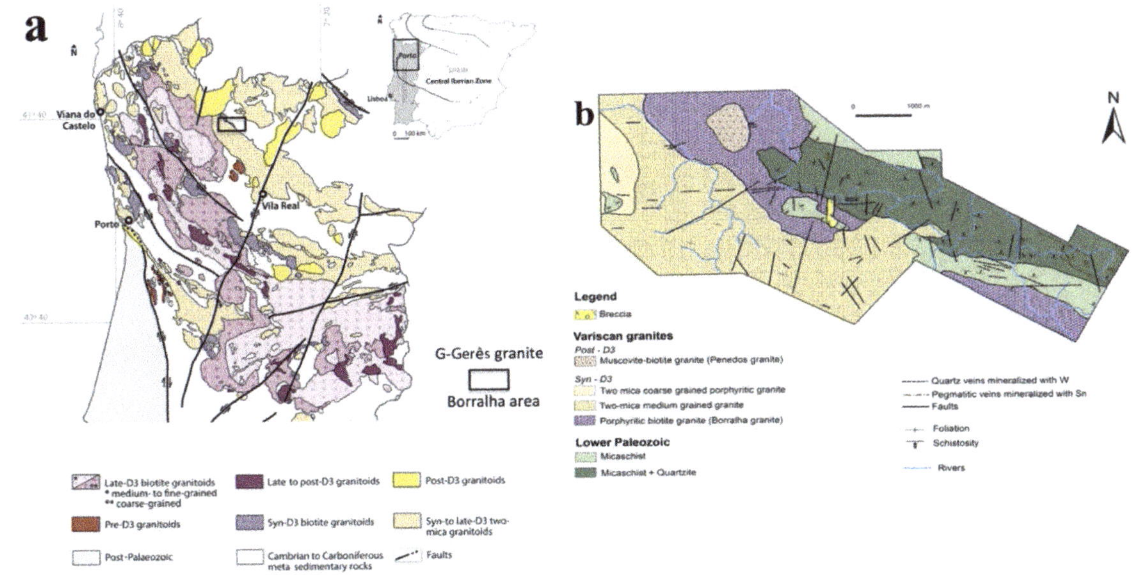

Figure 1. (**a**) Geological map of the northern part of Portugal and location of the W, (Cu,Mo) ore deposit of Borralha, Gerês Mountains [35]. (**b**) Geological map of the Borralha region [40]. (Both figures are reproduced from Bobos et al. [28] with permission of the Mineralogical Society of Great Britain and Ireland).

3. Materials and Methods

Sample collection. Wolframite to scheelite alteration and scheelite samples (SHS and 3N "vertical veins") with crystal sizes of a few centimeters associated with Fe,Mn-chlorite in large crystalline masses collected from the Santa Helena veins (SHS) and northern gallery (3N) from the exploitation level (−60 m underground) of the Borralha ore deposit were selected for mineralogy, FI, and oxygen isotope geochemistry analyses.

Optical microscopy. The petrography of gangue and ore minerals was conducted by optical microscopy (transmitted and reflected light) using a Nikon microscope ECLIPSE model E400POL (Minato, Tokyo, Japan), coupled with an AxioCam camera ZEISS model MRc, connected to a computer with an AxioVision imaging software (SE64 Rel.4.9).

Electron microprobe analysis. Mineral compositions were obtained with an electron probe micro-analyzer (EPMA) Jeol Hyperprobe JXA-8500F (Tokyo, Japan) operated at a 15 kV accelerating voltage and with a 10 nA beam current in the case of silicates; the operation mode for the tungstates involves 20 kV and 20 nA, with counting times of 10 s on the element peaks and 5 s on the background position. The major elements of the minerals were determined using detection limits that were acceptably (σ) above the mean background of 0.02 wt.% for most sulphides and silicates with mean counting times of 80 s. The standards used for the silicates were: Albite (Na$K\alpha$), orthoclase (Al$K\alpha$, Si$K\alpha$, K$K\alpha$), apatite (Ca$K\alpha$, P$K\alpha$), MgO (Mg$K\alpha$), $MnTiO_3$ (Mn$K\alpha$), TiO_2 (Ti$K\alpha$), and FeO (Fe$K\alpha$); for W minerals: Cassiterite (Sn $K\alpha$), pure Ta ($L\alpha$), Mo ($L\alpha$), Ni ($K\alpha$), Co ($K\alpha$), Cu ($K\alpha$), Ag ($L\alpha$), and Au ($M\alpha$) metals, $SrBaNbO_4O_{12}$ (Nb $L\alpha$), scheelite (W$L\alpha$), Sc_2O_3 (Sc $L\alpha$), UO_2 (U $M\beta$), ThO_2 (Th M_β), galena (Pb $M\alpha$), LaP_5O_{14} (La $L\alpha$), CeP_5O_{14} (Ce $L\alpha$), YAg (Y $L\alpha$), Cd ($L\alpha$), Sb_2S_3 (Sb$L\alpha$), AsGa (As $L\alpha$), and Bi_2Se_3 (Bi$M\alpha$). The calculation of the structural formula was based on four oxygens for tungstate minerals and 14 oxygens for chlorite.

Fluid inclusion microthermometry. Four bi-polished plate thin sections (two from SHS and the other two from 3N "vertical veins") were prepared and studied. Fluid petrography was performed on four scheelite (SHS1, SHS2, 3N1, and 3N2) bi-polished slide samples, where about 50 FIs were observed per bi-polished sample plate. The petrographic study was carried out by transmission microscopy, using a Leica DM/LSP microscope (Wetzlar, Germany) equipped with objectives up to 50× and a computer-controlled ICC50HD imaging system using the LAS EZ4 W software. The classification of FIs was performed according to the genetic criteria of Roedder [41], where the petrographic concepts of generations of FIs and fluid phase petrography were exposed by Van den Kerkhof and Hein [42], using Shepperd's et al. [43] abacuses. Microthermometric analyses were carried out using an Olympus microscope (Tokyo, Japan) with a maximum objective lens of 80× and a 12×, equipped with a Chaixmeca stage [44] for cryometry, and a Linkam TH600 stage, coupled to a NIKON POL-Optiphot microscope, for heating. The calibration of the equipment was done regularly using a set of SynFlinc standards. The estimated accuracy was −0.1 °C for the Chaixmeca stage and 1 °C in the Linkam stage. The salinity of the fluids was calculated with the help of the software developed by Bakker [45].

Raman spectroscopy. The study was conducted with a LabRam Dilor-JobinYvon-Spex (Lille, France) equipped with a He-Ne 632.8 nm laser source, coupled to an Olympus microscope with a 100× objective. The spectral decomposition of the bands and Raman parameters was carried out with Dilor LabSpec-JobinYvon-Spex Raman software (version 1.1). Raman spectrometry analysis of aqueous-carbonic inclusions allowed the estimation of the bulk composition and density of the fluids using the methods and equations of Touret [46] and Ramboz et al. [47]. The calibration and other technical details are described in Prieto et al. [48].

Oxygen isotopes. The scheelite, quartz, and chlorite samples were selected for oxygen isotope analysis carried out at the Laboratory for Stable Isotopes at the University of Salamanca (Salamanca, Spain). A conventional procedure was used for the extraction of oxygen from scheelite, chlorite, and quartz with BrF_5 and the quantitative conversion to CO_2 prior to the analysis by stable-gas mass spectrometry [49]. The $\delta^{18}O$ values were reported in ‰ relative to the Vienna-Standard Mean Ocean Water (V-SMOW) and normalized to the NBS-30 reference (+5.17 per mil vs. V-SMOW) and NBS-28 (+9.34 per mil vs. V-SMOW). The accuracy of $\delta^{18}O$ measurements was ±0.2‰ (1σ). The following isotope fractionation equations were used in this work between minerals and water in the system: Wolframite-H_2O [50], scheelite-H_2O [51,52], chlorite-H_2O [53], and quartz-H_2O [54].

4. Results

4.1. Mineralogy

The selected samples (3N and SHS) are characterized by the following mineral assemblages: (a) 3N: Wolframite, scheelite, Fe,Mn-chlorite, quartz, and (b) SHS: Fe,Mn-chlorite, scheelite. Fe,Mn-chlorite is represented by the following samples used for EPMA: FMN1, FMN2, FMN3, FMN8, FMN9, and FMN12.

Wolframite scheelitization. The wolframite crystals have been partially or totally replaced by scheelite in the samples analyzed. A net contact may be observed between wolframite (white field) replaced by scheelite (grey) (Figure 2a, NII). Small scheelite crystals occur along fissures or fractures and cleavages in wolframite, where the replacement of wolframite by scheelite developed a shredded texture (Figure 2b, NII) until complete replacement. Wolframite scheelitization implies a later circulation of a rich alkaline fluid (Ca^{2+}) (Figure 2c, NII).

In addition, small inclusions of sulphide (pyrite or chalcopyrite) were identified along fractures in wolframite (Figure 2d, N+). EPMA data of wolframite show a mixed composition between ferberite (F) and hübnerite (H), where the Fe amount of 0.53 atoms per formula unit (apfu) is higher than Mn (0.43 apfu) (Table 1).

The hübnerite and ferberite ratio (H/F) calculated shows values from 46.25% to 46.7%. Small amounts of Nb_2O_5 (0.18%), Ta_2O_5 (0.01%), and SnO_2 (0.01%) were also measured. EPMA data of scheelite show traces of Fe, Mn and also Nb, Ta, Mo, and Sn (Table 1). X-ray element maps of wolframite replaced by scheelite show the Fe, Mn, Ca, and W distributions in wolframite and scheelite crystals (Figure 3).

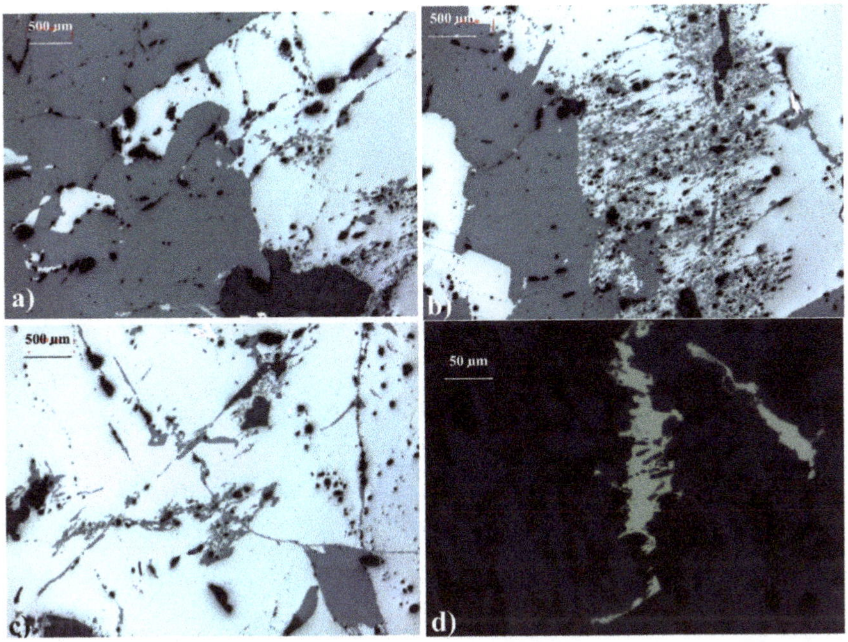

Figure 2. Photomicrographs showing textural relationships between wolframite and scheelite: (**a**) Wolframite (white field) replaced by scheelite (gray); (**b**,**c**) precipitation of scheelite crystals along to fissures or fractures of wolframite; and (**d**) inclusions of sulphide (pyrite) were identified along to fractures of wolframite.

Table 1. Electron microprobe data of wolframite and scheelite and calculus of structural formula based on four oxygens.

Oxides	10 Analyses Minimum	Fe,Mn-W Maximum	Fe,Mn-W Average	Sigma	14 Analyses Minimum	Scheelite Maximum	Scheelite Average	Sigma
WO_3	75.12	76.35	75.94	1.23	79.29	80.85	80.34	2.14
FeO		12.88	12.51	0.7		0.07	0.02	0.0738
MnO		11.03	10.62	0.66		0.13	0.03	0.00994
CaO		0.01	0	0.0024		19.32	19.18	0.68
Sc_2O_3		0.04	0.01	0.0274		0.02	0.01	0.02
Nb_2O_5		0.26	0.18	0.13		0.2	0.04	0.164
SnO_2		0.03	0.01	0.031		0.13	0.01	0.126
Ta_2O_5		0.14	0.04	0.012		0.3	0.06	0.2951
TiO_2		0	0	0		0.04	0.01	0.0172
V_2O_3		0.04	0.02	0.02		0.08	0.02	0.0581
Bi_2O_3		0.03	0.01	0.018		0.06	0.02	0.0453
MoO_3						0.241	0.09	0.0321
Sb_2O_5		0.05	0.02	0.02		0	0	0
Total		100.86	99.36			100.47	99.81	
W		0.99	1			0.99	1	

Table 1. *Cont.*

Oxides	10 Analyses Minimum	Fe,Mn-W Maximum	Fe,Mn-W Average	Sigma	14 Analyses Minimum	Scheelite Maximum	Scheelite Average	Sigma
Fe		0.54	0.53			0	0	
Mn		0.47	0.46			0	0	
Ca		0	0			0.98	0.99	
Sc		0	0			0	0	
Nb		0.01	0			0	0	
Sn		0	0			0	0	
Ta		0	0			0	0	
Ti		0	0			0	0	
V		0	0			0	0	
Bi		0	0			0	0	
Mo						0.02	0	
Sb		0	0			0	0	
Total		2.007	1.99			1.994	1.994	

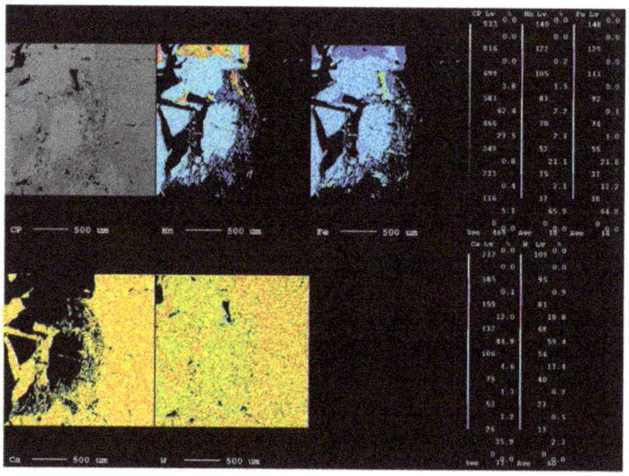

Figure 3. X-ray maps of Fe, Mn, and Ca distribution during wolframite scheelitization.

Scheelite. Scheelite ($CaWO_4$) aggregates adjacent to wolframite or Fe,Mn-chlorite were identified. The size of the scheelite aggregates reaches from 1 to 1.5 × 0.5 to 0.75 cm^2. The color is orange to yellow-brown and fluoresces with a blue-white color, indicating low MoO_3 content (average <0.09%). Traces of Sn, Pb, Bi, and Mo were measured in scheelite by EPMA (Table 1).

The average chemical composition of the scheelite associated with Fe,Mn-chlorite corresponds to $W_{1.00}Ca_{0.99}O_4$. The X-ray element maps show the W and Ca distributions in scheelite (Figure 4). Scheelite has a structure with tetrahedral $[WO_4]^{2-}$ and irregular dodecahedral $[CaO_8]^{14-}$ groups, where the hexavalent Mo can substitute for W, forming a complete solid solution series extending up to powellite ($CaMoO_4$) [55].

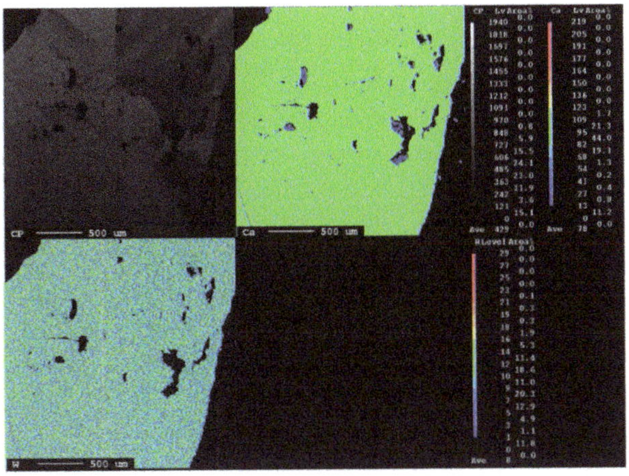

Figure 4. X-ray maps of W and Ca distribution in scheelite (reproduced from Bobos et al. [28] with permission of the Mineralogical Society of Great Britain and Ireland).

Fe,Mn-chlorite. Fe,Mn-chlorite aggregates observed in transmitted light under the microscope can be well identified by their green color pleochroism, varying from pale green for some massive aggregates to a dark pleochroism, a basal cleavage, and anomalous birefringence in cross-polarized light.

A strong pleochroism in green colors with an atypical dark orange interference color was observed (Figure 5), which is in accordance with the octahedral occupancy of heavy atoms (e.g., Mn^{2+}) [56] and the Fe + Mn/Fe + Mn + Mg ratio [57]. The elongation sign is opposite to the optical one and the optical plane is almost parallel to (010). The Fe,Mn-chlorite aggregates are radial with a radius size ranging from 500 μm to 1000 μm.

Figure 5. Photomicrograph showing radial aggregates of Fe,Mn-chlorite (NX).

The crystal chemistry of Fe,Mn-chlorite (Table 2) studied by EPMA shows an average crystal chemical composition (140 points analysed) corresponding to: $(Al_{1.385}Fe_{3.76}Mn_{0.59}Mg_{0.24}Ca_{0.01})_{5.99}(Si_{2.625}Al_{1.375})_4O_{10}(OH)_8$. There is no tetrahedral substitution [Si⇔Al(IV)], confirmed also by the calculated vacancy site (0.005) [58,59]. The R^{2+} exchanges (Mn^{2+} for Fe^{2+}) are evidenced in the octahedral sheet, also confirmed by the vacancy site calculated after Inoue et al. [59]. Small amounts of Zn^{2+}, Ca^{2+}, Na^{+}, and K^{+} were identified in Fe,Mn-chlorite. The ratio of Fe/(Fe + Mg) shows values from 0.92 to 0.94, whereas the Mg/(Mg + Fe) ratio is about 0.06. We also calculated the vacancy sites (Al^{VI}-Al^{IV}-Na-K)/2 [58], (Al^{VI}-Al^{IV})/2 [59], and 6-/Fe^{2+} + Fe^{3+}-Mg + Al^{VI} [60] (see Table 2).

Table 2. Electron microprobe data of Fe,Mn-chlorite and calculus of structural formula based on 14 oxygens (Fe^{2+}/Fe^{3+} and OH calculated assuming full site occupancy).

Oxides	**FMN 1**	**FMN2**	**FMN 3**	**FMN 12**	**FMN 8**	**FMN 9**	**Averages 146 Analyses**
	N = 31	**N = 29**	**N = 19**	**N = 22**	**N = 34**	**N = 11**	
SiO_2	23.18	22.7	23.2	23.11	23.73	22.46	22.82
TiO_2	0.08	0	0.02	0.01	0.04	0.04	0.06
Al_2O_3	19.88	19.95	19.82	19.82	20.14	20.86	20.37
FeO	39.52	39.86	39.53	39.13	39.55	38.59	39.055
MnO	6.23	5.03	5.82	6.21	5.28	5.81	6.02
MgO	1.87	1.44	1.88	1.67	1.75	1.01	1.44
CaO	0.04	0.01	0.03	0.05	0.05	0.06	0.05
Na_2O	0.03	0.03	0	0.01	0	0.01	0.02
K_2O	0.01	0	0	0	0	0.01	0.01
Total corrected by Cl + F	90.84	89.02	90.3	90.02	90.54	88.85	89.845
Si	2.64	2.64	2.66	2.66	2.69	2.61	2.62
Al IV	1.36	1.36	1.34	1.34	1.31	1.39	1.37
Total T site	4	4	4	4	4	4	4
Al VI	1.31	1.37	1.33	1.34	1.39	1.46	1.385
Ti	0.01	0	0	0	0	0	0.005
Fe2+	3.77	3.87	3.78	3.76	3.75	3.75	3.76
Mn	0.6	0.5	0.56	0.6	0.51	0.57	0.585
Mg	0.32	0.25	0.32	0.29	0.3	0.17	0.245
Ca	0	0	0	0.01	0.01	0.01	0.005
Na	0.01	0.01	0	0	0	0	0.005
K	0	0	0	0	0	0	0
Total O site	6.02	6	6.01	6	5.96	5.97	5.995
Total Cations	10.02	10	10.01	10	9.96	9.97	9.995
OH^-	8	8	8	8	8	8	8
Mg/(Fe + Mg)	0.08	0.06	0.08	0.07	0.07	0.04	0.06
Wt% H_2O calculated	10.52	10.32	10.48	10.43	10.57	10.33	10.425
Total Wt% (plus H_2O)	101.36	99.34	100.78	100.45	101.11	99.18	100.27
Vacancies							
(Al^{VI}-Al^{IV})/2 [58]	−0.02	0	−0.01	0	0.04	0.03	0.005
(Al^{VI} Al^{IV}-Na-K)/2 [59]	−0.02	0	−0.01	0	0.04	0.03	0.005
Calculation with Fe^{3+} estimated	0	0	0	0	0.08	0.06	0.02
Mg/(Fe^2 + Mg)	0.08	0.06	0.08	0.07	0.07	0.04	0.06
Vacancies							
6-(Fe_2 + Fe^3 + Mg + A^{VI}) [60]	0.6	0.5	0.56	0.61	0.56	0.62	0.61
(Al^{VI}-Al^{IV} + Fe^3-Na-K)/2 [59]	−0.02	0	−0.01	0	0.04	0.03	0.005

Compositionally, the Fe,Mn-chlorite falls in the middle field between daphnite to amesite (Figure 6) in the diagram of R^{2+} vs. Si^{4+} [61].

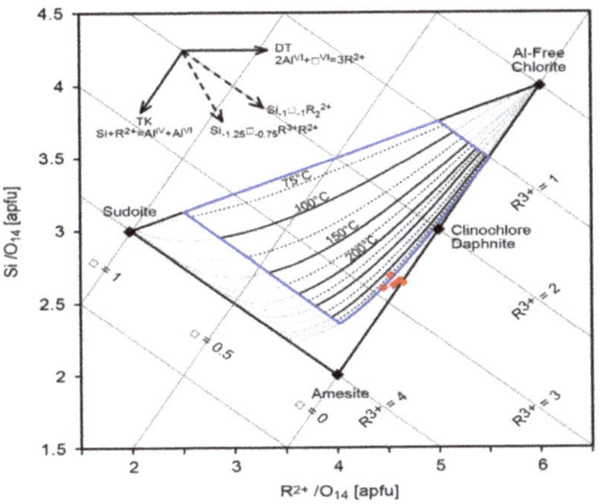

Figure 6. Chemical compositions of Fe,Mn-chlorite plotted in the R^{2+}-Si diagram [61]. Temperature estimation after Bourdelle [58].

The oxidation state of Fe,Mn-chlorite has previously been estimated [62], where the Fe^{3+} exhibited values between 0.00 and 0.08, with an average value of 0.02 for 145 spots analyzed.

4.2. Fluid Phase Petrography and Microthermometry

Petrography. The scheelite crystals exhibited fractures, making difficult the distinction between primary and secondary and pseudo-secondary inclusions.

The FIs occur in three preferred modes, namely: (i) In trails; (ii) isolated inclusions, and (iii) zonate ones (Figure 7; Table 3). Isolated inclusions are generally larger than 20 μm, whereas trails and zoning plane inclusions are smaller (average ~10 μm). The biggest measured FIs has a size about 160 μm with a negative crystal or oval shape, making the crystallographic form of the tetragonal crystal easy to identify. The FIs appear mostly dark due to internal reflections and occur in random clusters (primary inclusions) or in intergranular arrangements (pseudo-secondary inclusions).

Generally, the distinguished FIs are bi-phasic. Isolated tri-phasic primary FIs with a size exceeding 100 μm (sample SHS1), primary isolated FI, and intra-granular trails of pseudo-secondary FIs are shown in Figure 8a. Isolated FIs with a negative-crystal shape (sample SHS1), primary ones with a negative-crystal shape, and pseudo-secondary oval-shaped FIs in intra-granular trails are shown in Figure 8b. Large variations in the degree of fill (Φ_w) were also observed, especially in the 3N1-3N2 primary FIs and in both primary and pseudosecondary FIs of SHS1 and SHS2 (Table 3).

Freezing temperature, microthermometry, and salinity. The freezing temperature data include the first melting point, freezing depression, and melting of clathrate. Three different types of fluids in the FIs hosted by scheelite were identified: (i) Aqueous-carbonic fluids with CO_2 as the dominant carbonic phase, sometimes containing CH_4 and N_2; (ii) aqueous-carbonic fluids with CH_4 as the dominant carbonic phase, containing small amounts of CO_2; and (iii) aqueous fluids generally with a low to medium salinity. Aqueous-carbonic FIs with a dominant CH_4 phases (sample 3N1) are shown in Figure 9a,b; those with dominantly CO_2 phases (samples 3N2.1c, SHS1.3a, and SHS2.5b) are shown in Figure 9c,d.

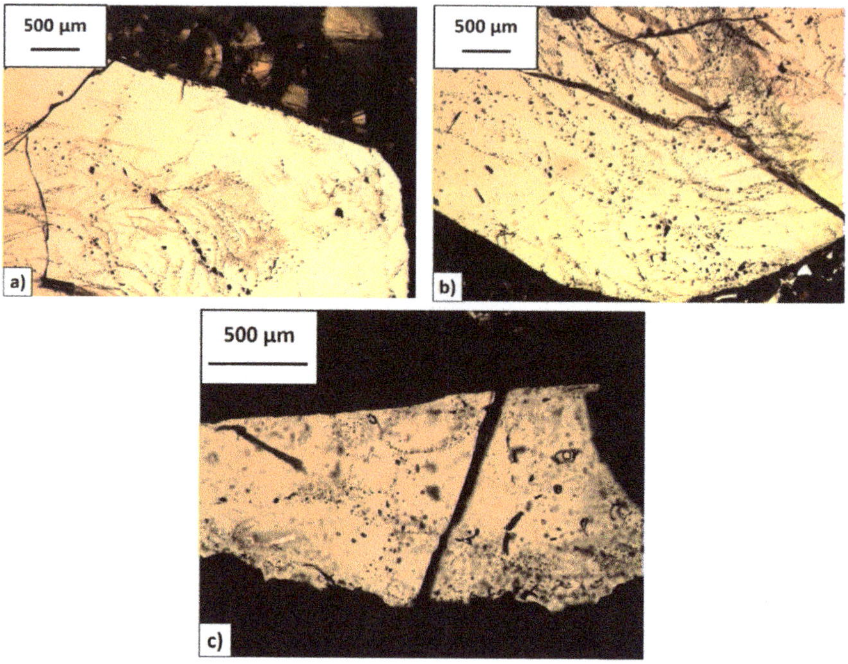

Figure 7. Photomicrographs showing: zonings and trails of fluid inclusions (**a**,**b**) and isolated fluid inclusions in sample SHS2 (**c**).

Table 3. Petrographic characteristics of fluid inclusions observed in scheelite. Degree of fill by liquid water is an approximate range. Size is generalized, the biggest fluid inclusions is 160 µm in the 3N1 sample (see photo Figure 8a). Total number of studied fluid inclusions was n = 190.

Samples	Type	Mode of Ocurrences	Shape	Number of Phases	Size (µm)	Φ_w (%)
3N1 and 3N2	Primary	Isolated	Irregular or rectangular	2	>20	50–80
3N1 and 3N2	Pseudosecondary	Intragranular trails and clusters	Negative crystal, rectangular or oval	2	<30	70–80
SHS1 and SHS2	Primary	Isolated	Negative crystal and irregular	2	>20	50–70
SHS1 and SHS2	Pseudosecondary	Intragranular trails and clusters	Rectangular and negative crystal	2	<30	70–95

Microthermometric data of aqueous-carbonic fluids dominated by CO_2 are shown in Table 4. The T_{mice} of aqueous-carbonic FIs consisting mainly of CO_2 ranges from −7.2 °C to 0 °C; T_{clat} from 0.2 °C to 11.5 °C; T_h from 380 °C to 200 °C; salinity from 3.5 wt.% to 5.5 wt.% NaCl equiv.; and 80 to 100 mol% XCO_2. N_2 was only identified in six of the CO_2-dominant aqueous-carbonic inclusions and always with $XN_2 < 8$ mol%. The T_{mice} of aqueous-carbonic FIs with dominant CH_4 ranges from −6.3 °C to 0 °C; T_{clat} from 7 °C to 10.4 °C; T_h from 323 °C to 220 °C; salinity from 4.3 to 5.1 wt.% NaCl equiv.; and 100 mol% CH_4 (Table 4). The data for aqueous FIs are: T_{mIce} from −5.7 °C to 0 °C; T_h from 265 °C to 130 °C; and salinity from 0 to 6.2 wt.% NaCl equiv. (Table 4).

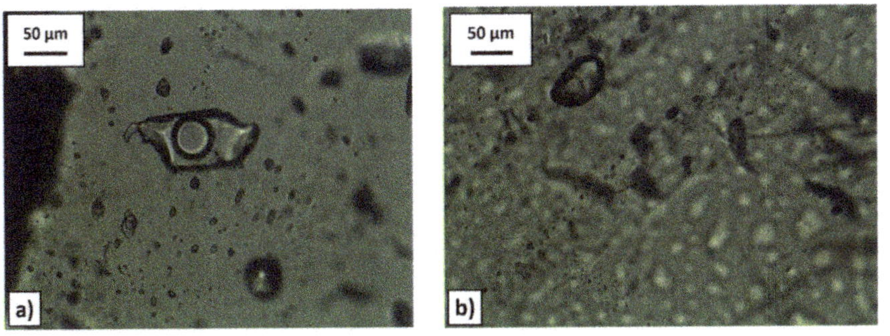

Figure 8. (**a**) Isolated bi-phasic primary fluid inclusions with a size of over 100 µm (sample SHS2, objective 5×). Other primary isolated fluid inclusions (back and front of the big isolated fluid inclusions) and intra-granular trails of pseudo-secondary fluid inclusions. (**b**) Isolated fluid inclusions in a negative-crystal shape (sample SHS1, objective 50×). Primary negative-crystal shape fluid inclusions and pseudo-secondary oval-shaped fluid inclusions in intra-granular trails are observed in the back side.

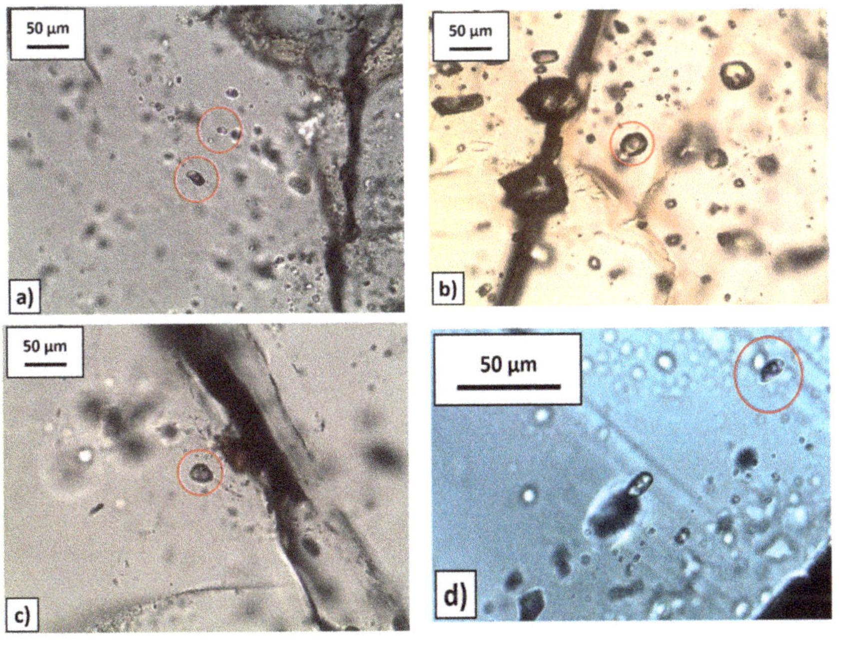

Figure 9. Aqueous-carbonic fluid inclusions: (**a**) Sample 3N1.6a and b predominantly CH_4 phase. (**b**) Sample 3N2.1c predominantly CO_2 phase. (**c**) Sample SHS1.3a predominantly CO_2 phase. (**d**) Sample SHS2.5b predominantly CO_2 phase.

Table 4. Microthermometric data for aqueous-carbonic fluid. $AqCO_2$—aqueous CO_2 FI; $AqCH_4$—aqueous CH_4 FI; N—number of analysis; Φw—degree of fill liquid water; T_e—eutectic temperature; T_{mice}—ice melting temperature; T_{clat}—clathrate melting temperature; T_h—homogenization temperature; Sal.—salinity in weight percent equivalents of NaCl; Xgas—amount of CO_2 or CH_4 in mol%; D—density in g/cc; Min-Max Av.—minimum, maximum, average.

Type		Φw (%)	T_e (°C)	T_{mice} (°C)	T_{clat} (°C)	T_h (°C)	Sal (wt.eq. NaCl)	Xgas (mol%)	D (g/cc)
$AqCO_2$ (n = 79)	Min-max	50 to 80	−75.7 to −50	−7.2 to 0	0.2 to 11.5	200 to 380	3.5 to 5.5	80 to 100	0.55 to 1
	Av.	70	−59	−2.3	5.9	284	4.8	94.2	0.77
$AqCH_4$ (n = 11)	Min-max	60 to 80	−68.7 to −52	−6.3 to 0	7 to 10.4	220 to 323	4.3 to 5.1	100	0.6 to 0.8
	Av.	75	−59	−2.5	9.4	262	4.7	100	0.8
Aqueous (n = 96)	Min-max	50 to 95	−67 to −23.1	−5 to 0		130 to 265	0 to 6.2		1 to 1.0
	Av.	82	−54.8	−0.75		218.5	1.18		1.01

Salinity (wt.% NaCl equiv.) measured for each type of fluid was plotted against T_h (°C) where three distinct evolution stages were distinguished (Figure 10). The first stage corresponds to the evolution of the aqueous-carbonic fluid with CO_2 dominating at high temperatures (<400 °C). The separation of carbonic phases shown in the T_h histogram occurred at about 330 °C through boiling.

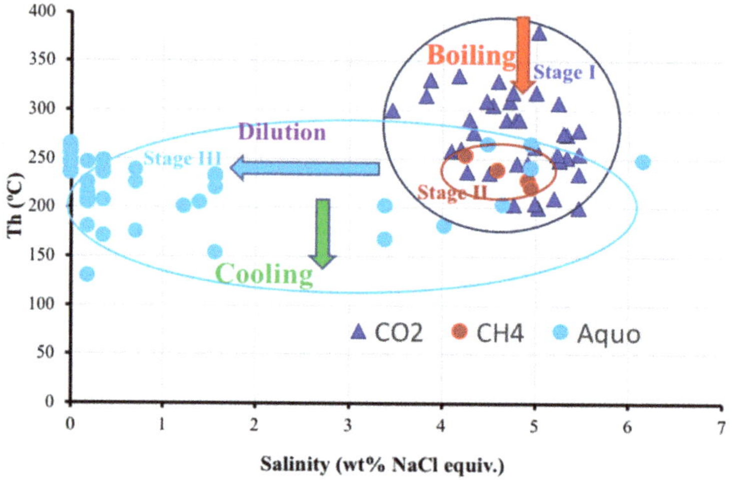

Figure 10. The diagram of salinity (wt.% NaCl eq.) vs. T_h (°C) corresponding to fluids identified in scheelite.

At this stage, the fluids lose their volatile components and start to precipitate their metals at the boiling point. Evidence for the boiling is also shown by the variations of liquid/vapor ratios observed in the FI and homogenization in liquid and vapor phase. The CH_4 and water were separated during decompression at temperatures below 300 °C and then mixed with an aqueous fluid of lower salinity. The T_h, T_{mice}, and salinity (wt.% NaCl equiv.) histograms corresponding to the whole set of FIs are shown in Figure 11.

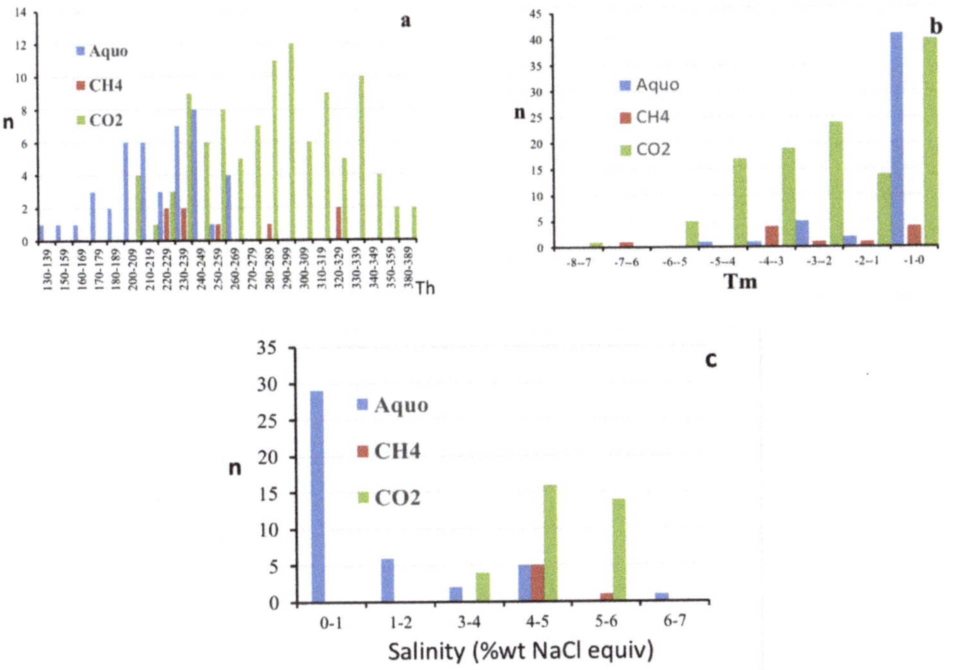

Figure 11. Histograms of T_h (**a**), T_m (**b**), and salinity of fluid inclusions (**c**) in scheelite from Borralha.

The T_{mi} and salinity histograms show that the hotter fluids were saltier (with about 5 wt.% NaCl equiv.) than the aqueous colder fluids which mixed at the end of the process. In addition, averages of the bulk compositions and densities of the FIs studied were calculated (see Table 4). No distinction between the primary and pseudosecondary inclusions could be established using the microthermometry of the fluid inclusion assemblages, but rather on the composition of the fluids. Raman spectrometry analyses were performed on representative aqueous-carbonic type (CO_2 or CH_4 dominant) FIs, where differences in the CO_2/CH_4 ratio were identified (Figure 12 and Table S1 in the Supplementary Materials).

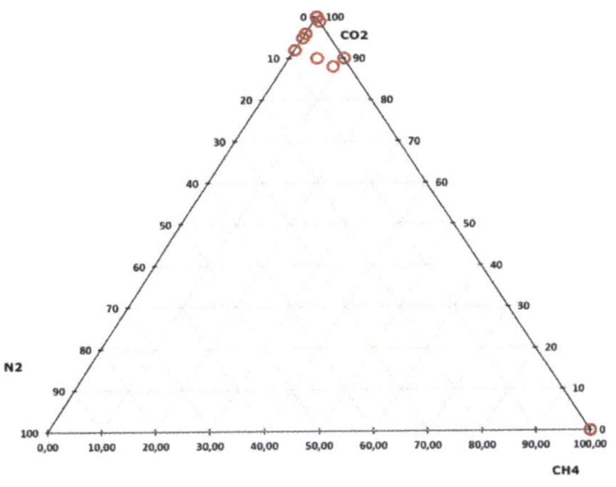

Figure 12. The ternary diagram of CO_2-CH_4-N_2.

4.3. Oxygen Isotope Analysis

Oxygen isotope measurements were obtained for quartz, wolframite, scheelite, and Fe,Mn-chlorite (Table 5), and previously published data on Fe,Mn chlorite [28] were re-analysed. The $\delta^{18}O$ isotope (V-SMOW) compositions of quartz (Q_{wf}) associated with wolframite shows a mean value of 12.6‰ (1σ) and a mean value of 11.5‰ (1σ) for the quartz join to scheelite.

Table 5. Oxygen isotope data of quartz, wolframite, scheelite and Fe,Mn-chlorite minerals.

Samples	Location	$\delta^{18}O$ (0) μεασυρεδ	Quartz-Water Pair			Wolframite-Water Pair			Scheelite-Water Pair			Chlorite-Water Pair		
			300 °C	350 °C	400 °C	300 °C	350 °C	400 °C	300 °C	350 °C	400 °C	300 °C	350 °C	400 °C
Quartz (wf)	SHS 48 m	12.6	5.71	7.3	8.54	-	-	-	-	-	-	-	-	-
Quartz (sch)	3N	11.5	4.61	6.2	7.44	-	-	-	-	-	-	-	-	-
Wolframite	SHS-48 m	3.8	-	-	-	6.11	-	6.31	-	-	-	-	-	-
Scheelite	3N	2.8	-	-	-	-	-	-	4.44	5.09	5.6	-	-	-
Scheelite	3N	3	-	-	-	-	-	-	4.64	5.29	5.8	-	-	-
Scheelite	SHS 48 m	3.8	-	-	-	-	-	-	5.44	6.09	6.6	-	-	-
Fe,Mn-chlorite	3N	3.1	-	-	-	-	-	-	-	-	-	3.09	3.17	4.36
Fe,Mn-chlorite	SHS	3.2	-	-	-	-	-	-	-	-	-	3.15	3.27	4.46

The $\delta^{18}O_{Fluid}$ of the quartz-water pair ($10^3\ln\alpha_{Quartz-water} = 3.38 \times 10^6T^{-2}-3.40$; [49]) yielded the following values for the precipitated quartz (Q_{wf}) associated with wolframite: +5.71‰ (1σ) at 300 °C, +7.30‰ (1σ) at 350 °C, and 8.54 (1σ) at 400 °C. For the precipitated quartz (Q_{sch}) associated with scheelite yielded: +4.61‰ (1σ) at 300 °C, +6.20‰ (1σ) at 350 °C, and +7.44‰ (1σ) at 400 °C (Table 5). The $\delta^{18}O$ (V-SMOW) measured for wolframite yielded a value of 3.8‰ (1σ) (i.e., much lighter than quartz). The $\delta^{18}O_{Fluid}$ of the wolframite-water pair ($10^3\ln\alpha_{Wf-water} = 1.03 \times 10^6T^{-2}-4.96$; and $10^3\ln\alpha_{Wf-water} = 0.21 \times 10^6T^{-2}-2.91$ for temperature intervals between 250–370 °C and 370–420 °C [50]) yielded a value of 6.11‰ (1σ) for an estimated temperature of 300 °C and 6.31‰ (1σ) for 400 °C.

Combining the quartz-water fractionation equation [54] with the wolframite-water pair [50], the following oxygen isotope fractionation equations were proposed for the quartz-wolframite pair [63]: (i) 250–370 °C: $10^3\ln\alpha_{q-Wf} = 2.35 \times 10^6T^{-2} + 1.56$ and (ii) 370–420 °C: $10^3\ln\alpha_{q-Wf} = 3.17 \times 10^6T^{-2}-0.49$. The quartz-wolframite oxygen isotope fractionation estimated for temperatures ranging from 300 °C to 400 °C yielded 9 and 7.11‰ (1σ). The data plotted (not shown) in the $10^3\ln\alpha$ vs. T °C [63] fall between 300 °C and 400 °C.

The $\delta^{18}O$ isotope (V-SMOW) composition measured on scheelite yielded values of 2.8‰, 3‰, and 3.8‰ (1σ). The $\delta^{18}O_{Fluid}$ of the scheelite-water pair ($10^3\ln\alpha_{scheelite-water} = 1.39 \times 10^6T^{-2}-5.87$; [52]) yielded 4.44‰, 4.64‰, and 5.44‰ (1σ) for 300 °C, 5.09‰, 5.29‰, and 6.09‰ (1σ) for 350 °C, and 5.60‰, 5.80‰, and 6.60‰ (1σ) for 400 °C (Table 5. The oxygen isotope fractionation in the quartz-scheelite and quartz-wolframite pairs is large, offering a much more sensitive geothermometer [64].

Combining the quartz-water fractionation equation [54] with the scheelite-water equation [52], the following fractionation equation was obtained for quartz-scheelite fractionation: $10^3\ln\alpha_{q-scheelite} = 1.99 \times 10^6T^{-2} + 2.47$. The $\Delta_{quartz-scheelite}$ calculated was 6.86‰ (1σ) for a T °C of 400 °C and 7.54‰ (1σ) for 350 °C. The $\Delta_{quartz-scheelite}$ values plotted in the diagram of $10^3\ln\alpha$ vs. T °C (not shown) intersected the quartz-scheelite fractionation curve at about 350–400 °C.

The $\delta^{18}O$ isotope (V-SMOW) signatures measured for Fe,Mn-chlorite yielded values of +3.1 and +3.2‰ ± 1.2‰, where the $\delta^{18}O_{Fluid}$ calculated for the chlorite-water pair ($10^3\ln\alpha_{chlorite-water} = 1.56 \times 10^6T^{-2} - 4.70$; [53]) yielded +3.27‰ for a T °C of 350 °C, +4.46‰ for 400 °C (Table 5), and +4.92‰ for 450 °C. The $\Delta_{chlorite-water}$ fractionation factor calculated was about +0.05‰ for 330 °C and values of −0.68‰ and −1.26‰ corresponded to temperatures of 380 °C and 450 °C, respectively. After plotting the data in the $10^6 \times T^{-2}$

(°K) vs. $10^3 \ln\alpha_{chlorite\text{-}water}$ diagram (Figure 13) [65], an average crystallization temperature of 400 °C was obtained.

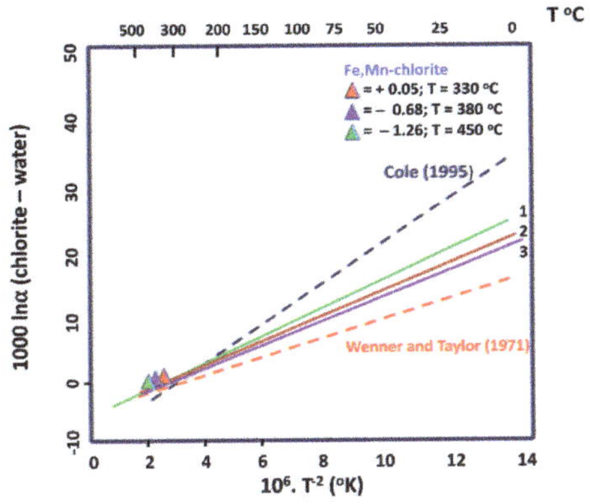

Figure 13. Estimation the oxygen isotope fractionation ratio between Fe,Mn-chlorite and water using the diagram of $10^6 \times T^{-2}$ (°K) vs. $10^3 \ln\alpha_{chlorite\text{-}water}$ [53,66].

4.4. Chlorite Geothermometry

Several chlorite thermometers based on empirical calibrations (linking T to Al^{IV} content) or semi-empirical models (linking T to the chlorite + quartz + water equilibrium constant) have been developed for Al-, Fe-, and Mg-rich chlorites over the last decade [58–60,67–69]. In a recent review on chlorite geothermometry, Bourdelle [29] noted that it could be worth applying these models for Mn-rich or Al-free chlorites to Mn-redox behavior, like Fe-redox.

In this study, 146 chemical analyses of Fe,Mn-chlorite (Table 2) were used for evaluating both empirical and semi-empirical models known in the literature. The average amount for FeO was about 39.05% and 6.05% for MnO, where the estimated Fe^{3+} was about 0.02 apfu. High contents of Fe [Fe/(Fe + Mg + Mn) = 0.4–0.95] and Mn [Mn/(Fe* + Mg) > 0.05] in chlorite minerals are also diagnostic of chemical environments related to ore-forming processes [70]. The Fe,Mn-chlorite temperature values obtained using semi-empirical geothermometers [58–60,71] are shown in Table 6.

Table 6. Calculated temperatures using semi-empirical thermometry.

Samples Calculated Temperatures (°C)	FMN1	FMN2	FMN3	FMN12	FMN8	FMN9
Semi-Empirical Thermometry	**T °C**	**T °C**	**T °C**	**T °C**	**T °C**	**T °C**
Inoue et al. [60,71]. Fe_{Total} = FeO + Fe_2O_3. Quadratic equation thermometer (valid for <350°, may be <400 °C)	229	246	230	220	212	228
Inoue et al. [71]. Fe_{Total} = FeO (valid for <350 °C, may be <400 °C)	197	210	197	190	183	196
Bourdelle et al. [58] Fe_{Total} = FeO. (Valid for <350 °C and <4 kb)	-	-	-	-	435	519
Lanari et al. [59]. Chl1: Fe_{Total} = FeO + Fe_2O_3 (valid between 100–500 °C)	-	-	-	-	718	984
Lanari et al. [59] Chl2: Fe_{Total} = FeO (valid between 100–500 °C and 1–20 kbar)	-	-	-	-	373	458

The average calculated temperature of Fe,Mn-chlorite [59,70] was estimated at about 228 °C (Fe_{total} = FeO + Fe_2O_3) and 196 °C (Fe_{total} = FeO). The temperatures estimated for samples FMN 8 and FMN 9 were respectively about 435 °C and 519 °C (Fe_{total} = FeO) after Bourdelle [58] and 373 °C and 458 °C (Fe_{total} = FeO) after Lanari [59]. Here, the geothermometers of Bourdelle and Lanari [58,59] are not applicable (Table 7). The average temperature calculated for Fe,Mn-chlorite crystallization was about 373 °C, after Cathelineau [67], and 392 °C after Jowett [68]. The chemical composition of Fe,Mn-chlorite plotted in the Si^{4+} vs. R^{2+} diagram [61] falls into a field with temperatures ranging from 400 °C to 550 °C (Figure 6).

Table 7. Calculated temperatures using empirical thermometry *.

Samples Empirical Thermometry (Fe_{Total} = FeO)	**FMN1 T °C**	**FMN2 T °C**	**FMN3 T °C**	**FMN12 T °C**	**FMN8 T °C**	**FMN9 T °C**
Cathelineau [66]. (valid for <350 °C)	375	376	371	371	359	387
Jowett [67]. For Si < 3.3 apfu and Ca < 0.07 apfu. Fe/(Fe + Mg) < 0.6, and T° < 325 °C	393	395	389	389	377	406
Restrictions of each thermometer calibration (1 = Ok; 0 = not valid)						
Bourdelle et al. [58].	NOT APPLICABLE	NOT APPLICABLE	NOT APPLICABLE	NOT APPLICABLE	NOT APPLICABLE	NOT APPLICABLE
Uncertainty of 30 °C for T < 300 °C and 50–60 °C for T > 300 °C						
$Na_2O + K_2O + CaO$ <1? (wt.%)	1	1	1	1	1	1
Si (apfu) between 2 to 4 apfu	1	1	1	1	1	1
Vacancies >0.05 apfu?	0	0	0	0	0	0
Lanari et al. [59].	NOT APPLICABLE	NOT APPLICABLE	NOT APPLICABLE	NOT APPLICABLE	NOT APPLICABLE	NOT APPLICABLE
Uncertainty of 50 °C.						
Si < 3 apfu	1	1	1	1	1	1
$Na_2O + K_2O + CaO$ <1? (wt.%)	1	1	1	1	1	1
Vacancies (Fe Total = FeOT) > 0.05 (for <0.05 = Thermometer Chl2 not applicable)	T° Chl2 should be not used	T° Chl2 should be not used	T° Chl2 should be not used	T° Chl2 should be not used	T° Chl2 should be not used	T° Chl2 should be not used
Inoue et al. [60,71].	Composition Valid	Composition Valid	Composition Valid	Composition Valid	Composition Valid	Composition Valid
Vacancies apfu < 1	1	1	1	1	1	1
$Na_2O + K_2O + CaO$ < 0.5	1	1	1	1	1	1

* Thermometry calculations was carried out using the database of Verdecchia et al. [72].

5. Discussion

The successful application of chlorite minerals and stable isotope geothermometry to ore deposits requires several conditions to satisfy any pair of minerals: (i) Fractionation factors between them are well-calibrated and strongly temperature dependent; (ii) the mineral(s) species were in isotopic equilibrium at the time of deposition; and (iii) they have retained their isotopic composition since formation. Therefore, the sequence mineral pairs identified in the W, (Cu,Mo) ore deposit of Borralha satisfy all the requirements to evaluate the isotope geochemistry relationship between tungstates and neoformed silicates (quartz and Fe,Mn-chlorite) and empirical chlorite geothermometers, based also on FI microthermometry data.

5.1. Wolframite to Scheelite and Fe,Mn-Chlorite Crystallization

The scheelite samples collected are excellent samples due to their size, purity, and their genetic relationship with Mn-bearing wolframite and Fe,Mn-chlorite. Wolframite crystallized from fluids with slightly acidic to neutral pH, where WO_4^{2-} may be easily complexed with Fe^{2+}, Mn^{2+}, and Ca^{2+}. It should be noted that the stability of the Fe^{2+} end-member (ferberite) is much more sensitive to the high variations of f_{O2} than the Mn^{2+}

end-member wolframite (hübnerite) since the composition of ferberite might be strongly affected by the oxidation state [26]. Only Mn-bearing wolframite is associated with the late evolving W mineralization, possibly triggered by an alkalinity rising in the fluid as a result of an increasing S activity and the availability of Ca^{2+} in the system, which is correlated with a late alkaline metasomatic reaction common in geochemical settings [73]. Several authors have suggested that Mn-rich wolframites precipitated at higher temperatures than Fe-rich wolframites [73–76]. Nevertheless, the H/F ratio used as a geothermometer [77] failed, because temperature is not the only factor controlling the composition of wolframite [27]. The H/F ratio can be used as an indicator of contrasting W deposition environments in perigranitic ore-forming systems, where hübnerite can be viewed as a relatively simple indicator of a strong magmatic control on W deposition [76]. The H/F ratio measured in wolframite samples from Borralha ranged from 46.25% to 46.70%, as a consequence of the high Mn/Fe ratio in the magmatic-hydrothemal fluid. By contrast, a H/F ratio < 40 and down to 0 indicates a W deposition environment where a significant non-magmatic contribution to the ore fluid and W deposition should be necessary [78].

Experimental studies on the stability of $H_4[Si(W_3O_{10})_4]2H_2O$ at high P-T conditions and variable pH in the presence of Ca^{2+} have confirmed the separation of silico-tungstic acid under the effect of solution neutralization [79], where $[WO_4]^{2-}$ is preferentially combined with Ca^{2+} to form scheelite, favoring the incorporation of the available Fe^{2+} and Mn^{2+} from wolframite and SiO_2 from the fluid into the chlorite structure. The CO_2 loss from the hydrothermal fluid during boiling caused pH changes from acidic to alkaline conditions [80–83], accompanying the wolframite dissolution and scheelite deposition in the presence of Ca^{2+}. Calcium for scheelite crystallization was supplied from the albitization and/or sericitization of plagioclase from granitic and pegmatite host rocks. Optical microscopy showed that the scheelitization of Mn-bearing wolframite occurred along fissures and fractures. This implies that high-pressure fluids are required to trigger fracture initiation and propagation, where the fluctuations of fluid pressure were recorded by FIs. Hence, Mn-bearing wolframite supported a dissolution process where $[WO_4]^{2-}$ separation served for scheelite crystallization, whereas Mn and Fe for chlorite. It is important to note that the term "metamorphic fluid" is used here to describe a fluid that was partially equilibrated with metamorphic rocks and this does not necessarily imply that the fluid components were produced from prograde metamorphic reactions.

5.2. Characterization of the Mineralizing Fluid

Previous FIs studies on scheelite from Borralha revealed two generations of FIs (primary and pseudosecondary) characterized by the presence of an aqueous, low salinity fluid [38]. The fluids entrapped in scheelites showed a complex history with different fluids and boiling during W precipitation, where boiling (immiscibility), mixing, cooling, and fluid-rock interactions (e.g., Ca^{2+} needed for scheelite precipitation) played an important role in the scheelite precipitation. The primary fluid stage was characterized by a complex aqueous-carbonic fluids of low salinity, indicating a typical metamorphic fluid. The CO_2 ubiquitous recorded in FIs from a variety of mineralized systems underwent boiling/unmixing presumably leading to ore deposition. Scheelite was mainly precipitated during the early stage from an aqueous-carbonic fluid with a CO_2–dominated composition, where the T_h of the ore-forming fluids decreased from 380 °C to 200 °C.

Fluid boiling/immiscibility has two fundamental effects on the fate of metals. Indeed, it is known to induce the precipitation of some metals from the liquid phase owing to enhanced metal concentrations and to change the redox conditions and ligand contents in the residual liquid as a result of the removal of volatile components (H_2S, H_2, HCl, and CO_2) into the vapor [81–83]. The partitioning of CO_2 and CH_4 into the vapor phase occurred through boiling, where both participated in redox reactions. The coexistence of vapor- and liquid-rich inclusions indicates fluid boiling/immiscibility. The boiling resulted in the separation of the aqueous-carbonic fluid with predominant CH_4 and mixing with an aqueous fluid of low salinity at about 330 °C, deduced from the histograms (Figure 8) and the diagram of salinity vs. T_h (Figure 9). The salinity (wt.% NaCl equiv.) vs. T_h

(°C) diagram shows two distinct fields corresponding to a magmatic-hydrothermal vs. metamorphic fluid (Stages I and II) and a hydrothermal-meteoric fluid (Stage III) (Figure 9).

Stage I occurred around the maximum T_h measured, corresponding to the beginning of separation between CO_2-CH_4 fluids during boiling (immiscibility). The L-rich FIs appeared to be in equilibrium with the CO_2-CH_4 inclusions (see Figures 10 and 11a), which is another indication of boiling. Boiling also increased oxygen fugacity, leading to the immiscibility of CH_4 where a disequilibrium isotopic fractionation occurred between CO_2 and CH_4 below 400 °C [83]. The process ended during the final of Stage II, when the aqueous saline FIs were generated. Changes in the aqueous-carbonic fluid composition were recognized as temperature decreased, where CH_4 became dominant (Stage II) and the T_h decreased (average of 262 °C). Stage III corresponded to the dilution of metamorphic fluids and cooling by the circulation of meteoric fluids. Exhaustion of deposition led to low salinity FIs. This stage corresponded to an aqueous fluid of lower temperature (average T_h = 218.5 °C). In addition, the salinity gradually decreased as temperature decreased (Figures 10 and 11c). The large dispersal of T_h values in the histogram and the reduced salinity observed for the aqueous inclusions was indicative of mixing.

Experimental studies have demonstrated that the solubility of scheelite decreases substantially with decreasing temperatures [73], where tungstate destabilization was influenced by the pH changes of the fluid [84]. This process removed CO_2 from the hydrothermal fluid, a component that can not only be a buffer fluid pH [73,85], but also favors W transport [86]. The trace amount of N_2 addition to the mineralizing fluids derived from the devolatilization of metasediments [87].

5.3. Oxygen Isotopes

The $\delta^{18}O_{Fluid}$ calculated for the quartz-water pair (sample Qwf) yielded values of + 5.7‰ at about 300 °C, + 7.30‰ at 350 °C, and + 8.54‰ at 400 °C, where the calculated values fall within the magmatic water field [88,89] from 5.5‰ to 10.0‰ [90–92]. In fact, from the isotopic point of view, quartz (Q_{wf}) and wolframite (+ 6.11 to + 6.31‰ at 300 and 400 °C, respectively) crystallized directly from a magmatic-hydrothermal fluid, which may represent primary magmatic water or water from any source whose isotopic composition is controlled by an exchange with a large volume of igneous rock. Nevertheless, the $\delta^{18}O_{Fluid}$ calculated for quartz (Q_{sch}) and scheelite yielded values from 5.55‰ to 6.6‰ for sample SHS, for temperatures ranging from 300 °C to 400 °C; for sample 3N values of 4.44‰ to 5.8‰ were obtained in the same temperature range. The samples analyzed suggest several scheelite generations crystallized from a magmatic-hydrothermal vs. metamorphic fluid (sample SHS) at temperatures below 400 °C, which were later diluted by a hydrothermal meteoric fluid (sample 3N). In fact, the differentiation between magmatic and metamorphic fluids in terms of $\delta^{18}O_{Fluid}$ is limited, creating difficult interpretations concerning the mixing of these two fluid phases. According with the FIs data, scheelite crystallization could be largely influenced by metamorphic fluids which equilibrated with the magmatic-hydrothermal fluid during the first wolframite crystallization. $\delta^{18}O_{Fluid}$ data support an input of hydrothermal-meteoric waters, confirming an exchange and re-equilibration with the primary fluid at elevated temperatures and low water/rock ratios. This assumption is supported by FIs data obtained from the studied scheelite samples, where an input of aqueous FIs after boiling was observed (Figure 10). Based on $\delta^{18}O$ data, it seems that Fe,Mn-chlorite crystallized in equilibrium with scheelite, but from a diluted hydrothermal-meteoric fluid contribution at a high temperature. The crystallization temperatures estimated from the $10^6 \times T^{-2}$ (°K) vs. $10^3\ln\alpha_{chlorite\text{-}water}$ diagram (Figure 13) [65,66] were between 350 °C to 450 °C, where the chlorite-water pair fractionation yielded values from +3.17‰ at 350 °C to +4.46‰ at 400 °C, confirming the presence of a hydrothermal-meteoric fluid in equilibrium with scheelite crystallization.

The mixing of magmatic and metamorphic fluids with different origin in granite systems has been inferred in the literature, where the external fluids (i.e., metamorphic, hydrothermal-meteoric) can mix and dilute the primary magmatic fluid, even when enriched by metals extracted from wall-rocks in the frame of the resulting convective circulation [93–95].

5.4. Chlorite Geothermometer

Chlorite consists of a 2:1 layer with a negative charge $[(R^{2+}, R^{2+})_3(_xSi_{4x}R^{2+}{}_y)O_{10}OH_2]^-$ balanced by a positively charged interlayer octahedral sheet $[(R^{2+}, R^{3+})_3(OH)_6]^+$. $R^{2+,3+}$ is represented by Mg, Fe(II), and Fe(III) with Mg-rich chlorite (clinochlore) generally being metamorphic (high temperature), while the Fe-rich chlorite (chamosite) is typically diagenetic (low temperature). Additionally, Ni-rich chlorite (nimite) and Mn-rich chlorite (pennantite) are the other two less common varieties of chlorite. Most chlorites are trioctahedral in both sheets, i.e., the ferric iron content is low. Furthermore, it is well known that the ubiquitous and large thermodynamic stability of chlorite is in metamorphic, diagenetic, and hydrothermal systems, where its chemical composition is influenced by temperature and pressure. Thus, chlorite may crystallize over a large range of temperatures from 80 °C to more than 700 °C and pressures from 1 Kbar to 12 Kbar [96]. For example, cookeite, a Li-di,trioctaedral chlorite, is an example of a polymorph that has crystallized at low to high pressures [97,98].

The Fe,Mn-chlorite associated with W mineralizations from Borralha crystallized directly from the hydrothermal fluid and isotopically re-equilibrated with respect to scheelite. Hydrothermal fluids in the crust generally have a low oxidation potential and dissolved Fe predominantly occurs in its +2 oxidation state [99]. Under these reduced conditions, the amount of Fe^{3+} in hydrothermal fluids is negligible compared to that of Fe^{2+} [71]. Both Fe and Mn were available in the hydrothermal fluid from wolframite replaced by scheelite. Considering the textural relationships observed between wolframite to scheelite and Fe,Mn-chlorite, and the high amount of Fe and Mn in the hydrothermal fluid, chlorite crystallized after scheelite preserving the redox conditions during the fluid migration. The redox conditions are supported by the obtained FI data, where CO_2 and CH_4 subsequently reduced oxidation. By contrast, adequate oxidizing conditions were needed to buffer the oxygen activity in the case of a hydrothermal fluid with a Fe- or Mn-poor or -rich character and a low fluid/rock ratio. In this case, both Fe^{3+} and Mn^{3+} had a limited stability field in aqueous solution, being rapidly trapped during chlorite crystal growth.

Several authors have suggested that empirical or semi-empirical geothermometers have only been developed for the most common chlorites (i.e., Al-, Fe-, or Mg-rich chlorites) and it could worth extending to rarer chlorites, like Mn-rich chlorite [29], because the actual geothermometers do not include a Mn-rich end-member, such as pennantite. Nevertheless, the systematic presence of measurable amounts of Mn has been tested as a new pathfinder for chlorites developed in the course of ore-forming processes. A valid example is related to Fe,Mg,Mn-chlorite identified in Pb-Zn, Bi, and Ag ore veins from the Toyoha geothermal system, Japan [70]. In this case, the structural formula calculation, similar to ferric iron, will arise without the quantification of the total Mn^{3+}/Mn ratio, but the ratio of Mn^{3+}/Mn^{2+} would probably theoretically modify the chlorite's structural formula [29]. Nevertheless, the presence of Mn^{3+} in chlorite, which would eventually be substituted with Al^{3+} in the octahedral site, is expected to lead to the structural distortion of chlorite due to the different ionic radii of Al^{3+} (0.535 Å) and Mn^{3+} (0.645 Å) [100].

The chemical compositions of Fe,Mn-chlorite are plotted in the field of daphnite (average Si = 5.30 apfu and Fe 7.51 apfu) in the Si vs. Fe_{tot} diagram (not shown) or in the field of daphnite-amesite according to the Si vs. R^{2+} diagram (Figure 6; [61]), which makes it valid for the four-thermodynamic-component solid-solution model (Mg-sudoite, Mg-amesite, clinochlore, and daphnite) used for semi-empirical geothermometers.

The estimated Fe,Mn-chlorite temperatures using the semi-empirical thermometer of Lanari [59] were about 373 °C and 458 °C (Fe_{total} = FeO) and 435 °C and 519 °C (Fe_{total} = FeO) (Bourdelle [58]). This represents average values of 415 and 477 °C. If we consider the uncertainly of 50–60 °C for T °C > 300 °C suggested by Bourdelle [58], we propose a temperature between 375 °C to 410 °C. The estimated temperatures calculated from semi-empirical thermometers seem to be supported by the $\delta^{18}O_{Fluid}$ calculated using the fractionation equation for the chlorite-water pair [53] which confirmed a temperature of chlorite crystallization estimated at around 400 °C.

The magmatic-hydrothermal vs. metamorphic fluid circulation evidences was proposed in this work based on FI data and the oxygen stable isotope fingerprints obtained on tungstates, quartz, and Fe,Mn-chlorite (Figure 14). The presence of a metamorphic fluid with an aqueous-carbonic composition and low salinity was contributed to wolframite destabilization and scheelite precipitation in the presence of Ca^{2+} derived from plagioclase alteration. During boiling, the CO_2 was removed from the aqueous-carbonic fluid, favoring W transport [84], where Fe,Mn-chlorite crystallized after scheelite crystallization at a temperature close to the crystallization of scheelite.

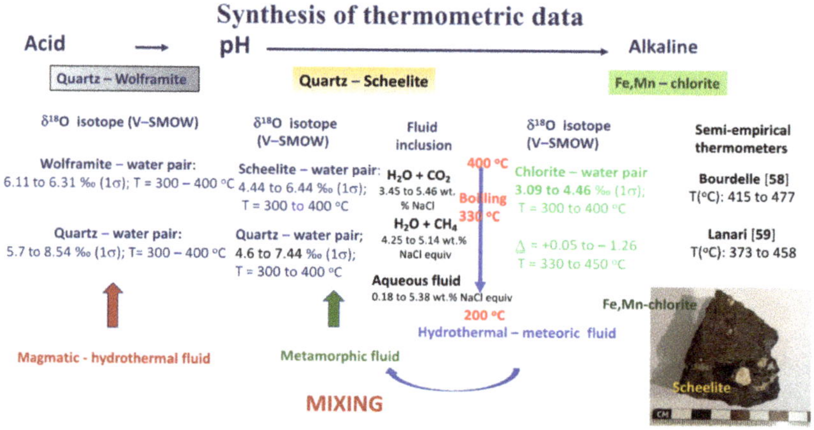

Figure 14. Synthesis of microthermometry and oxygen isotope data correlated with semi-empirical chlorite thermometers [58,59].

6. Conclusions on Fe,Mn-Chlorite Thermometry

The diversity of chlorite minerals in fossils or active hydrothermal systems is connected in most cases with the alteration types of magmatic rocks, where chlorite may crystallize directly from a solution or replace the primary minerals. The sequence of mineral pairs identified in the W, (Cu,Mo) ore deposit of Borralha satisfies all conditions to evaluate the oxygen isotope signature between tungstates and silicates. Besides, the question arises on Mn-oxidation, the effect of fluid chemistry on the mineral-water fractionation factor in chlorite may represent an uncertainty however, fractionation factors between minerals and aqueous solutions are independent of solution chemistry.

Our multiproxy study based on FIs, stable isotopes, and semi-empirical chlorite thermometers showed good consistency between the results obtained. Combining FIs and oxygen isotope geochemistry data, it was revealed that a metamorphic fluid supported the crystallization of scheelite and quartz at temperatures below 400 °C. The geothermometers applied according to Lanari and Bourdelle for Fe,Mn-chlorite produced temperature estimates between 373 °C and 458 °C, and 435 °C and 519 °C, respectively. The $\Delta_{chlorite\text{-}water}$ calculated fractionation factor corresponded to temperatures of 380 °C and 450 °C, respectively. An average chlorite crystallization temperature of 400 °C was assumed considering an uncertainly of 50–60 °C for T °C > 300 °C, as suggested by the authors of the geothermometers.

Supplementary Materials: The following are available online: https://www.mdpi.com/article/10.3390/min12010024/s1. Table S1. The CO_2, CH_4 and N_2 (mol%) of aqueous-carbonic fluid measured by Raman spectroscopy analysis (50 analyses).

Author Contributions: Conceptualization, I.B., C.M.d.S. and F.N.; Formal analysis, I.B.; Investigation, I.B. and C.M.d.S.; Methodology, I.B., C.M.d.S. and F.N.; Writing—original draft, I.B. and C.M.d.S.; Writing—review and editing, I.B., C.M.d.S. and F.N. All authors have read and agreed to the published version of the manuscript.

Funding: POCTEP-ESMIMET 2018–2020 Interreg Spain-Portugal Financed by POCTEP.

Acknowledgments: This work was developed in the frame of projects POCTEP-ESMIMET 2018–2020 Interreg Spain-Portugal Financed by POCTEP and UIDB/04683/2020 ICTerra, FCT -Lisboa. Carlos Sá acknowledge the post-doc scholarship attributed in the frame of this project during 2019–2020. Thanks also to ICTerra-Polo Porto for using the fluid inclusions lab. The first author is greatly indebted to F. Nieto and Franck Bourdelle for their thorough reading and helpful suggestions carried out on an early manuscript. The authors are grateful to the Academic Editor and to all three reviewers who contributed with many suggestions and greatly improved the final manuscript version. Also thanks to Mark Ryan who reviewed the English presentation.

Conflicts of Interest: The authors declare no conflict of interest.

References

1. Ren, Y.S.; Zhao, H.L.; Lei, E.; Wang, H.; Ju, N.; Wu, C.Z. Trace element and rare earth element geochemistry of the scheelite and ore genesis of the Yangjingou large scheelite deposit in Yanbian area, northeastern China, China. *Acta Petrol. Sin.* **2010**, *26*, 3720–3726.
2. Song, G.; Qin, K.; Li, G.; Evans, N.G.; Chen, L. Scheelite elemental and isotopic signatures: Implications for the genesis of skarn-type W-Mo deposits in the Chizhou Area, Anhui Province, Eastern China. *Am. Miner.* **2014**, *99*, 303–317. [CrossRef]
3. Poulin, R.S.; McDonald, A.M.; Kontak, D.J.; McClenaghan, M.B. On the relationship between cathodoluminescence and the chemical composition of scheelite from geologically diverse ore-deposit environments. *Can. Miner.* **2016**, *54*, 1147–1173. [CrossRef]
4. Poulin, R.S.; Kontak, D.J.; McDonald, A.; McClenaghan, M.B. Assessing scheelite as an ore-deposit discriminator using its trace-element and REE chemistry. *Can. Miner.* **2018**, *56*, 265–302. [CrossRef]
5. Guo, S.; Chen, Y.; Liu, C.-Z.; Wang, J.-G.; Su, B.; Gao, Y.-J.; Wu, F.-Y.; Sein, K.; Yang, Y.-H.; Mao, Q. Scheelite and coexisting F-rich zoned garnet, vesuvianite, fluorite, and apatite in calc-silicate rocks from the Mogok metamorphic belt, Myanmar: Implications for metasomatism in marble and the role of halogens in W mobilization and mineralization. *J. Asian Earth Sci.* **2016**, *117*, 82–106. [CrossRef]
6. Wang, H.; Feng, C.; Zhao, Y.; Zhang, M.; Chen, R.; Chen, J. Ore genesis of the Lunwei granite-related scheelite deposit in the Wuyi Metallogenic belt, southeast China: Constraints from geochronology, fluid inclusions, and H-O-S isotopes. *Res. Geol.* **2016**, *66*, 240–258. [CrossRef]
7. Wang, Y.C.; Wang, K.Y.; Konare, Y. N_2-rich fluid in the vein-type Yangjingou scheelite deposit, Yanbian, NE China. *Sci. Rep.* **2018**, *8*, 5662. [CrossRef]
8. Chen, J.; Sheng, D.; Shao, Y.; Zhang, J.; Liu, Z.; Wei, H.; Yang, Q.; Luo, X.; Du, Y. Silurian S-type granite-related W-(Mo) mineralization in the Nanling Range, south China: A case study of the Pingtan W-(Mo) deposit. *Ore Geol. Rev.* **2019**, *107*, 186–200. [CrossRef]
9. Kilias, S.P.; Konnerup-Madsen, J. Fluid inclusion and stable isotope evidence for the genesis of quartz-scheelite veins, Metaggitsi area, central Chalkidiki Peninsula, N. Greece. *Miner. Depos.* **1997**, *32*, 581–595. [CrossRef]
10. Sciuba, M.; Beaudoin, G.; Grzela, D.; Makvandi, S. Trace element composition of scheelite in orogenic gold deposits. *Miner. Depos.* **2019**, *55*, 1149–1172. [CrossRef]
11. Meinert, L.D.; Dipple, G.M.; Nicolescu, S. World Skarn Deposits. In *Economic Geology 100th Anniversary Volume*; Hedenquist, J.W., Thompson, J.F.H., Goldfarb, R.J., Richards, J.P., Eds.; Society of Economic Geologists, Inc.: Littleton, CO, USA, 2005; pp. 299–336.
12. Li, W.; Xie, G.Q.; Cook, N.J.; Mao, J.W.; Li, C.; Ciobanu, C.L.; Zhang, Z.Y. Tracking dynamic hydrothermal processes: Textures, in-situ Sr-Nd isotopes and trace element analysis of scheelite from the Yangjiashan vein-type W deposit, South China. *Am. Miner.* **2021**, *106*, 1987–2002. [CrossRef]
13. Wang, G.W.; Li, R.X.; Carranza, E.J.M.; Zhang, S.T.; Yan, C.H.; Zhu, Y.Y.; Qu, J.N.; Hong, D.M.; Song, Y.W.; Han, J.W. 3D geological modeling for prediction of subsurface Mo targets in the Luanchuan district, China. *Ore Geol. Rev.* **2015**, *71*, 270–274. [CrossRef]
14. McClenaghan, M.B.; Cabri, L.J. Review of gold and platinum group element (PGE) indicator minerals methods for surficial sediment sampling. *Geochem. Explor. Environ. Anal.* **2011**, *11*, 251–263. [CrossRef]
15. Mishra, B.; Pruseth, K.L.; Hazarika, P.; Chinnasamy, S.S. Nature and source of the ore-forming associated with orogenic gold-deposits in the Dharwar Craton. *Geosci. Front.* **2018**, *9*, 715–726. [CrossRef]
16. Singoyi, B.; Zaw, K. A petrological and fluid inclusion study of magnetite–scheelite skarn mineralization at Kara, Northwestern Tasmania: Implications for ore genesis. *Chem. Geol.* **2001**, *173*, 239–253. [CrossRef]
17. Zaw, K. The CanTung E-Zone Orebody, Northwest Territories: A Major Scheelite Skarn Deposit. Unpublished Master's Thesis, Queen's University, Kingston, ON, Canada, 1976.

18. Zaw, K.; Clark, A.H. Fluoride–hydroxyl ratios of skarn silicates, CanTung E-zone scheelite orebody, Tungsten, Northwest Territories. *Can. Miner.* **1978**, *16*, 207–221.
19. Song, S.W.; Mao, J.W.; Xie, G.Q.; Jian, W.; Chen, G.H.; Rao, J.F.; Ouyang, Y.P. Petrogenesis of scheelite-bearing albitite as an indicator for the formation of a world-class scheelite skarn deposit: A case study of the Zhuxi tungsten deposit. *Econ. Geol.* **2021**, *116*, 91–121. [CrossRef]
20. Yuvan, J.; Shelton, K.; Falck, H. Geochemical investigations of the high-grade quartz-scheelite veins of the Cantung mine, Northwest Territories. *Geol. Surv. Can. Open File* **2007**, *5344*, 177–190.
21. Xie, G.; Mao, J.; Li, W.; Fu, B.; Zhang, Z. Granite-related Yangjiashan tungsten deposit, southern China. *Miner. Depos.* **2019**, *54*, 67–80. [CrossRef]
22. Wilkinson, J.J.; Zhaoshan Chang, Z.; Cooke, D.R.; Baker, M.J.; Wilkinson, C.C.; Inglis, S.; Chen, H.; Gemmell, J.B. The chlorite proximitor: A new tool for detecting porphyry ore deposits. *J. Geochem. Explor.* **2015**, *152*, 10–26. [CrossRef]
23. Walshe, J.L. A six-component chlorite solid solution model and the conditions of chlorite formation in hydrothermal and geothermal systems. *Econ. Geol.* **1986**, *81*, 681–703. [CrossRef]
24. Cathelineau, M.; Nieva, D. A chlorite slid solution geothermometer the Los Azufres (Mexico) geothermal system. *Contrib. Miner. Petrol.* **1985**, *91*, 235–244. [CrossRef]
25. Sillitoe, R.H. Porphyry copper systems. *Econ. Geol.* **2010**, *105*, 3–41. [CrossRef]
26. Hsu, L.C. The stability relations of the wolframite series. *Am. Miner.* **1976**, *61*, 944–955.
27. Amossé, J. Physicochemical Study of the Hubnerite-Ferberite ($MnWO_4$-$FeWO_4$) zonal distribution in wolframite ($Mn_xFe_{(1-x)}WO_4$) Deposits. Application to the Borralha Mine. *Phys. Chem. Miner.* **1978**, *3*, 331–341. (In Portuguese) [CrossRef]
28. Bobos, I.; Noronha, F.; Mateus, A. Fe-, Fe,Mn- and Fe,Mg-chlorite: A genetic linkage to W, (Cu,Mo)-mineralization in the magmatic-hydrothermal system of Borralha, Northern Portugal. *Miner. Mag.* **2018**, *82*, 259–279. [CrossRef]
29. Bourdelle, F. Low-Temperature chlorite geothermometry and related recent analytical advances: A review. *Minerals* **2021**, *11*, 130. [CrossRef]
30. Ribeiro, A.; Pereira, E.; Dias, R. Structure in the Iberian Peninsula. In *Pre-Mesozoic Geology of Iberia*; Dallmeyer, R.D., Martinez, G., Eds.; Springer: Berlin, Germany, 1990; pp. 220–236.
31. Ribeiro, A.; Munhá, J.; Dias, R.; Mateus, A.; Peirera, E.; Ribeiro, L.; Fonseca, P.; Araújo, A.; Oliveira, T.; Romão, J.; et al. Geodynamic evolution of the SW Europe Variscides. *Tectonics* **2007**, *26*, 1–24. [CrossRef]
32. Matte, P. Accretionary history and crustal evolution of the Variscan Belt in Western Europe. *Tectonophysics* **1991**, *196*, 309–337. [CrossRef]
33. Ribeiro, A. Contribuition à l'étude tectonique de Trás-os-Montes Oriental. *Comunic. Serv. Geológicos Port.* **1974**, *24*, 1–168.
34. Noronha, F.; Ramos, J.M.F.; Rebelo, J.; Ribeiro, A.; Ribeiro, M.L. Essai de corrélation des phases de déformation hercyniennes dans le NW de la Péninsule Ibérique. *Leidse Geol. Meded.* **1981**, *52*, 89–91.
35. Ferreira, N.; Iglesias, M.; Noronha, F.; Pereira, E.; Ribeiro, A.; Ribeiro, M.L. Granitoides da Zona Centro Ibérica e seu enquadramento geodinâmico. In *Geologia de Los Granitoides y Rocas Asociadas del Macizo Hesperico*; Bea, F., Carnicero, A., Gonzalo, J., Lopez Plaza, M., Rodriguez Alonso, M., Eds.; Libro de Homenaje a García de Figuerola L.C.; Rueda: Madrid, Spain, 1987; pp. 37–51.
36. Dias, R.; Ribeiro, A. The Ibero-Armorican Arc: A collision effect against an irregular continent? *Tectonophysics* **1995**, *246*, 113–128. [CrossRef]
37. Noronha, F. Les brèches du gisement de tungstène de Borralha (Nord du Portugal). Quelques données pour leur classification. *Public. Domus. Laboratório Mineral. Geológico Fac. Ciências Porto* **1979**, *4*, 187–212.
38. Noronha, F. Caractéristiques physico-chimiques des fluides associés à la génese du gisement de tungsten de Borralha (Nord Portugal). *Bull. Miner.* **1984**, *107*, 273–284. [CrossRef]
39. Noronha, F.; Vindel, E.; López, J.A.; Dória, A.; García, E.; Boiron, M.C.; Cathelineau, M. Fluids related to tungsten ore deposits in Northern Portugal and Spanish Central system: A comparative study. *Rev. Soc. Geol. España* **1999**, *12*, 397–403.
40. Noronha, F. Metalogenic Study of the Wore Deposit of Borralha. Ph.D. Thesis, Universidade do Porto, Porto, Portugal, 1983. Unpublished work (In Portuguese).
41. Roedder, E. Fluid inclusions. *Rev. Miner. Geochem.* **1984**, *12*, 644.
42. Van den Kerkhof, A.M.; Hein, U.F. Fluid Inclusion Petrography. *Lithos* **2001**, *55*, 27–47. [CrossRef]
43. Shepherd, T.J.; Rankin, A.H.; Alderton, D.H.M. *A Practical Guide to Fluid Inclusion Studies*; Blackie & Son Ltd.: Glasgow, UK, 1985; 239p.
44. Poty, B.; Leroy, J.; Jachimowicz, L. Un nouveil appareil pour la mesure des temperatures sous le microscope: I: Installation de microthermometrie Chaixmeca. *Bull. Miner.* **1976**, *99*, 182–186.
45. Bakker, R.J. Package FLUIDS Computer programs for analysis of fluid inclusion data and modeling bulk fluid properties. *Chem. Geol.* **2003**, *194*, 3–23. [CrossRef]
46. Touret, J. The significance of fluid inclusionsin metamorphic rocks. In *Thermodynamics in Geology*; Fraser, D.G., Ed.; Amsterdam Reidel Pub. Co.: Dordrecht, The Netherlands, 1977; pp. 203–227.
47. Ramboz, C.; Schnapper, D.; Dubessy, J. The P-V-T-X-FO_2 evolution of H_2O-CO_2-CH_4 bearing fluid in a wolframite vein: Reconstruction from fluid inclusion studies. *Geochim. Cosmochim. Acta* **1985**, *49*, 205–219. [CrossRef]
48. Prieto, A.C.; Guedes, A.; Doria, A.; Noronha, F.; Jimenez, J. Quantitative determination of gaseous phases in fluid inclusions by Raman microspectrometry. *Spectr. Lett.* **2012**, *45*, 1–5. [CrossRef]

49. Clayton, R.N.; Mayeda, T.K. The use of bromide pentafluoride in the extraction of oxygen from oxides and silicates for isotopic analysis. *Geochim. Cosmochim. Acta* **1963**, *27*, 43–52. [CrossRef]
50. Zhang, L.; Liu, J.; Zhou, H.; Chen, Z. Oxygen isotope fractionation in the quartz-water-salt system. *Econ. Geol.* **1989**, *84*, 1643–1650. [CrossRef]
51. Landis, G.P.; Rye, R.O. Geologic, fluid inclusion, and stable isotope studies of the Pasto Bueno tungsten-base metal ore deposit northern Peru. *Econ. Geol.* **1974**, *69*, 1025–1059. [CrossRef]
52. Wesolowski, D.; Ohmoto, H. Calculated oxygen isotope fractionation factors between water and the minerals scheelite and powellite. *Econ. Geol.* **1986**, *81*, 471–477. [CrossRef]
53. Wenner, D.B.; Taylor, H.R., Jr. Temperatures of serpentinization of ultramafic rocks based on $^{18}O/^{16}O$ fractionation between coexisting serpentine and magnetite. *Contrib. Miner. Petrol.* **1971**, *32*, 165–168. [CrossRef]
54. Clayton, R.N.; O'Neil, J.R.; Mayeda, T.K. Oxygen isotope exchange between quartz and water. *Geophys. Res.* **1972**, *77*, 3057–3067. [CrossRef]
55. Brugger, J.; Giere, R.; Grobe, B.; Uspenski, E. Scheelite -powellite and paraniite-(Y) from the Fe-Mn deposit at Fianel, Eastern Swiss Alps. *Am. Miner.* **1998**, *83*, 1100–1110. [CrossRef]
56. Bailey, S.W. Clinochlores: Structures and crystal chemistry, In: Hydrous Phyllosilicates (S.W. Bailey, editor), Mineralogical Society of America, Washington. *Rev. Miner.* **1988**, *19*, 398–404.
57. Nieto, F. Chemical composition of metapelitic chlorites: X-ray diffraction and optical property approach. *Eur. J. Miner.* **1997**, *9*, 829–841. [CrossRef]
58. Bourdelle, F.; Parra, T.; Chopin, C.; Beyssac, O. A new chlorite geothermometer for diagenetic to low-grade metamorphic conditions. *Contrib. Miner. Petrol.* **2013**, *165*, 723–735. [CrossRef]
59. Lanari, P.; Wagner, T.; Vidal, O. A thermodynamic model for di-trioctahedral chlorite from experimental and natural data in the system MgO–FeO–Al_2O_3–SiO_3–H_2O: Applications to P–T sections and geothermometry. *Contrib. Miner. Petrol.* **2014**, *167*, 968–984. [CrossRef]
60. Inoue, A.; Meunier, A.; Patrier-Mas, P.; Rigault, C.; Beaufort, D.; Vieillard, P. Application of chemical geothermometry to low-temperature trioctahedral chlorites. *Clays Clay Miner.* **2009**, *57*, 371–382. [CrossRef]
61. Wiewióra, A.; Weiss, Z. Crystallochemical classifications of phyllosilicates based on the unified system of projection of chemical composition: II. The chlorite group. *Clay Miner.* **1990**, *25*, 83–92. [CrossRef]
62. Jacobson, C.E. Estimation of Fe^{3+} from electron microprobe analyses: Observations on calcic amphibole and chlorite. *J. Metamorf. Geol.* **1989**, *7*, 507–513. [CrossRef]
63. Shieh, Y.N.; Zhang, G.X. Stable isotope studies of quartz-vein type tungsten deposits in Dajistan mine, Jiangxi Province, Southeast China. In *Stable Isotope Geochemistry: A Tribute to Samuel Epstein*; Taylor, H.P., O'Neil, J.R., Kaplan, I.R., Eds.; SPE Publ.: London, UK, 1991; Volume 3, pp. 425–435.
64. Zheng, Y.-F. Oxygen isotope fractionation in wolframite. *Eur. J. Miner.* **1992**, *4*, 1133–1135. [CrossRef]
65. Wenner, D.B.; Taylor, H.P., Jr. Oxygen and hydrogen isotope studies of the serpentinization of ultramafic rocks in oceanic environments and continental complexes. *Am. J. Sci.* **1973**, *273*, 207–239. [CrossRef]
66. Cole, D.R. A preliminary evaluation of oxygen isotopic exchange between chlorite–water. *Geol. Soc. Am. Annu. Meet.* **1985**, *17*, 550.
67. Cathelineau, M. Cation site occupancy in chlorites and illites as a function of temperature. *Clay Miner.* **1988**, *23*, 471–485. [CrossRef]
68. Jowett, E.C. Fitting Iron and Magnesium into the Hydrothermal Chlorite Geothermometer. In Proceedings of the GAC/MAC/SEG Joint Annual Meeting, Toronto, ON, Canada, 27–29 May 1991.
69. Bourdelle, F.; Cathelineau, M. Low-temperature chlorite geothermometry: A graphical representation based on a T–R^{2+}–Si diagram. *Eur. J. Miner.* **2015**, *27*, 617–626. [CrossRef]
70. Inoue, A.; Kurokawa, K.; Hatta, T. Application of chlorite geothermometry to hydrothermal alteration in Toyoha geothermal system, southwestern Hokkaido, Japan. *Res. Geol.* **2010**, *60*, 52–70. [CrossRef]
71. Inoue, A.; Inoue, S.; Utada, M. Application of chlorite thermometry to estimation of formation temperature and redox conditions. *Clay Miner.* **2018**, *53*, 143–158. [CrossRef]
72. Verdecchia, S.O.; Collo, G.; Zandomeni, P.S.; Wunderlin, C.; Fernhrmann, M. Crystallochemical index and geothermobarometric calcultions as a multiproxy approach to P-T condition of the low-grade metamorphism: The case of he San Luis Formation, eastern Sierras Pampeanas. *Lithos* **2019**, *324–325*, 385–401. [CrossRef]
73. Wood, S.A.; Samson, I.M. The hydrothermal geochemistry of tungsten in granitoid environments: I. Relative solubilities of ferberite and scheelite as a function of T, P, pH and m_{HCL}. *Econ. Geol.* **2000**, *95*, 143–182. [CrossRef]
74. Oelsner, O. Über Erzgebirgische Wolframite. *Ber. Freib. Geol. Ges.* **1944**, *20*, 44–49.
75. Leutwein, F. Die chemische Zusammensetzung der Wolframite und ihre lagerstttenkundliche Bedeutung. *Acta Geol. Akad. Sci. Hungary* **1952**, *1*, 133–141.
76. Taylor, R.G.; Hosking, K.F.G. Manganese-iron ratios in wolframites, South Crofty Mine, Cornwall. *Econ. Geol.* **1970**, *65*, 47–53. [CrossRef]
77. Moore, F.; Howie, R.A. On the application of the hubnerite ferberite ratio as a Geothermometer. *Miner. Depos.* **1978**, *13*, 391–397. [CrossRef]

78. Michaud, J.A.; Pichavant, M. The H/F ratio as an indicator of contrasted wolframite deposition mechanisms. *Ore Geol. Rev.* **2019**, *104*, 266–272. [CrossRef]
79. Lecumberri-Sanchez, P.; Vieira, R.; Heinrich, C.A.; Pinto, F.; Wälle, M. Fluid-rock interaction is decisive for the formation of tungsten deposits. *Geology* **2017**, *45*, 579–582. [CrossRef]
80. Baumer, A.; Caruba, R.; Guy, B. Experimental study of hydrothermal transformation scheelite in ferberite: Preliminary results. *Bull. Miner.* **1985**, *108*, 15–20. [CrossRef]
81. Reed, M.H.; Spycher, N. Calculation of high temperature pH and mineral equilibria in hydrothermal waters, with application to geothermometry and studies of boiling and dilution. *Geochim. Cosmochim. Acta* **1984**, *48*, 1479–1492. [CrossRef]
82. Reed, M.H.; Spycher, N. Boiling, cooling, and oxidation in epithermal systems: A numerical modeling approach. In *Geology and Geochemistry of Epithermal Systems*; Berger, B.R., Bethke, P.M., Eds.; Soc. Econ. Geologists: Littleton, CO, USA, 1985; Volume 2, pp. 249–272.
83. Drummond, S.E.; Ohmoto, H. Chemical evolution and mineral depositionin boiling hydrothermal systems. *Econ. Geol.* **1985**, *80*, 126–147. [CrossRef]
84. Wood, S.A.; Samson, I.M. Solubility of ore minerals and complexation of ore metals in hydrothermal solutions. *Rev. Econ. Geol.* **1998**, *10*, 33–80.
85. Phillips, G.N.; Evans, K.A. Role of CO_2 in the formation of gold deposits. *Nature* **2004**, *429*, 860–863. [CrossRef] [PubMed]
86. Higgins, N.C. Fluid inclusion evidence for the transportof tungsten by carbonate complexes in hydrothermal solutions. *Can. J. Earth Sci.* **1980**, *17*, 823–830. [CrossRef]
87. Kreulen, R.; Schuiling, R.D. N_2- CH_4-CO_2 during formation of the Dome de l'Agoul, France. *Geochim. Cosmochim. Acta* **1981**, *46*, 193–203. [CrossRef]
88. Taylor, H.P., Jr. The application of oxygen and hydrogen isotope studies to problems of hydrothermal alteration and ore deposition. *Econ. Geol.* **1974**, *69*, 843–883. [CrossRef]
89. Taylor, H.P., Jr. Oxygen and hydrogen isotope relationships in hydrothermal mineral deposits. In *Geochemistry of Hydrothermal Ore Deposits*, 2nd ed.; Barnes, H.L., Ed.; Wiley: New York, NY, USA, 1979; pp. 236–277.
90. Ohmoto, H. Stable isotope geochemistry of ore deposits. In *Stable Isotopes in High Temperature Geological Processes*; Valley, J.W., Taylor, H.P., Jr., O'Neil, J.R., Eds.; Mineralogical Society of America: Chantilly, VA, USA, 1986; Volume 16, pp. 491–559.
91. Sheppard, S.M.F. Characterization and isotopic variations in natural water. In *Stable Isotopes in High Temperature Geological Processes*; Valley, J.W., Taylor, H.P., Jr., O'Neil, J.R., Eds.; De Gruyter: Berlin, Germany, 1986; Volume 16, pp. 165–183.
92. Taylor, H.P., Jr. Oxygen and hydrogen isotope relationships in hydrothermal mineral deposits. In *Geochemistry of Hydrothermal Ore Deposits*, 3rd ed.; Barnes, H.L., Ed.; John Wiley & Sons: New York, NY, USA, 1997; pp. 229–302.
93. Marignac, C.; Cathelineau, M. *The Nature Ore-Forming Fluids in Peri-Batholitic Sn-W Deposis*; James Cook University: Townsville, Australia, 2018.
94. Polya, D.A.; Foxford, K.A.; Stuart, F.; Boyce, A.; Fallick, A.E. Evolution and paragenetic content of low D hydrothermal fluids from Panasqueira W-Sn deposit, Portugal: New evidence from microthermometric, stable isotope, nobe gas and halogen analyses of primary fluid inclusions. *Geochim. Cosmochim. Acta* **2000**, *64*, 3357–3371. [CrossRef]
95. Wilkinson, J.J. The role of metamorphic fluids in the development of the Cornubian prefield: Fluid inclusion evidence from south Cornwall. *Miner. Mag.* **1990**, *54*, 219–230. [CrossRef]
96. Vidal, O.; Goffeé, B. Cookeite experimental study and thermodynamic analysis of its compatibility relations in the Li_2O-Al_2O_3-SiO_2-H_2O system. *Contrib. Miner. Petrol.* **1991**, *108*, 72–81. [CrossRef]
97. Bobos, I.; Vieillard, P.; Charoy, B.; Noronha, F. Alteration of spodumene to cookeite and its thermodynamic stability. *Clays Clay Miner.* **2007**, *57*, 296–304.
98. Goffeé, B.; Azanõn, J.M.; Bouybaouene, M.L.; Jullien, M. Metamorphic cookeite in Alpine metapelites from the Rif (Morocco) and Betic chain (Spain). *Eur. J. Miner.* **1996**, *8*, 335–348. [CrossRef]
99. Heinrich, C.A.; Seward, T.M. A spectrophotometric study of aqueous iron (II) chloride complexing from 25 to 200 °C. *Geochim. Cosmochim. Acta* **1990**, *54*, 2207–2221. [CrossRef]
100. Burns, R.G. *Mineralogical Applications of Crystal Field Theory*; Cambridge University Press: New York, NY, USA, 1993.

minerals

MDPI

Article

Genesis of the Baiyun Gold Deposit in Northeast Hubei Province, China: Insights from In Situ Trace Elements and S-Fe Isotopes of Sulfide

Weifang Song [1,2], Jianzhong Liu [3,4,5,*], Yuanbing Zou [1,2,*], Xingping Liu [1,2], Taocheng Long [1,2], Jiandong Zhu [1,2], Shengbo Fu [1,2], Song Chen [1,2], Yangfu Xiong [1,2], Runjie Zhou [2,6], Jingjing You [1,2], Xinqi Zhou [1,2], Zaixi Yang [1,2] and Jie Fang [1,2]

1 NO. Six Geological Team of Hubei Geological Bureau, Xiaogan 432000, China; gs.wfsong19@gzu.edu.cn (W.S.); dragontc@163.com (T.L.); 18963975955@163.com (J.Z.); 13669029269@163.com (S.F.); fengkuang2352006@126.com (S.C.); xf_139@163.com (Y.X.); youjing19921110@163.com (J.Y.); xinqizhou2024@163.com (X.Z.); chinayangzaixi@163.com (Z.Y.); 13477623737@163.com (J.F.)
2 Hubei Key Laboratory of Resources and Eco-Environment Geology, Xiaogan 432000, China; rjzhou@foxmail.com
3 Guizhou Bureau of Geology and Mineral Exploration & Development, Guiyang 550004, China
4 Innovation Center Ore Resources Exploration Technology in the Region of Bedrock, Ministry of Natural Resources of the People's Republic of China, Guiyang 550018, China
5 College of Resource and Environmental Engineering, Guizhou University, Guiyang 550025, China
6 Hubei Geological Survey, Wuhan 430034, China
* Correspondence: m13765115603@163.com (J.L.); zouyuanbing@163.com (Y.Z.)

Abstract: The Baiyun gold deposit is a medium-sized deposit in northeastern Hubei around the southern margin of the Tongbai-Dabie metallogenic belt. However, its genesis has not been determined. The metallogenic process of the Baiyun gold deposit can be divided into three stages: quartz + feldspar, quartz + native gold + electrum + polymetallic sulfides, and quartz + pyrite + calcite + iron dolomite + illite. In this study, LA-ICP-MS was used for in situ trace element and isotope analyses in the main and late ore stage hydrothermal sulfides to evaluate the genesis and evolution of ore-forming fluids. Gold is positively correlated with Ag, Cu, Pb, Zn, and Te and the Co/Ni ratio is greater than 1. The S isotope values of Py1 and Py2 are −0.23–3.04‰ and 1.27–6.09‰, respectively. As mineralization progressed, S isotope values increased. In situ S isotope values of the two types of galena symbiotic with pyrite in the main metallogenic stage are 2.97–3.47‰. In situ Fe isotopic values of pyrite are −0.05–0.82‰; values in the two stages are similar without significant fractionation. We inferred that the Baiyun gold deposit formed via magmatic mineralization related to the subduction of the Pacific Plate during the Yanshanian.

Keywords: pyrite; galena; in situ analysis; trace elements; S-Fe isotopes; Baiyun gold deposit

1. Introduction

The Tongbai-Dabe metallogenic belt is an important site of precious metals, rare metals, and strategic non-metals in China [1–9], and the Baiyun gold deposit in northeastern Hubei Province is the largest gold deposit in the southern margin of the metallogenic belt. This deposit has important typicality and representative characteristics for understanding gold mineralization in the area [10–13]. Previous studies of the Baiyun gold deposit have mainly focused on the geological characteristics [10–13], trace elements in the ore and surrounding rock [10,11], occurrence of gold [10–13], alteration of the surrounding rock [10–13], sulfur isotopes of sulfides [10–13], and fluid inclusions and hydrogen and oxygen isotopes [10]. It has been preliminarily identified as a magmatic-hydrothermal deposit [10–13]. However, data related to the genesis of the deposit are limited. Furthermore, previous research was mainly conducted at the end of the last century, with few recent studies of this deposit. Thus, advanced in situ analysis technology and more accurate geochemical data are required.

Pyrite and galena are important hydrothermal sulfides in the paragenetic sequence with the main ore minerals (natural gold and electrum). Therefore, the study of these sulfides is of great significance for understanding the geochemical characteristics of deposits. However, the structural characteristics, trace elements, and isotopic compositions of sulfides differ significantly with respect to environmental conditions. Previous studies have suggested that in situ trace elements and isotopic compositions of hydrothermal sulfides are important for determining the source of ore-forming fluids in medium- to low-temperature hydrothermal gold deposits [14–22]. Moreover, the surrounding rock of the Baiyun gold deposit is Neoproterozoic granodiorite, Early Cretaceous lamprophyre, Huangmailing Formation strata, and Qijiaoshan Formation strata of the Neoproterozoic Hongan Rock Group and the stratigraphic lithology is mainly dolomite albitite gneiss, quartz schist, albitite hornblende schist, albitite granulite, light granulite, manganese marble, graphite schist, and epidotite albitite hornblende schist, and part of the surrounding rock is of magmatic origin [10–13]. Therefore, sulfur isotope values of sulfides may be similar to those of magmatic sulfur, indicating that the S isotopes of sulfide powder cannot provide reliable data for understanding the genesis of the Baiyun gold deposit. In this study, in situ trace element analyses and isotope testing of the hydrothermal sulfides in ores by LA-(MC)-ICP-MS (laser ablation multi collector *inductively* coupled plasma *mass spectrometry*) were performed to reveal the source and evolution of ore-forming fluids. Moreover, characteristics of geochemical data were analyzed to further discuss the genesis of ore deposits from the perspective of microzones.

2. Geological Background

2.1. Regional Geology

2.1.1. Tongbai-Dabie Metallogenic Belt

The Tongbai-Dabie metallogenic belt is located between the North China Block and the Yangtze Block. It represents the eastern extension of the Qinling composite orogenic belt (an important part of the Central Orogenic Belt of China) and is one of 26 important metallogenic belts in China [23–31]. It belongs to the southern Qinling-Huaiyang fold belt of the Qinling fold system and can be divided into three secondary tectonic units: southern Qinling, Tongbai-Dabie, and northern Huaiyang metallogenic belts [24,30]. The Precambrian metamorphic rock series is largely exposed in the belt, and magmatic activity is active, with long invasion and eruption times. Magmatic rocks of various types are widely distributed in the belt. Intermediate-acid intrusive rocks are mainly exposed in the Mesozoic, followed by basic–ultrabasic rocks, while neutral rocks are rare [13]. Medium-low metamorphism occurred in the Wudang-Suizhou and northern Huaiyang areas and medium-high metamorphism occurred in the Tongbai-Dabie area. The Dabie area is one of the largest and best-preserved ultrahigh-pressure metamorphic bodies in the world [23,26,27,29]. The northern margin of the metallogenic belt of Henan Province hosts the Qianechong, Doupo, and other large molybdenum deposits and the Laowan and Bodaoling large-super to large gold and silver deposits [6,32–35]. The North Huaiyang metallogenic belt of Anhui Province hosts the Shapinggou and other super large molybdenum deposits [1,5,7,9,36]. The southern margin of the metallogenic belt in northeastern Hubei Province hosts the Baiyun gold deposit, Niangniangding beryllium deposit, Wangjiawuji tungsten deposit, Chenlingou gold deposit, and Huahe large fluorite deposit [10–13,37] (Figure 1). The output of these deposits is closely related to the Yanshan intrusion in time and space.

Our research group and others have studied the typical deposits produced by the above intrusions and their peripheries extensively, revealing several key results. (1) Highly differentiated granitic intrusions related to regional rare metal mineralization were mostly formed in the Yanshan period [38,39]. (2) The main genetic types of rare metal deposits produced in the study area include porphyry, altered rock, and quartz veins. There is general support for a magmatic-hydrothermal genesis with mineralization in the late Yanshan period; the mineralizing rock mass is a highly differentiated intermediate-acid granitic rock mass [40–50]. (3) In terms of metallogenic series, the study area mainly shows an Early Cretaceous Mo-Be-W ore-forming subseries, an Early Cretaceous Au-Ag ore-forming subseries, and a late Early Cretaceous Mo-Pb-Zn ore-forming subseries. The above metallogenic subseries belong to a large metallogenic series and are related to Pacific Plate subduction and retreat in metallogenic dynamics in the Yanshan period [1–5,51–53]. (4) Some researchers believe that the material source of molybdenum deposits in the Dabie region was the mantle, while the metallogenic fluid was sourced from the lower crust, rather than from the fluid metallogenic differentiation of a large rock base and rock mass produced near the surface. The contact belt of the rock base, rock mass, and veins exposed on the surface only provided a channel and space for mineralization [54]. (5) From the rock base to the inner and outer contact belts and then to the distal end, the evolution trend was from high-temperature deposits (Mo, Be, W, and Li) to medium-temperature (Au and Ag) and low-temperature (fluorite and barite) deposits, which were the product of the same metallogenic system in different locations. The northern and southern margins of the Tongbai-Dabie metallogenic belt had similar metallogenic tectonic backgrounds. However, the types of deposits produced are different, and there are differences in the ore-controlling factors, mineralization forms, and metallogenic ages. Among these, gold, molybdenum, tungsten, beryllium, fluorite, and other strategic minerals as the dominant minerals in the metallogenic belt have been the focus of research.

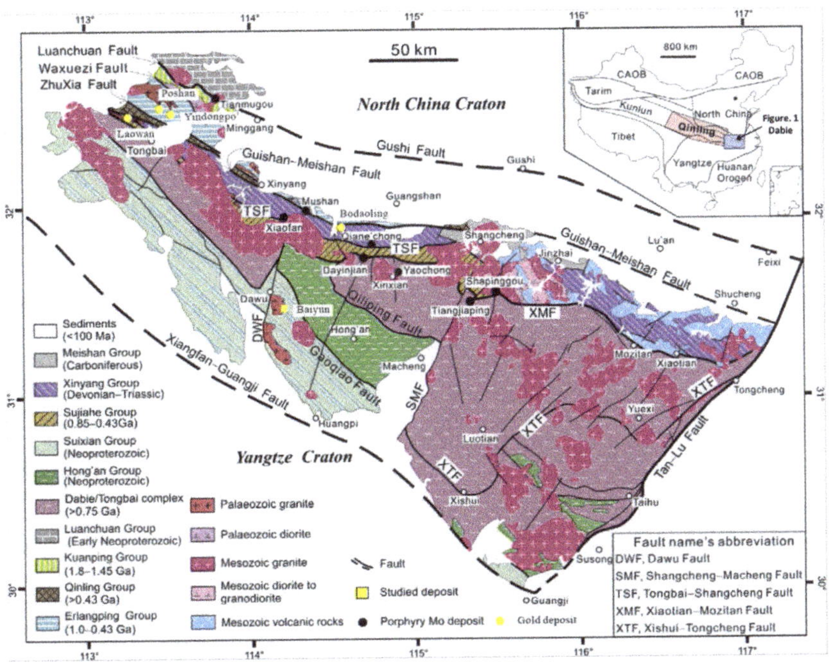

Figure 1. Diagram of geology and deposits in the Tongbai-Dabei metallogenic belt (modified from [2]).

The Baiyun gold deposit in Dawu County is accompanied by silver mineralization and is representative of the area. The Baiyun gold deposit is controlled by the NW and NE fault fracture zones and contact zones between different rock masses (veins), the orebodies extend to a deep level without a sealing edge, and the depth still has prospecting potential. Gold and silver orebodies have been found in the secondary structures parallel to the main ore-bearing structures in the NW and NE directions of the Baiyun gold deposit, indicating that the gold mineralization intensity was high and the geological conditions for gold mineralization were good in the area [10–13]. Partial studies on trace elements in ores and surrounding rocks, sulfur isotopes of sulfide powder and fluid inclusions in quartz veins have yielded results. For example, the formation of the Baiyun gold deposit may be related to granitic magmatic activity during the Yanshanian [10,11]. However, systematic and reliable geochemical data are lacking, limiting analyses of the source of ore-forming materials, the genesis of the deposits, and the potential for deep-edge prospecting in mining areas.

2.1.2. Baiyun Gold Deposit

The area of the Baiyun gold deposit underwent complex tectonic movement and magmatic emplacement, forming a large area of emerging Neoproterozoic gneissic granitic rocks and Mesozoic large granite bases, rock masses, and dikes. Strata in the metallogenic belt have exhibited different degrees of regional metamorphism, among which ultrahigh-pressure metamorphic strata in the Dabie area are a major site for studies of intracontinental orogeny. The exposed metamorphic strata in the region mainly include a set of Neoproterozoic Hongan Rock Group schist and gneiss formations formed by intense metamorphism, such as plagio-amphibolite, biotite hornblende plagioclase gneiss, monzonitic gneiss, and granulite. The original rocks belong to the calc-alkaline basalt-dacite-rhyolitic volcanic series and siliceous rocks [13]. The large faults in the Tongbai-Dabie metallogenic belt are mainly the SEE-trending Xiaotian-Mozitan, Gueshan-Meishan, and Tongbai-Chengcheng faults and the NEE-trending Tanlu, Tuanfeng-Macheng, and Huanshui faults. A large number of secondary faults are also distributed in the NW, NE, and SW directions, forming the integral structural framework of the Tongbai-Dabie metallogenic belt. The Baiyun gold mining area is located east of the intersection of the NW-trending Xincheng Huangpi shear zone and the NNE-trending Huanshui fault [10–13]. Strong structural and magmatic activities gave rise to the unique structural characteristics of the Daleishan dome. The orebody occurs in a brittle fault zone with a steep dip in the NW and NW directions [12,13]. The ore-bearing surrounding rocks are Neoproterozoic Daleishan gneissic granodiorite, Mesozoic lamprophyre, and Huangmailing Formation of the Neoproterozoic Hongan Rock Group, and the rock types are relatively simple (Figure 2).

2.2. Deposit Geology

The Baiyun gold deposit is a medium-sized gold deposit (gold resources of 5.995 tons) located east of the Daleishan Dome within Dawu County. The exposed strata of the mine are mainly the Huangmailing and Qijiaoshan formations of the Neoproterozoic Hongan Rock Group, and the stratigraphic lithology is mainly dolomite albitite gneiss, quartz schist, albitite hornblende schist, albitite granulite, light granulite, manganese marble, graphite schist, and epidotite albitite hornblende schist [10–13]. The fault structure in the mining area mainly consists of four groups: NW-, NNW-, NE-, and NNE-trending faults [12,13]. The brittle-ductile shear zone near the EW- or NW-trending faults is the main ore-hosting structure, and it shows multi-stage activity characteristics. The orebodies mainly exist in late superimposed brittle faults, and rock breakage and alteration in the fault zone tend to contain high-grade ore. At the intersection of the NW and NE faults, mineralization is often more significant, and a small number of ore veins are produced in the NE faults, suggesting that the gold mineralization in this area is closely related to the formation time of the NE-trending fault (Figure 2). The magmatic rocks exposed in the area are mainly Neoproterozoic Daleishan intrusion rocks, which are batholith-shaped and have been transformed by deformation and metamorphism. The main lithology is a set of granitic and granodioric gneisses, and their petrochemical characteristics show that they belong to deep-source synfusion calc-alkaline granites. Lamprophyre intrusions (dikes) in different directions are widely distributed in the mining area, most of which are filled along the NE- and NW-trending faults. The scale of the dikes ranges from several meters to hundreds or even thousands of meters, and the thickness is generally 0.50–10 m. Gold orebodies are produced in the contact area between the dike and surrounding rock.

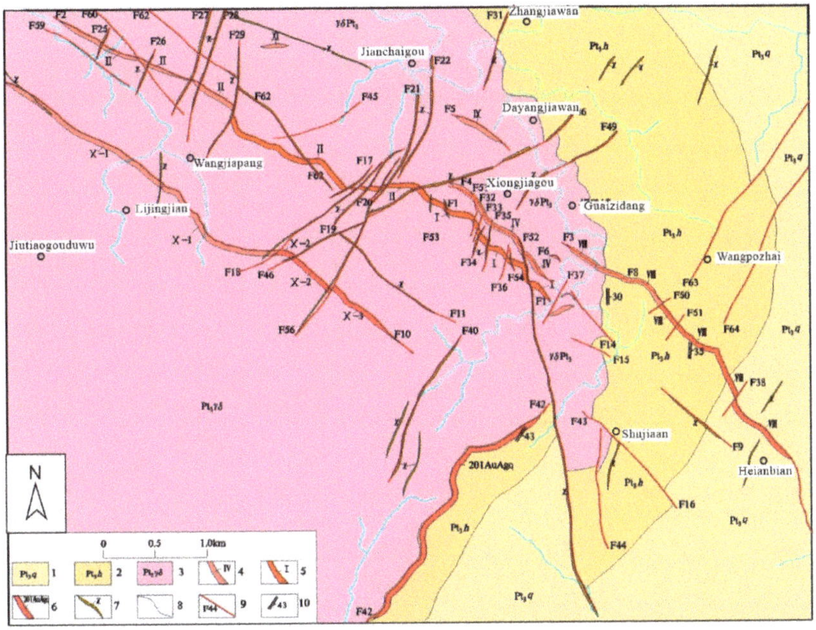

Figure 2. Geological map of the Baiyun gold deposit in NE Hubei Province (modified from [13]). 1—Qijiangshan Formation of Neoproterozoic Hongan group; 2—Huangmaling Formation of Neoproterozoic Hongan group; 3—Neoproterozoic granite diorite; 4—Au-bearing mineralized quartz veins and their No.; 5—Au-bearing quartz veins and their No.; 6—Au-Ag-bearing quartz veins and their No.; 7—Lamprophyre veins; 8—Geological boundary; 9—Faults and their number; 10—Foliation occurrence.

The entire ore-forming length is over 9 km, the width is approximately 2 km, and more than 10 veins are observed [10–13]. Five veins are found in the industrial orebody with a certain scale; the No. 201 vein is deposited in the NNE-trending interlayer fracture zone, and the No. I, No. II, No. VIII, and No. X veins are deposited in the quartz vein of the NW-trending fracture zone (Figures 2 and 3). The overall trend of the orebody is 290–325°, with a dip of 35–75° from west to east, and the dip angle changes from steep to slow and then to steep. The length of a single orebody is 1140–5300 m, with a thickness of 0.13–4.76 m, and the average grade is $1.27–9.79 \times 10^{-6}$. The average grade of associated silver is $10.91–235.85 \times 10^{-6}$. A single orebody is vein-shaped, plate-shaped, or plate-like, consists of multiple small orebodies, and is produced in a roughly parallel lateral form in space. The lateral direction is 260°–280°, and the lateral angle is 30°–40°. The mineralization is characterized by an uneven to extremely uneven distribution along the trend and the dip (i.e., from west to east), the gold content generally changes from low to high to low, the silver content gradually increases from low to high, and the Au/Ag ratio changes from 1:3.2 in vein II to 1:59 in vein VIII [10]. The change along the trend is greater than that along the dip. The ore types in the mining area are mainly quartz vein type and altered rock type. The ore minerals are mainly native gold, electrum, pyrite, galena, sphalerite, and chalcopyrite, whereas the gangue minerals are mainly quartz, sericite, Fe-dolomite, and calcite. The gold in the veins of the Baiyun gold deposit is mainly produced in the form of native gold and electrum, which are mostly symbiotic with pyrite, quartz, galena, and sphalerite. The mineral particle size is between 5 and 100 μm (i.e., micro-gold). The electrum particles are liquid droplets and irregular strips filled in the cracks or gaps of pyrite and quartz particles, and a small amount is wrapped in sulfide minerals and quartz particles. The ore textures are mainly granular, disseminated, semi-automorphic–heteromorphic, contained, and granular metamorphosed. The ores are characterized by isolated granular, disseminated, zonal, veinlet, and cataclastic structures. The main wall-rock alterations in the ore district are silicification, potassium feldspathization, sulfidation, carbonatization, and illitization. Silicification is commonly seen on both sides of the gold-bearing quartz veins, with a width of 5–30 cm and locally up to about 1 m. Sulfidation is commonly seen on the side of the wall-rock near the gold-bearing quartz vein, and it generally shows a scattered, disseminated, and strip-shaped output. Moreover, areas closer to the gold-bearing quartz vein show stronger mineralization. Carbonatization is a type of alteration during late mineralization that mainly occurs in gelatinous-protocrystalline carbonate veins, which are produced in the fissures of the wall-rock, and gold-bearing quartz veins (Figure 4E).

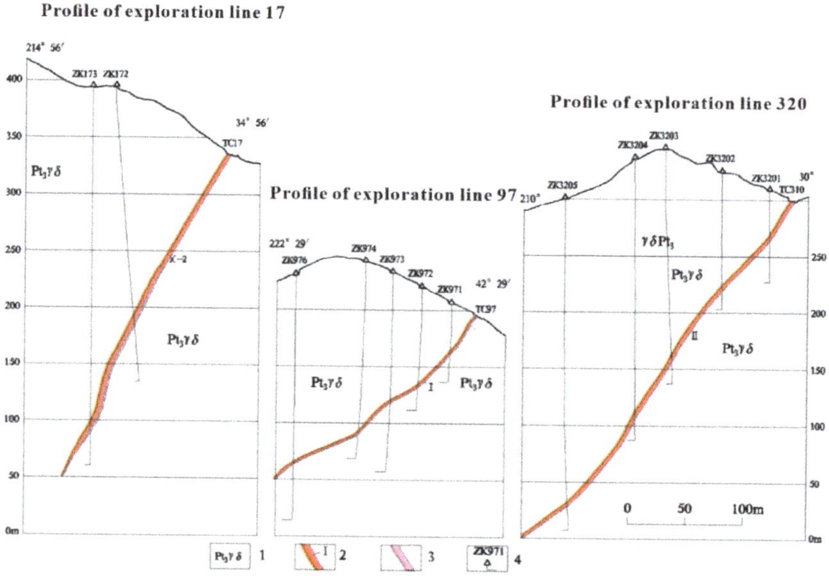

Figure 3. Profile of exploration lines 17, 97 and 320 in the Baiyun gold deposit (modified from [13]). 1—Neoproterozoic granite diorite; 2—Orebodies and their number; 3—Mineralized body and their number; 4—Drilling and their number.

Based on the mineral combination, formation sequence, ore texture and structure, relationship between wall-rock alteration and mineralization, and occurrence and interpenetration of ore veins, the Baiyun gold deposit can be divided into two metallogenic periods: hydrothermal metallogenic period and epigenetic metallogenic period. The hydrothermal mineralization period can be divided into three mineralization stages: early quartz-potassium feldspar stage, main quartz-native gold-electrum-porous pyrite-fine polymetallic sulfides mineralization stage (Figure 4A,B,D,E), and late quartz-rim-pyrite-coarse polymetallic sulfides-native silver-carbonate-illite mineralization stage (Figure 4A,C,D,F). During the supergene period, pyrite, chalcopyrite, and other sulfides were mainly oxidized to limonite and copper blue.

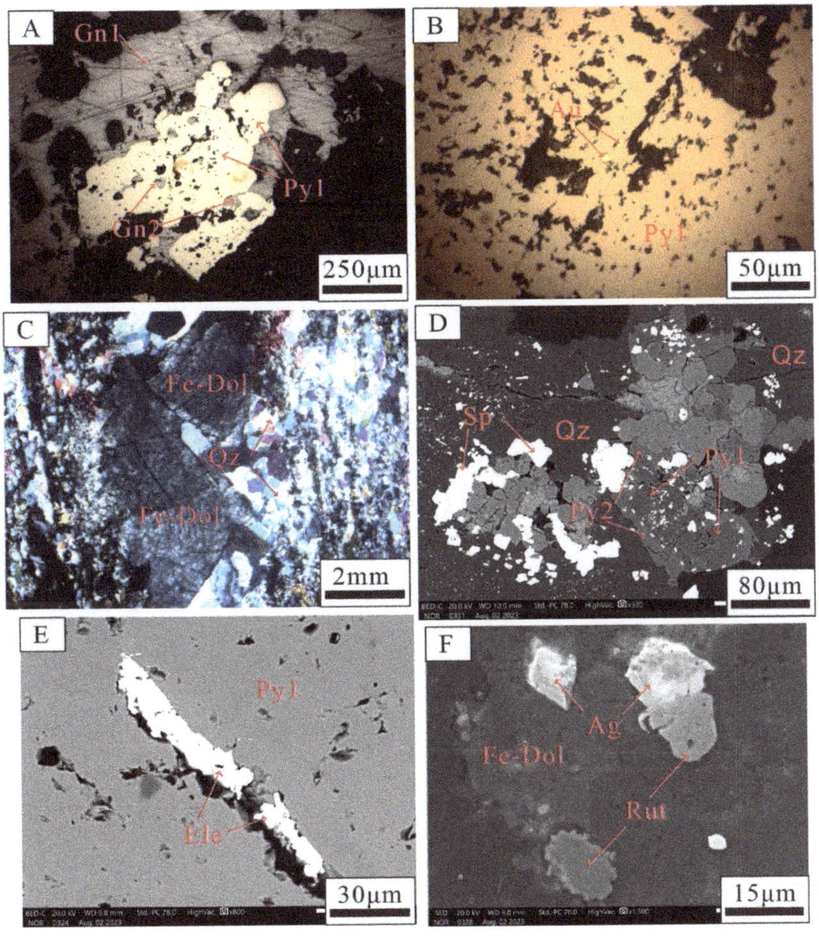

Figure 4. Morphological and textural features of different minerals in the Baiyun gold deposit including the (**A**) Porous Py1 contains Gn1 and symbiotic with Gn2 (RPPL image), (**B**) Porous Py1 contains natural gold particles (RPPL image), (**C**) The Fe-dolomite in the late ore-forming stage is filled along the fissures (RPPL image), (**D**) Porous Py1 contains fine-grained sphalerite and associated with coarse-grained sphalerite; Py2 overgrown by Py1. (BSE image), (**E**) Electrum is produced in Py fractures (BSE image), (**F**) Fine natural silver and rutile particles are produced in dolomite and quartz (BSE image). Abbreviations. Py: pyrite; Qz: quartz; Rut: rutile; Gn: galena; Sp: sphalerite; Au: natural gold; Ele: electrum; Ag: natural silver; Fe-Dol: Fe-dolomite. BSE: backscattered electron; and RPPL: reflected plane-polarized light.

3. Samples and Analytical Methods

3.1. Sample Preparation

Twelve ore samples were collected from the cross-section of an outcrop of the No.202 orebody (Figure 2), and 18 laser sections were prepared and polished to a thickness of approximately 100 μm for mineralographic and geochemical analyses. All sections were observed using optical microscopy and scanning electron microscopy (SEM) to identify the minerals, symbiotic relationships, and structural characteristics. Based on the generated data and comprehensive evaluation, coarse-grained sulfide minerals assigned to different ore stages were selected for in situ analyses using LA-ICP-MS. In situ analyses also involved

Fe and S isotopes of sulfides located adjacent to the associated sample used in the screening analyses, and the results are presented in Tables 1–3.

Table 1. Trace element data (ppm) for two types of pyrite samples from the Baiyun gold deposit obtained using LA-ICP-MS.

Pyrite Type	Spot No.	Au	Co	Ni	Cu	Ti	Zn	As	Mo	Ag	Te	Bi	Pb
Py1	BY-3-1-8	0.54	0.30	1.43	0.38	0.07	0.00	0.06	0.56	0.00	0.04	0.00	1.92
	BY-5-1-1	0.55	0.36	1.57	1.58	0.61	0.00	0.07	0.82	0.02	0.09	0.01	3.42
	BY-5-1-3	0.97	0.39	1.91	2.38	0.76	0.00	0.07	0.83	0.43	0.10	0.03	4.57
	BY-5-1-4	1.12	0.68	2.27	4.26	0.97	0.00	0.08	1.43	1.49	0.19	0.04	7.71
	BY-5-1-12	1.85	0.97	4.50	4.36	0.98	2.29	0.09	1.54	1.65	0.31	0.08	13.22
	BY-2-1-1	2.30	2.07	7.33	4.50	1.03	3.25	0.14	3.15	1.78	0.45	0.14	15.99
	BY-2-1-2	2.36	2.78	7.33	7.72	1.39	3.85	0.15	5.71	2.74	0.56	0.27	16.15
	BY-2-1-3	2.46	4.80	8.49	9.92	2.04	3.90	0.16	7.38	4.22	0.63	0.30	19.93
	BY-2-1-4	2.61	11.65	11.20	13.09	3.47	4.35	0.18	7.44	8.06	0.71	0.55	31.99
	BY-2-1-5	2.94	21.90	24.09	41.57	26.60	5.38	0.19	19.57	12.04	0.81	1.21	41.70
	BY-2-1-6	3.04	24.34	36.46	45.93	61.24	6.49	0.22	21.13	15.34	1.60	1.24	216.07
	BY-2-1-7	3.60	51.45	41.48	46.30	221.48	7.47	0.36	23.79	15.71	1.82	1.64	326.25
	BY-1-1-12	3.94	57.13	50.70	55.86	330.71	7.54	10.02	83.48	36.81	3.18	4.08	921.45
	BY-1-1-13	8.29	59.67	53.14	86.77	366.47	7.95	32.40	90.06	57.53	3.36	12.16	6379.76
	BY-1-1-14	33.56	68.46	86.66	490.07	1938.54	13.86	142.31	698.28	491.68	3.57	126.20	74353.59
Py2	BY-3-1-1	0.00	0.00	0.00	0.00	0.00	0.00	0.03	0.00	0.00	0.00	0.00	0.00
	BY-3-1-2	0.49	0.00	0.00	0.00	0.00	0.00	0.05	0.00	0.15	0.00	0.00	0.00
	BY-3-1-3	0.58	0.00	0.00	0.00	0.00	0.00	0.07	0.00	0.22	0.00	0.00	0.03
	BY-3-1-4	0.65	0.51	0.00	0.01	0.00	0.00	0.09	0.00	0.30	0.00	0.00	0.03
	BY-3-1-5	0.74	0.54	0.46	0.02	0.00	0.00	0.09	0.00	0.37	0.00	0.00	0.04
	BY-3-1-6	0.83	0.68	0.52	0.13	0.00	0.00	0.10	0.01	0.40	0.00	0.00	0.07
	BY-3-1-7	0.92	0.79	0.83	0.20	0.18	0.00	0.11	0.01	0.56	0.00	0.00	0.08
	BY-5-1-2	0.96	0.88	0.90	0.30	0.40	0.05	0.13	0.02	0.60	0.01	0.00	0.09
	BY-5-1-5	1.03	1.01	1.16	0.41	0.41	0.86	0.13	0.02	0.63	0.01	0.00	0.10
	BY-5-1-6	1.35	1.58	1.19	0.43	0.44	1.30	0.13	0.02	0.69	0.01	0.00	0.13
	BY-5-1-7	1.37	1.64	1.30	0.45	0.45	1.44	0.13	0.03	0.69	0.01	0.00	0.13
	BY-5-1-8	1.55	1.66	1.30	0.51	0.55	1.49	0.14	0.04	0.70	0.01	0.00	0.23
	BY-5-1-9	1.55	3.39	1.44	0.55	0.61	1.69	0.14	0.04	0.80	0.02	0.01	0.41
	BY-5-1-10	1.56	3.64	1.45	0.60	0.62	2.12	0.16	0.06	0.91	0.02	0.01	0.58
	BY-5-1-11	1.86	8.80	2.04	0.87	0.62	2.38	0.16	0.12	1.05	0.02	0.02	1.85
	BY-2-1-8	1.86	11.27	2.62	0.88	0.69	2.58	0.16	0.24	1.15	0.02	0.02	2.36
	BY-1-1-1	1.95	14.65	2.92	1.65	0.70	3.63	0.17	0.32	1.41	0.03	0.02	2.38
	BY-1-1-2	2.04	18.62	3.01	1.67	0.71	4.57	0.17	0.39	1.60	0.03	0.02	2.41
	BY-1-1-3	2.25	20.78	3.71	1.78	0.71	5.30	0.18	0.40	1.70	0.03	0.03	3.44
	BY-1-1-4	2.25	22.33	6.67	2.03	0.73	5.39	0.18	0.69	3.50	0.05	0.03	5.15
	BY-1-1-5	2.26	23.13	8.38	2.55	0.74	5.84	0.18	0.88	4.42	0.07	0.03	10.77
	BY-1-1-6	2.29	36.99	9.28	3.33	0.77	7.85	0.18	0.94	5.78	0.08	0.04	15.79
	BY-1-1-7	2.34	41.10	10.49	3.38	0.79	8.71	0.18	1.08	6.12	0.09	0.09	19.78
	BY-1-1-8	2.55	56.15	12.16	4.30	0.90	8.87	0.19	3.27	6.43	0.11	0.13	21.82
	BY-1-1-9	2.62	70.03	12.46	5.37	0.99	9.05	0.21	3.28	10.05	0.12	0.17	22.15
	BY-1-1-10	2.68	78.42	17.72	5.43	1.04	9.99	0.21	5.43	11.35	0.12	0.18	24.44
	BY-1-1-11	3.03	78.68	18.01	5.47	1.07	18.62	0.22	5.76	13.04	0.25	0.21	50.61
	BY-1-1-15	3.84	160.83	29.51	8.70	1.14	36.80	0.24	6.90	16.59	0.64	0.46	67.65

Abbreviations: 0.00 = below detection limit.

Table 2. S isotope data (‰) for different types of pyrite and galena samples in the Baiyun gold deposit determined using LA-MC-ICP-MS.

Types	Stages	Spot No.	$\delta^{34}S_{V-CDT}$ (‰)
Py1	Main ore stage	BY-2-1-PY-1	0.79
		BY-2-1-PY-2	2.07
		BY-2-1-PY-3	0.41
		BY-2-1-PY-4	−0.23
		BY-2-1-PY-5	2.19
		BY-2-1-PY-6	0.05
		BY-2-1-PY-7	1.81
		BY-2-1-PY-8	0.63
		BY-2-1-PY-9	−0.16
		BY-1-1-PY-10	1.42
		BY-5-1-PY-11	1.16
		BY-5-1-PY-12	0.92
		BY-1-1-PY-4	0.15
		BY-1-1-PY-5	2.73
		BY-1-1-PY-6	2.79
		BY-1-1-PY-7	2.42
		BY-1-1-PY-8	1.06
		BY-1-1-PY-9	1.89
		BY-1-1-PY-10	1.57
		BY-1-1-PY-11	1.70
		BY-1-1-PY-12	3.04
		BY-1-1-PY-13	1.91
		BY-1-1-PY-14	0.94
		BY-1-1-PY-15	1.79
		BY-1-1-PY-16	1.26
Py2	Late ore stage	BY-1-1-PY-1	5.89
		BY-1-1-PY-2	4.80
		BY-1-1-PY-3	5.30
		BY-1-1-PY-4	6.09
		BY-1-1-PY-5	1.27
		BY-1-1-PY-6	3.51
		BY-1-1-PY-7	1.80
		BY-1-1-PY-8	5.45
		BY-1-1-PY-9	4.93
Gn1	Main ore stage	BY-1-1-GN-1	2.97
		BY-1-1-GN-2	3.31
		BY-1-1-GN-3	3.22
		BY-1-1-GN-4	3.08
		BY-1-1-GN-5	3.24
		BY-1-1-GN-6	3.34
		BY-1-1-GN-7	3.17
		BY-1-1-GN-8	3.47
		BY-1-1-GN-9	3.08
		BY-1-1-GN-10	3.16
		BY-1-1-GN-11	3.15
		BY-1-1-GN-12	3.30
Gn2	Main ore stage	BY-1-1-GN-13	3.16
		BY-1-1-GN-14	3.16
		BY-1-1-GN-15	3.30
		BY-1-1-GN-16	3.00

Table 3. Fe isotope data (‰) for two different types of pyrite samples in the Baiyun gold deposit determined using LA-MC-ICP-MS.

Types	Stages	Spot No.	$\delta^{56}Fe_{IRMM014}$‰
Py1	Main ore stage	BY-1-1-1	0.23
		BY-1-1-2	0.18
		BY-1-1-3	−0.05
		BY-1-1-4	0.38
		BY-5-1-1	0.45
		BY-5-1-2	0.43
		BY-5-1-3	0.05
		BY-5-1-4	0.31
		BY-5-1-5	0.37
		BY-5-1-6	0.34
		BY-2-1-1	0.33
		BY-2-1-2	0.47
		BY-2-1-3	0.33
		BY-2-1-4	0.36
		BY-2-1-5	0.21
		BY-2-1-6	0.33
		BY-2-1-7	0.36
		BY-3-1-1	0.50
		BY-3-1-2	0.26
		BY-3-1-3	0.11
		BY-3-1-4	0.21
		BY-1-1-1	0.23
Py2	Late ore stage	BY-1-1-5	0.42
		BY-1-1-6	0.82
		BY-1-1-7	0.60
		BY-1-1-8	0.41
		BY-1-1-9	0.27
		BY-1-1-10	0.27
		BY-1-1-11	0.50
		BY-1-1-12	0.49
		BY-1-1-13	0.45
		BY-1-1-14	0.47
		BY-5-1-7	0.61
		BY-5-1-8	0.59
		BY-5-1-9	0.57
		BY-5-1-10	0.52
		BY-5-1-11	0.59
		BY-5-1-12	0.45
		BY-2-1-8	0.66
		BY-2-1-9	0.65
		BY-3-1-5	0.41
		BY-3-1-6	0.36
		BY-3-1-7	0.25

3.2. LA-ICP-MS Trace Element Analyses

Analyses of the pyrite samples were conducted by Beijing Createch Texting Technology Co., Ltd. (Beijing, China) using a RESOlution 193 nm laser ablation instrument equipped with a COMPex Pro 102 ArF excimer laser and a MicroLas optical system. Details on the analytical instruments and procedures employed in the sample analysis have been reported previously [55]. An Analytik Jena PQMS Elite ICP-MS instrument was used to determine the signal intensities of different ions. He and Ar served as the ion carrier and makeup gas, respectively. They were mixed before their input into the T-connector, and the mixture was then released into the ICP. A signal smoothing component was installed in the LA system to produce a stable signal at a low laser repetition rate of 1 Hz [56]. The diametrical laser beams for the sulfide and calcite veins had laser beam sizes of 24 and

50 μm, respectively, and the laser repetition rate was 8 Hz. The compositional data for the sulfides were corrected using data from the USGS MASS-1 sulfide standard [57,58]. Each LA point was associated with 20 s and 45 s of background and sample–signal acquisition, respectively. Offline calculations of the concentrations were conducted using the ICP-MS Data Cal program [59,60].

3.3. In Situ S Isotope Analysis

The S isotope analyses of pyrite and galena were conducted using a Nu Plasma II MC-ICP-MS equipped with a Resolution-S155 excimer ArF LA system at Beijing Createch Texting Technology Co., Ltd. (Beijing, China). The 193 nm ArF excimer laser, which was homogenized via a set of beam delivery systems, was focused on the surface of the sample at a fluence of 2.5 J/cm^2. Each acquisition incorporated 30 s of background and 40 s of a spot with a diameter of 24 μm, with a repetition rate of 8 Hz. He (800 mL/min) was used as the carrier gas to efficiently purge aerosols from the ablation cell and was mixed with Ar (~0.8 L/min) in a T-connector before its entry into the ICP torch. The integration time of the Nu Plasma II system was 0.3 s. The Wenshan natural pyrite crystal (GBW07267) prepared by the Chinese Academy of Geological Sciences, Beijing, China [55] was utilized as a standard, and the reproducibility of the δ^{34}S measurements was better than 0.6‰.

3.4. In Situ Fe Isotope Analysis

The LA-MC-ICP-MS in situ Fe isotope analysis was completed by Tianjin Kehui Test Technology Co., Ltd. (Tianjin, China). A Neptune Plus multi-receiver plasma mass spectrometer (Thermo Scientific, Waltham, MA, USA) with a RESOLution SE 193 nm solid laser was used. The detection environment required a temperature of 18–22 °C and a relative humidity of <65%. According to images of the sample after scanning, the appropriate area was selected and the sulfide was stripped using a laser stripping system. Stripping was performed by point stripping, with an adjustable stripping diameter of 10–150 μm, an energy density of 3 J/cm^2, and a frequency of 5 Hz. High-purity He was used as the carrier gas to blow out the aerosol produced by stripping, and the aerosol was sent to MC-ICP-MS for mass spectrometry. The Fe isotopes were simultaneously received by the static Faraday cup, and its structural parameters are listed in Table 1. The integration time was 0.131 s and 200 sets of data were collected, requiring approximately 27 s in total. Before the formal test, the instrument parameters were debugged with sulfide standard samples, including the plasma part (plasma tube position and carrier gas flow rate) and ion lens parameters. A highly sensitive injection cone combination ('X' skimmer cone + 'Jet' sample cone) was used to achieve a signal intensity for ^{56}Fe of approximately 10 V in high-resolution mode. To minimize the matrix effect, sulfide standards similar to the sample matrix were used as standard samples (pyrite standard sample Tianyu-Py, pyrrhotite standard sample JC-Po, niobium pyrite standard sample JC-Pn, and chalcopyrite standard sample Tianyu-Ccp), and the standard sample–standard crossover method was used for quality discrimination correction [61–63].

4. Results

4.1. Occurrence and Texture of Pyrites

Pyrite is one of the main ore minerals in the Baiyun gold deposit. Two main pyrite types were identified based on chemical and textural analyses. The types and origins of the pyrites in the studied samples are depicted in Figure 4. The pyrite types, structures, compositions, and physical characteristics are described below.

4.1.1. Py1

Py1 formed during the main stage of mineralization. SEM revealed an irregular shape and obvious dissolution pores and fissures. Some of the pores were filled with galena, sphalerite, and argon-gold formed during the same period, and the crystals were relatively coarse and often wrapped in pyrite formed in the late stage of mineralization.

4.1.2. Py2

Py2 formed in the late ore-forming stage, forming a typical girdling structure. This pyrite was irregular in shape and often grew around it during the main ore-forming stage. SEM revealed that Py2 was bright and smooth and that some pyrite particles were associated with fine sphalerite.

4.2. Occurrence and Texture of Galenas

According to the symbiosis with pyrite and mineral structure characteristics, galena can be divided into two types: fine galena coated with Py1 and disseminated coarse-grained galena grown around Py1.

4.2.1. Gn1

Gn1 is fine-grained and coated with Py1 in the main mineralization stage. It shows a subhedral-anhedral crystal structure, isolated distribution, and a particle size of less than approximately 100 μm.

4.2.2. Gn2

Gn2 particles are disseminated and closely symbiotic with Py1. The mineral particles are coarse and have a clear internal structure, and the pores inside the particles are developed. These pores are often filled with quartz, and they precipitate with Py1 at the same time in the main ore-forming stage.

4.3. Trace Elements of Pyrites Samples Determined Using LA-ICP-MS

The diameter of the beam spot of LA-ICP-MS used to test the in situ trace elements of pyrite was 24 μm. In total, 15 points were analyzed for Py1 and 29 points were tested for Py2. A correlation diagram of the data and mineralization elements is shown in Table 1 and Figure 5.

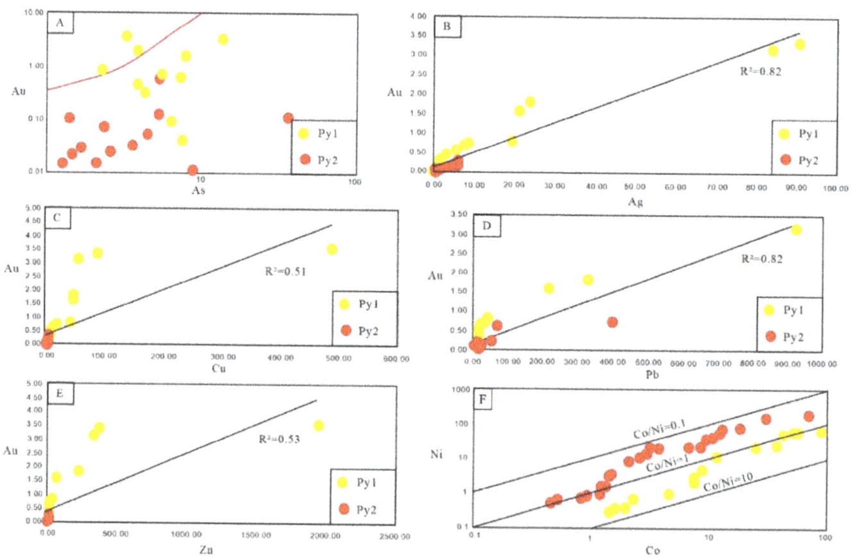

Figure 5. Elemental compositions of different types of pyrite in the Baiyun gold deposit and correlations including (**A**) As vs. Au, (**B**) Ag vs. Au, (**C**) Cu vs. Au, (**D**) Pb vs. Au, (**E**) Zn vs. Au, and (**F**) Co vs. Ni. The red line in Figure (**A**) means the maximum solubility of Au in pyrites. The black lines in the figures mean R2 correlation factors. The Au saturation line in (**A**) was obtained from Reich et al. [64]. The measuring unit of element contents: ppm.

The measured concentrations of Co, Ni, Cu, Ti, Pb, Zn, Ag, Au, As, Mo, Te, and Bi in the 44 pyrite samples are listed in Table 1. The trace elements of the main ore stage Py1 were Au (0.04–3.57 ppm), Co (0.30–68.46 ppm), Ni (1.43–86.66 ppm), Cu (0.38–490.07 ppm), Ti (0.54–33.56 ppm), Zn (0.07–1938.54 ppm), As (0.00–13.86 ppm), Mo (0.06–142.31 ppm), Ag (0.56–698.28 ppm), Te (0.00–491.68 ppm), Bi (0.00–126.20 ppm), and Pb (1.92–74353.59 ppm) (Table 1). The trace elements of the main ore stage Py2 were Au (0.00–0.73 ppm), Co (0.00–199.52 ppm), Ni (0.00–67.30 ppm), Cu (0.00–16.94 ppm), Ti (0.00–11.93 ppm), Zn (0.00–1.19 ppm), As (0.00–66.56 ppm), Mo (0.03–0.37 ppm), Ag (0.00–8.49 ppm), Te (0.00–21.20 ppm), Bi (0.00–0.82 ppm), and Pb (0.00–399.20 ppm) (Table 1). Compared with those for Py2 in the late ore-forming stage, Py1 in the main ore-forming stage had higher contents of Ni, Cu, Ti, Pb, Zn, Ag, Au, Mo, Te, and Bi but lower contents of Co and As and a lower Co/Ni ratio.

4.4. Sulfur Isotope Ratios of Pyrite and Galena Samples

In situ S isotope tests were conducted for two types of pyrite and galena samples using LA-MC-ICP-MS. The results are shown in Table 2 and Figure 6. The Py1, Py2, Gn1, and Gn2 samples yielded $\delta^{34}S$ values varying from −0.23‰ to 3.04‰ (mean = 1.37‰), 1.27‰ to 6.09‰ (mean = 4.34‰), 2.97‰ to 3.47‰ (mean = 3.21‰), and 3.00‰ to 3.30‰ (mean = 3.15‰), respectively. These results show that the in-situ S isotope data for the two galena types were concentrated, and no obvious fractionation was observed. Py2 in the late ore-forming stage has a larger sulfur isotope value than that for Py1 in the main ore-forming stage.

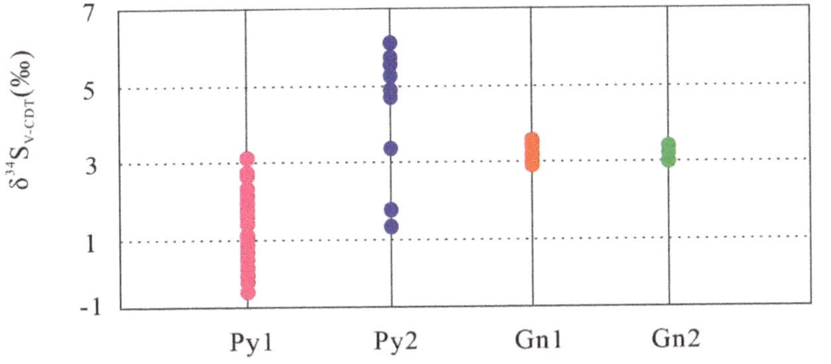

Figure 6. Plot displaying sulfur isotope ratios (‰) of different types of pyrite and galena from the Baiyun gold deposit.

4.5. Iron Isotope Ratios of Pyrite Samples

In situ Fe isotope tests were conducted for the two types of pyrite using LA-MC-ICP-MS. The results are shown in Table 3 and Figure 7. Py1 and Py2 samples yielded $\delta^{56}Fe$ values varying from −0.05‰ to 0.50‰ (mean = 0.29‰) and from 0.25‰ to 0.82‰ (mean = 0.49‰), respectively. The in situ Fe isotope values of Py1 and Py2 were both within the range of Fe isotopes of magmatic-hydrothermal pyrite calculated previously. However, compared with that for Py1 in the main ore-forming stage, Py2 in the late ore-forming stage had a relatively high Fe isotope value.

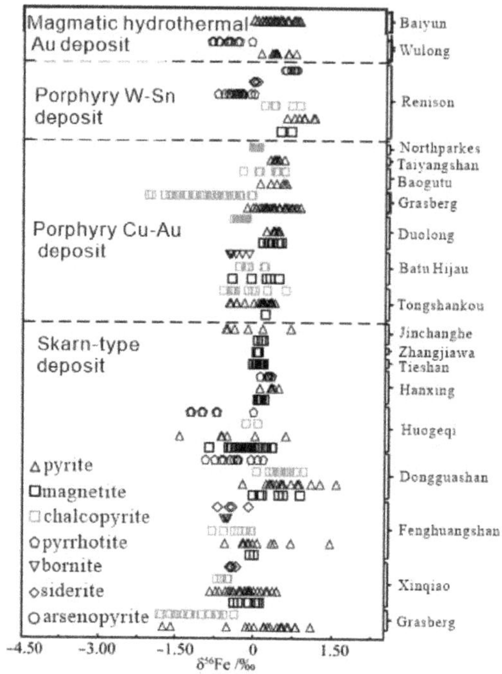

Figure 7. Iron isotope composition of iron-bearing minerals in different magmatic-hydrothermal deposits (modified from [65]).

5. Discussion

5.1. Genesis of Pyrite and Mechanism of Gold Precipitation

Pyrite is common in hydrothermal mineralization systems. During the formation of ore deposits, other elements can be precipitated simultaneously [66,67], which can effectively control the distribution of a series of economic trace elements [68,69], such as Au, Ag, As, Pb, Zn, and Cu, which can enter the pyrite lattice (solid solution, micron/nano-level mineral inclusions) [64,70–72]. Thus, research on pyrite can provide a better understanding of the characteristics and zoning of metals and can clarify the mechanism of superimposed mineralization events driven by physical and chemical processes (hydrothermal alteration, metamorphism, deformation, etc.) and changes in the composition and distribution of trace elements in ore-forming fluids [70,73–75]. Thus, the composition and microstructure of pyrite are useful complements to other geochemical information and provide insight into the evolutionary process of hydrothermal systems [57,74,76–78].

The Co and Ni contents in pyrite can reflect changes in the fluid temperature; higher temperatures are favorable for the replacement of Fe with Co and Ni in the pyrite lattice [67,70–82]. In addition, the Te content in pyrite shows a negative correlation with the fO_2 of fluids [83,84]. The Te content in stage 2 (average value 3.88×10^{-6}) was significantly lower than that in stage 1 (average value 44.30×10^{-6}), indicating that the fO_2 of fluids in stage 2 was higher than that in stage 1. Furthermore, the trace element content of pyrite can reflect the physical and chemical processes occurring in the fluids. For example, pyrite formed during the rapid boiling process has high Cu, Pb, Ag, and Au concentrations and low Co and Ni concentrations (Table 1) [67]. In addition, in terms of morphology, the porous structure and abundant mineral micro-inclusions (mainly galena and sphalerite) in pyrite (Py1) particles may indicate fluid boiling [68,85]. Pyrite (Py2) formed during the fluid mixing process was mostly the product of slow crystallization that occurred under relatively stable physical and chemical conditions, and it is characterized by a well-formed proliferation zone that generally did not show bubbles and mineral inclusions.

The ratios of different elements can also be used to determine the source of pyrite and infer the metallogenic evolution and genesis of the deposit. The Co/Ni ratio of pyrite is an important marker for ensuring changes in the metallogenic environment, particularly as a source of hydrothermal gold deposits [69,86,87]. In degassing deposits related to basic intrusions, the Co and Ni contents in pyrite are relatively high. The Co/Ni ratio of the magmatic source is generally higher than that of pyrite formed in a sedimentary environment [69,87–90], with magmatic-hydrothermal pyrite exhibiting a high Co/Ni ratio (>1) but sedimentary pyrite exhibiting a low Co/Ni ratio (<1) [69,89]. The Co/Ni ratios of pyrite in the two metallogenic stages obtained in this study were mostly greater than 1, indicating that the source of the pyrites is magmatic-hydrothermal fluids with potential metasomatism with the old strata.

The Co, Ni, Cu, Ti, Pb, Zn, Ag, Au, As, Mo, Te, and Bi contents in the pyrites in the main metallogenic stage (Py1) were higher than those in the late metallogenic stage. Pyrites in this stage had an obvious porous structure, and a large number of metal sulfides (such as chalcopyrite, galena, and sphalerite), native gold, and electrum inclusions were found, indicating that fluid boiling may have occurred in this stage and a large number of minerals precipitated. The high Te content, Co/Ni ratio, and $\delta^{34}S$ value showed that the mineralizing substances were mainly derived from magma, which may react with the surrounding rocks. Pyrites in the late metallogenic stage (Py2) had relatively low mineralized element contents and ring-like structures. There were fewer sulfide inclusions, such as sphalerite and galena, in the pyrites in this stage, indicating that the physical and chemical conditions of the fluids in this stage are relatively stable. The relatively higher Te content, Co/Ni ratio, and $\delta^{34}S$ value show that the mineralizing substances also mainly come from magmatic-hydrothermal fluids. In addition, in the two stages, the Au content was positively correlated with the Cu, Pb, Zn, Ag, and As contents, and the gold content in some pyrites was above the Au-As solid solution dissolution line, indicating that gold in the deposit mainly occurred in the form of granular gold (natural gold and electrum), although structural gold or solid solution gold was also observed.

5.2. Sources of Sulfur and Iron

The $\delta^{34}S$ value of pyrite is often used to track the source of minerals and to discuss the mineralization process [91–93]. Typical $\delta^{34}S$ values for magma sulfur are −5‰ to +5‰, for sedimentary sulfur are <0‰, and for marine sulfur are +20‰ [94–96]. The S isotopic composition of hydrothermal sulfides is greatly affected by fO_2, pH, temperature, and primary fluid composition but is less affected by pressure [91,95]. Increasing data have revealed the Fe isotopic composition and variation of various geological reservoirs, magma minerals, rocks, and hydrothermal fluids/minerals [97–99]. Theoretical calculations and experimental research have also yielded some mineral–melt, mineral–fluid, and mineral–mineral Fe isotopic equilibrium fractionation coefficients [100–105]. These results provide a basis for the development of geological applications of Fe isotopes. Currently, Fe isotope data are primarily used in analyses of biological processes, environmental changes, and magma evolution, among other areas of research [106–111]. The application of Fe isotopes in mineral deposits mainly focuses on stripe iron formation; submarine hydrothermal mineralization, magmatic mineralization, magmatic-hydrothermal mineralization, and other hydrothermal mineralization systems; and the supergene oxidation of mineral deposits [112–122]. These studies show that Fe isotopes have great potential for tracing the source of mineralization materials and constraining the genesis of mineral deposits. The Fe isotopic composition of the main Fe-containing minerals (pyrite, magnetite, pyrrhotite, chalcopyrite, etc.) has been evaluated in magmatic-hydrothermal deposits.

The $\delta^{56}Fe$ values of Fe-containing minerals in magmatic-hydrothermal deposits vary greatly (−2.07‰ to +1.58‰) [97–99]. The $\delta^{34}S$ of pyrite in the main stage of mineralization was −0.23‰ to +3.04‰, with an average value of +1.37‰, which was within the range for magmatic sulfur. The $\delta^{34}S$ of pyrite in the late stage of mineralization was +1.27‰ to +6.09‰, with an average value of +4.34‰, which was generally consistent with the range for magmatic sulfur, and some values exceeded the range for magmatic sulfur slightly, indicating that the sulfur in this stage mainly comes from magma, and some stratigraphic sulfur is added, indicating that fluid mixing occurs in the late stage of mineralization. The $\delta^{56}Fe$ of pyrite in the main metallogenic stage was −0.05‰ to +0.50‰, with an average value of +0.29‰, and the $\delta^{56}Fe$ of pyrite in the late metallogenic stage was +0.25‰ to +0.82‰, with an average value of +0.49‰, which was within the range for magmatic iron, indicating that the iron in this stage was mainly derived from magma.

5.3. Genesis of the Deposit

Systematic studies on the mineralogy, mineragraphy, in situ trace elements, and isotopes of polymetallic sulfides showed that the main ore minerals of the Baiyun gold deposit are native gold, electrum, pyrite, sphalerite, and galena (Figure 8). Based on the morphology, Co/Ni ratio, and geochemical characteristics of the pyrite, it is believed that the pyrite in the ores is mainly divided into two stages: the main metallogenic stage Py1 and the late metallogenic stage Py2. Py1 had a highly porous structure and contained a large amount of native gold, electrum, and metal sulfides (mainly sphalerite and galena), indicating that the pyrite in this stage was formed under the action of fluid boiling, which is the main factor controlling gold precipitation. Py2 was formed under a relatively stable fluid condition, and pyrite in this stage develops a typical annular structure with high brightness and a smooth surface under a SEM. This is the late metallogenic stage, and the contents of Co, Ni, Cu, Ti, Pb, Zn, Ag, Au, As, Mo, Te, Bi, and other mineralization elements were lower than those of Py1. The in situ S and Fe isotope tests of the two-stage pyrite showed that the ore-forming material was derived from the magma fluid. Previous researchers have used quartz veins of the Baiyun gold deposit as the research object, revealing that the metallogenic temperature of the Baiyun gold deposit was 170–350 °C and thus it belonged to the medium-low temperature hydrothermal gold deposit [10,11]. H-O isotope mapping revealed that most of the points were distributed in the range of magma water and a small part of the transition area between magma water and meteoric water, indicating that the ore-forming fluid mainly came from magma hydrothermal fluid, and meteoric water participated in the mineralization process [10,11]. Previous researchers have applied the quartz inclusion Rb-Sr dating method to the Baiyun gold deposit to confirm that the metallogenic age is 131.0 ± 10.0 Ma [123]. As mentioned previously, lamprophyre is an important ore-bearing rock in mining areas. According to the phenomenon that the quartz veins fill the cracks of lamprophyre, U-Pb dating of zircon in lamprophyre was carried out, which limited the metallogenic age of the Baiyun gold deposit to 126.8 ± 2.0 Ma [124]. This age is basically consistent with that obtained using quartz inclusion dating, indicating that the metallogenic age is the Early Cretaceous, consistent with the Yanshan period Pacific Plate subduction and the regional large area chalk granitic magma intrusion time.

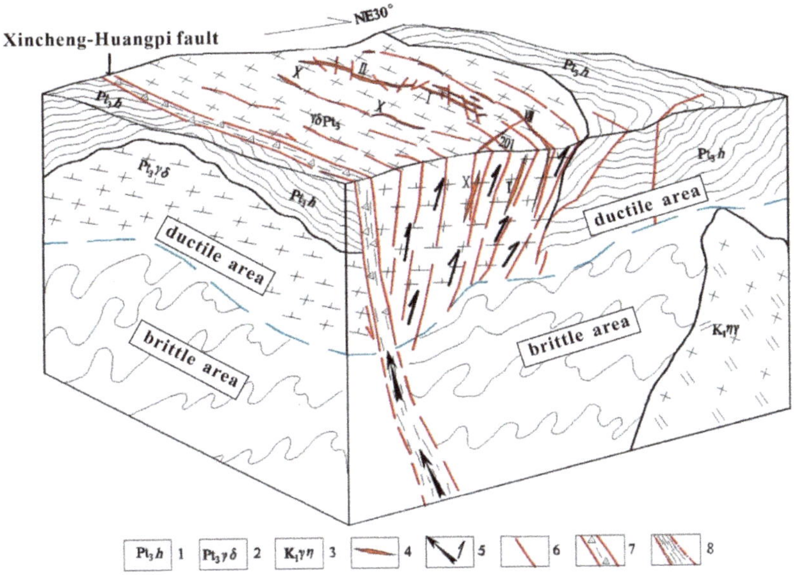

Figure 8. The metallogenic model of the Baiyun gold deposit (modified from [13]). 1—Neoproterozoic Huangmalingshan Formation; 2—Neoproterozoic Daleishan intrusive rock; 3—Cretaceous intrusive rock; 4—gold orebodies; 5—direction of ore-bearing hydrothermal migration; 6—brittle fault; 7—shallow brittle fracture zone; 8—deep ductile shear zone.

6. Conclusions

Two types of pyrite and galena were identified in the Baiyun gold deposit. Analyses of mineral morphology and mineral symbiosis suggest that fluid boiling occurred during the main stage of mineralization in the Baiyun gold deposit and was the main factor controlling gold precipitation. The gold content in some pyrites exceeded the Au-As solid solution dissolution line, indicating that gold in the deposit mainly exists in the form of grain gold (natural gold and electrum), with some in the form of structural or solid solution gold. Au was strongly positively correlated with Cu, Pb, Zn, Ag, As, Te, and other elements, indicating that these elements can be used as indicators of gold mineralization. The two types of pyrite had high Te contents and Co/Ni ratios. The in situ S and Fe isotopes of pyrite and the in situ S isotope of galena were all within the range of values for magma, indicating that the mineralization material originated from magmatic-hydrothermal processes. We concluded that the Baiyun gold deposit in Dawu County, Hubei Province is a magmatic-hydrothermal deposit formed in the Early Cretaceous.

Author Contributions: Author Contributions: Conceptualization, Investigation and Methodology, Data curation, Writing—original draft, Writing—review and editing, W.S.; Funding acquisition, W.S., J.L. and Y.Z.; Investigation and Methodology, X.L., T.L., J.Z., S.F. and S.C.; Project administration, W.S., J.L., Y.X., R.Z., J.Y., X.Z., Z.Y. and J.F.; Experimental data consolidation, W.S.; Writing—original draft, W.S.; and Writing—review and editing, W.S., J.L. and S.C.; All authors have read and agreed to the published version of the manuscript.

Funding: This study was jointly funded by Natural Science Foundation project of Hubei Province (2023AFD230), Science and technology project of Hubei Geological Bureau (KJ2024-28), and the Talent Team Program of Science and Technology Foundation of Guizhou Province (Qian Ke He Ping Tai Ren Cai CXTD [2021]007).

Data Availability Statement: Data are contained within the article.

Acknowledgments: We are grateful to E. Hongfang Chen for assistance during the BSE, SEM, and EDS analyses and E. Dewei Kong for the support provided when conducting the trace element and isotopic analyses of the sulfides. We would like to thank the anonymous reviewers for their constructive comments.

Conflicts of Interest: The authors declare no conflict of interest.

References

1. Chen, H.J.; Chen, Y.J.; Zhang, J.; Chen, X.Z.; Zhang, H.D. Zircon U–Pb ages and Hf isotope characteristics of the ore-bearing intrusion from the Shapinggou molybdenumdeposit, Jinzhai County, Anhui Province. *Acta Petrol. Sin.* **2013**, *29*, 131–145. (In Chinese with English Abstract)
2. Chen, Y.J.; Wang, P.; Li, N.; Yang, Y.F.; Pirajno, F. The collision-type porphyry Mo deposits in Dabie Shan, China. *Ore Geol. Rev.* **2017**, *81*, 405–430. [CrossRef]
3. Gao, Y.; Mao, J.W.; Ye, H.S.; Li, Y.F.; Luo, Z.Z.; Yang, Z.Q. Petrogenesis of ore-bearingporphyry from the Tangjiaping porphyry Mo deposit, Dabie orogen: Zircon U–Pb geochronology, geochemistry and Sr–Nd–Hf isotopic constraints. *Ore Geol. Rev.* **2016**, *79*, 288–300. [CrossRef]
4. Mao, J.W.; Xie, G.Q.; Bierlein, F.; Qu, W.J.; Du, A.D.; Ye, H.S.; Pirajno, F.; Li, H.M.; Guo, B.J.; Li, Y.F.; et al. Tectonic implications from Re–Os dating of Mesozoic molybdenum deposits in the East Qinling–Dabie orogenic belt. *Geochim. Cosmochim. Acta* **2008**, *72*, 4607–4626. [CrossRef]
5. Mao, J.W.; Pirajno, F.; Xiang, J.F.; Gao, J.J.; Ye, H.S.; Li, Y.F.; Guo, B.J. Mesozoic molybdenum deposits in the East Qinling–Dabie orogenic belt: Characteristics and tectonic settings. *Ore Geol. Rev.* **2011**, *43*, 264–293. [CrossRef]
6. Mi, M.; Chen, Y.J.; Yang, Y.F.; Wang, P.; Li, F.L.; Wan, S.Q.; Xu, Y.L. Geochronology and geochemistry of the giant Qian'echong Mo deposit, Dabie Shan, eastern China: Implications for ore genesis and tectonic setting. *Gondwana Res.* **2015**, *27*, 1217–1235. [CrossRef]
7. Wang, G.G.; Ni, P.; Yu, W.; Chen, H.; Jiang, L.L.; Wang, B.H.; Zhang, H.D.; Li, P.F. Petrogenesis of Early Cretaceous post-collisional granites at Shapinggou, Dabie Orogen: Implications for crustal architecture and porphyry Mo mineralization. *Lithos* **2014**, *184–187*, 393–415. [CrossRef]
8. Liu, Q.Q.; Li, B.; Shao, Y.J.; Lu, A.H.; Lai, J.Q.; Li, Y.F.; Luo, Z.Z. Molybdenum mineralization related to the Yangtze's lower crust and differentiation in the Dabie Orogen: Evidence from the geochemical features of the Yaochong porphyry Mo deposit. *Lithos* **2017**, *282–283*, 111–127. [CrossRef]
9. Chen, H.J.; Chen, Y.J. Zircon U-Pb ages and Hf isotope characteristics of granitic rocks from the Shapinggou molybdenum deposit, Dabie Shan, China. *Ore Geol. Rev.* **2022**, *150*, 105–151. [CrossRef]
10. Li, J.Z.; Xiao, Y.P. Preliminary researches on geological characteristics and genesis of Baiyun gold deposit. *Huibei Geol.* **1990**, *4*, 32–50. (In Chinese with English Abstract)
11. Liu, T.F. Geological characteristics and genesis of Baiyun gold deposit, Northeastern Hubei. *Gold Geol.* **1997**, *3*, 31–37. (In Chinese with English Abstract)
12. Xu, X.J.; Zou, Y.B.; Jiang, Z.F.; Meng, Z.P.; Chen, S.; Fang, J. Geological characteristics and genesis analysis of Dapoding gold deposit in northeast Hubei. *Gold* **2020**, *41*, 26–31. (In Chinese with English Abstract)
13. Fan, W.; Xu, X.J.; Wu, C.X.; Li, J.; Meng, Z.P.; Yang, N.H.; Li, B.B. Geological Characteristics and Metallogenic Model of Baiyun Gold Deposit in Dawu County, Hubei Province. *Resour. Environ. Eng.* **2020**, *34*, 13–17. (In Chinese with English Abstract)
14. Hu, X.L.; Gong, Y.J.; Zeng, G.P.; Zhang, Z.J.; Wang, J.; Yao, S.Z. Multistage pyrite in the Getang sediment-hosted disseminated gold deposit, southwestern Guizhou Province, China: Insights from textures and in situ chemical and sulfur isotopic analyses. *Ore Geol. Rev.* **2018**, *99*, 1–16. [CrossRef]
15. Hu, R.Z.; Su, W.C.; Bi, X.W.; Tu, G.Z.; Hofstra, A.H. Geology and geochemistry of Carlin-type gold deposits in China. *Mine. Depos.* **2002**, *37*, 378–392.
16. Su, W.C.; Heinrich, C.A.; Pettke, T.; Zhang, X.C.; Hu, R.Z.; Xia, B. Sediment-hosted gold deposits in Guizhou, China: Products of wall-rock sulfidation by deep crustal fluids. *Econ. Geol.* **2009**, *104*, 73–93. [CrossRef]
17. Su, W.C.; Hu, R.Z.; Xia, B.; Xia, Y.; Liu, Y.P. Calcite Sm-Nd isochron age of the Shuiyindong Carlin-type gold deposit, Guizhou, China. *Chem. Geol.* **2009**, *258*, 269–274. [CrossRef]
18. Chen, M.H.; Mao, J.W.; Li, C.; Zhang, Z.Q.; Dang, Y. Re-Os isochron ages for arsenopyrite from Carlin-like gold deposits in the Yunnan-Guizhou-Guangxi "golden triangle", southwestern China. *Ore Geol. Rev.* **2015**, *64*, 316–327. [CrossRef]
19. Su, W.C.; Xia, B.; Zhang, H.T.; Zhang, X.C.; Hu, R.Z. Visible gold in arsenian pyrite at the Shuiyindong Carlin-type gold deposit, Guizhou, China: Implications for the environment and processes of ore formation. *Ore Geol. Rev.* **2008**, *33*, 667–679. [CrossRef]
20. Xie, Z.J.; Xia, Y.; Cline, J.S.; Pribil, M.J.; Koenig, A.; Tan, Q.P.; Wei, D.T.; Wang, Z.P.; Yan, J. Magmatic origin for sediment-hosted Au deposits, Guizhou Province, China: In situ chemistry and sulfur isotope composition of pyrites, Shuiyindong and Jinfeng deposits. *Econ. Geol.* **2018**, *113*, 1627–1652. [CrossRef]
21. Large, R.R.; Maslennikon, V.V.; Robert, F.; Danyushevsky, L.V.; Chang, Z.S. Multistage sedimentary and metamorphic origin of pyrite and gold in the giant Sukhoi Log deposit, Lena gold province Russia. *Econ. Geol.* **2007**, *102*, 1233–1267. [CrossRef]

22. Xie, Z.J.; Xia, Y.; Cline, J.S.; Koenig, A.; Wei, D.T.; Tan, Q.P.; Wang, Z.P. Are there Carlin-type Au deposits in China? A comparison of the Guizhou, China, deposits with Nevada, USA. *Rev. Econ. Geol.* **2018**, *20*, 187–233.
23. Hacker, B.R.; Ratschbacher, L.; Webb, L.; Ireland, T.; Walker, D.; Dong, S.W. U–Pb zircon ages constrain the architecture of the ultrahigh-pressure Qinling–Dabie Orogen, China. *Earth Planet. Sci. Lett.* **1998**, *161*, 215–230. [CrossRef]
24. Hacker, B.R.; Ratshbacher, L.; Webb, L.E.; Ireland, T.R.; Calvert, A.; Dong, S.; Wenk, H.R.; Chateigner, D. Exhumation of ultrahigh-pressure continental crust in eastcentral China: Late Triassic–Early Jurassic tectonic unroofing. *J. Geophys. Res. Solid Earth* **2000**, *105*, 13339–13364. [CrossRef]
25. Li, S.G.; Huang, F.; Nie, Y.H.; Han, W.L.; Long, G.; Li, H.M.; Zhang, S.Q.; Zhang, Z.H. Geochemical and geochronological constraints on the suture location between the North and South China blocks in the Dabie orogen, central China. *Phys. Chem. Earth Part A Solid Earth Geod.* **2001**, *26*, 655–672. [CrossRef]
26. Ma, C.Q.; Ehlers, C.; Xu, C.H.; Li, Z.C.; Yang, K.G. The roots of the Dabieshan ultrahigh pressure metamorphic terrane: Constraints from geochemistry and Nd–Sr isotope systematics. *Precambrian Res.* **2000**, *102*, 279–301. [CrossRef]
27. You, Z.D.; Han, Y.J.; Yang, W.R.; Zhang, Z.M.; Wei, B.Z.; Liu, R. *The High-Pressure and Ultra High-Pressure Metamorphic Belt in the East Qinling and Dabie Mountains, China*; China University of Geosciences Press: Wuhan, China, 1996; pp. 1–150. (In Chinese with English Abstract)
28. Peng, S.G. Role and significance of "prospecting nearby the large deposits" theory in exploration practice of Wudang-Tongbai-Dabie metallogenic belt. *Resour. Environ. Eng.* **2024**, *38*, 16–25. (In Chinese with English Abstract)
29. Zhang, H.F.; Gao, S.; Zhong, Z. Geochemical and Sr-Nd-Pb isotopic compositions of Cretaceous granitoids: Constraints on tectonic framework and crustal structure of the Dabieshan ultrahigh-pressure metamorphic belt, China. *Chem. Geol.* **2002**, *186*, 281–299. [CrossRef]
30. Xu, X.C.; Lou, J.W.; Xie, Q.Q.; Xiao, Q.X.; Liang, J.F.; Lu, S.M. Geochronology and tectonic setting of Pb-Zn-Mo deposits and related igneous rocks in the Yinshan region, Jinzhai, Anhui province, China. *Ore Geol. Rev.* **2011**, *43*, 132–141. [CrossRef]
31. Zhang, H.F.; Zhong, Z.Q.; Gao, S.; Zhang, B.R.; Zhang, L.; Hu, S.H.; Hou, Q.Y. Pb and Nd isotopic composition of the Jigongshan granite: Constraints on crustal structure of Tongbaishan in the middle part of the Qinling Tongbai Dabie orogenic belt, Central China. *Lithos* **2004**, *73*, 215–227. [CrossRef]
32. Gao, Y.; Mao, L.W.; Ye, H.S.; Meng, F.; Li, Y.F. A review of the geological characteristics and geodynamic sitting of the late Early Cretaceous molybdenum deposits in the East Qinling-Dabie molybdenum belt, East China. *J. Asian Earth Sci.* **2015**, *108*, 81–96. [CrossRef]
33. Xie, Q.Q.; Xu, X.C.; Li, X.X.; Chen, T.H.; Lu, S.M. Rare earth elements geochemistry of Laowan gold deposit in Henan Province: Trace to source of ore-forming materials. *J. Rare Earths* **2006**, *24*, 115–120. (In Chinese with English Abstract) [CrossRef]
34. Li, F.L. Geological characteristics and metallogenic epoch of Qianechong large-size porphyry Mo deposit at the northern foot of Dabie Mountains Henan Province. *Miner. Depos.* **2011**, *30*, 457–468. (In Chinese with English Abstract)
35. Tang, X.W.; Yang, Z.Q.; Guo, Y.S. The study of fluid inclusion and metallogenic model of Xiaofan copper-molybdenum deposit in Henan Province. *Miner. Resour. Geol.* **2017**, *31*, 209–219. (In Chinese with English Abstract)
36. Zhang, H.D.; Wang, B.H.; Hao, Y.J.; Cheng, S.; Xiang, B. Geological characteristics and comprehensive ore-prospecting information of Shapinggou porphyry-type molybdenum deposit in Anhui Province. *Miner. Depos.* **2012**, *31*, 41–51. (In Chinese with English Abstract)
37. Zou, Y.B.; Liu, X.P.; Fan, W.; Li, J.; Cai, L.Y. Geological features and genesis of the Chenlingou gold deposit in Luotian county, Hubei Province. *Contrib. Geol. Miner. Resour. Res.* **2018**, *33*, 527–533. (In Chinese with English Abstract)
38. Zhang, H.D.; Shi, D.F.; Hao, Y.J.; Wang, B.H. Geological features of the Shapinggou porphyry molybdenum ore deposit, Jinzhai County, Anhui Province. *Geol. Anhui* **2010**, *20*, 104–108. (In Chinese with English Abstract)
39. Zhang, H.D.; Wang, G.J.; Shi, D.F. Metallogenic conditions of the Shapinggou porphyry molybdenumore deposit, Jinzhai County, Anhui Province and the prospecting direction. *Shanghai Geol.* **2010**, *31*, 203–205. (In Chinese with English Abstract)
40. Yu, W.; Ni, P.; Wang, G.G.; Shang, L.; Jiang, L.L.; Wang, B.H.; Zhang, H.D. Evolution of ore-forming fluids of the Shapinggou porphyry molybdenum deposit, Jinzhai, Anhui Province. *J. Nanjing Univ.* **2012**, *48*, 240–255. (In Chinese with English Abstract)
41. Zhang, H.; Sun, W.D.; Yang, X.Y.; Liang, H.Y.; Wang, B.H.; Wang, R.L.; Wang, Y.X. Geochronology and metallogenesis of the Shapinggou giant porphyry molybdenum deposit in the Dabie orogenic belt. *Acta Geol. Sin.* **2011**, *85*, 2039–2059. (In Chinese with English Abstract)
42. Zhang, H.D.; Wang, G.J.; Wang, B.H.; Hao, Y.J.; Xiang, B.; Peng, H.H. Shapinggou porphyry molybdenum ore-bearing intrusion and its alteration features. *Geol. Anhui* **2010**, *20*, 260–266. (In Chinese with English Abstract)
43. Yang, Z.Q. Geochemistry and tectonic setting of the Tangjiaping Mo-riched granodiorite porphyry in the northern Dabie Mountains. *Geol. Rev.* **2009**, *55*, 745–752. (In Chinese with English Abstract)
44. Wang, Y.X. Ore-forming features of explosion breccia pipe in Gaijing, Yinsha, Jinzhai county, Anhui Province. *Geol. Anhui* **2006**, *13*, 180–184. (In Chinese with English Abstract)
45. Wei, Q.G.; Gao, X.Y.; Zhao, T.P.; Chen, W.; Yang, Y.H. Petrogenesis of Tangjiaping granite porphyry in northern Dabie: Evidence from zircon LA-ICP-MS U-Pb dating and geochemical characteristics. *Acta Petrol. Sin.* **2010**, *26*, 1550–1562. (In Chinese with English Abstract)

46. Xu, X.C.; Lou, J.W.; Lu, S.M.; Xie, Q.Q.; Chu, P.L.; Yin, T. Re-Os ages of molybdenum-lead-zinc polymetallic deposits and Ar-Ar ages of related magmatic rocks in Yinshan area, Jinzhai, Anhui Province. *Miner. Depos.* **2009**, *28*, 621–632. (In Chinese with English Abstract)
47. Ni, P.; Wang, G.G.; Yu, W.; Chen, H.; Jiang, L.L.; Wang, B.H.; Zhang, H.D.; Xu, Y.F. Evidence of fluid inclusions for two stages of fluid boiling in the formation of the giant Shapinggou porphyry Mo deposit, Dabie Orogen, Central China. *Ore Geol. Rev.* **2015**, *65*, 1078–1094. [CrossRef]
48. Wang, B.H.; Zhang, H.D.; Wang, P.; Xu, X.C.; Hao, Y.J. Geological and geochemical characteristics and genesis of magmatic rocks related to porphyry Mo deposits in the Beihuaiyang region. *Earth Sci. Front.* **2016**, *23*, 46–62. (In Chinese with English Abstract)
49. Huang, F.; Wang, D.H.; Wang, C.H.; Chen, Z.H.; Yuan, Z.X.; Liu, X.X. Resources characteristics of molybdenum deposits and their regional metallogeny in China. *Acta Geol. Sin.* **2014**, *88*, 2296–2314. (In Chinese with English Abstract)
50. Dai, B.Z.; Jiang, S.Y.; Wang, X.L. Petrogenesis of the granitic porphyry related to the giant molybdenum deposit in Donggou, Henan Province, China: Constraints from petrogeochemistry, zircon U-Pb chronology and Sr-Nd-Hf isotopes. *Acta Petrol. Sin.* **2009**, *25*, 2889–2901. (In Chinese with English Abstract)
51. Tu, J.H.; You, J.J. Geological characteristics and prospecting direction of the Niangniangding tungsten-beryllium ore deposit in Hubei Province. *Acta Petrol. Mineral.* **2017**, *36*, 703–712. (In Chinese with English Abstract)
52. Zhong, S.Y.; Xiong, Y.L.; Yang, C.; Li, S.T.; Tu, J.H.; Shi, X.B.; Hu, T.P. Occurrence and Genesis of Beryllium Bearing Minerals in Yangxinyan Beryllium Ore Spot in Dawu County. *Resour. Environ. Eng.* **2020**, *34*, 518–524. (In Chinese with English Abstract)
53. Lu, S.M.; Ruan, L.S.; Zhao, L.L.; Wang, B.H.; Zhang, H.D.; Wang, G.G.; Chen, F. Two stages of diagenesis and metallogenesis of Shapinggou molybdenum-lead-zinc ore field in Jinzhai County, Anhui Province. *Acta Gelolgical Sin.* **2016**, *90*, 1167–1181. (In Chinese with English Abstract)
54. Lu, X.X.; Luo, Z.H.; Huang, F.; Liang, T.; Huang, D.F.; Han, N.; Gao, Y. "Small" magma and "big" fluid lead to form large scale deposit and transmagmatic fluid mineralization: Take for example of Mo deposits in eastern Qinling-Dabie mountain metallogenic belt. *Acta Petrol. Sin.* **2017**, *33*, 1554–1570. (In Chinese with English Abstract)
55. Zhu, Z.Y.; Jiang, S.Y.; Ciobanu, C.L.; Yang, T.; Cook, N.J. Sulfur isotope fractionation in pyrite during laser ablation: Implications for laser ablation multiple collector inductively coupled plasma mass spectrometry mapping. *Chem. Geol.* **2017**, *450*, 223–234. [CrossRef]
56. Hu, Z.C.; Zhang, W.; Liu, Y.S.; Gao, S.; Li, M.; Zong, K.Q.; Chen, H.H.; Hu, S.H. "Wave" signal-smoothing and mercury-removing device for laser ablation quadrupole and multiple collector ICPMS analysis: Application to lead isotope analysis. *Anal. Chem.* **2015**, *87*, 1152–1157. [CrossRef] [PubMed]
57. Wilson, S.A.; Ridley, W.I.; Koenig, A.E. Development of sulfide calibration standards for the laser ablation inductively-coupled plasma mass spectrometry technique. *J. Anal. At. Spectrom.* **2002**, *17*, 406–409. [CrossRef]
58. Liu, Y.S.; Hu, Z.C.; Gao, S.; Guenther, D.; Xu, J.; Gao, C.G.; Chen, H.H. In situ analysis of major and trace elements of anhydrous minerals by LA-ICP-MS without applying an internal standard. *Chem. Geol.* **2008**, *257*, 34–43. [CrossRef]
59. Hu, X.L.; Yao, S.Z.; Ding, Z.J.; He, M.C. Early Paleozoic magmatism and metallogeny in Northeast China: A record from the Tongshan porphyry Cu deposit. *Miner. Depos.* **2017**, *52*, 85–103. [CrossRef]
60. Hu, X.L.; Ding, Z.J.; Yao, S.Z.; He, M.C.; Shen, J.; Zhu, B.P.; Chen, B. Geochronology and Sr-Nd-Hf isotopes of the Mesozoic granitoids from the Great Xing'an and Lesser Xing'an ranges: Implications for petrogenesis and tectonic evolution in NE China. *Geol. J.* **2016**, *51*, 1–20. [CrossRef]
61. Chen, L.; Liu, Y.; Li, Y.; Li, Q.L.; Li, X.H. New potential pyrrhotite and pentlandite reference materials for sulfur and iron isotope microanalysis. *J. Anal. At. Spectrom.* **2021**, *36*, 1431–1440. [CrossRef]
62. Chen, L.; Feng, Y.T.; Yu, H.M.; Zhang, W.; Kang, J.T.; Huang, F.; Hu, Z.C.; Li, X.H. Tianyu-Py pyrite: A new natural reference material for micro-beam determination of iron isotopic ratios. *J. Anal. At. Spectrom.* **2022**, *37*, 2300–2308. [CrossRef]
63. Chen, L.; Feng, Y.T.; Yu, H.M.; Li, L.; Ding, X.; Zhang, W.; Kang, J.T.; Li, X.H. Mineralogical and Geochemical Characterization at a Micro-scale of a New Natural Chalcopyrite Reference Material for In-situ Fe Isotopic Ratio Analysis. *At. Spectrosc.* **2023**, *44*, 14–23. [CrossRef]
64. Deditius, A.P.; Utsunomiya, S.; Reich, M.; Kesler, S.E.; Ewing, R.C.; Hough, R.; Walshe, J. Trace metal nanoparticles in pyrite. *Ore Geol. Rev.* **2011**, *42*, 32–46. [CrossRef]
65. Wang, X.W.; Li, Y.X.; An, F. Status and progress on geochemical behavior of iron isotope in magmatichydrothermal system. *Miner. Depos.* **2023**, *42*, 1214–1228. (In Chinese with English Abstract)
66. Reich, M.; Deditius, A.; Chryssoulis, S.; Li, J.W.; Ma, C.Q.; Parada, M.A.; Barra, F.; Mittermayr, F. Pyrite as a record of hydrothermal fluid evolution in a porphyry copper system: A SIMS/EMPA trace element study. *Geochim. Cosmochim. Acta.* **2013**, *104*, 42–62. [CrossRef]
67. Román, N.; Reich, M.; Leisen, M.; Morata, D.; Barra, F.; Deditius, A.P. Geochemical and micro-textural fingerprints of boiling in pyrite. *Geochim. Cosmochim. Acta* **2019**, *246*, 60–85. [CrossRef]
68. Deditius, A.P.; Utsunomiya, S.; Renock, D.; Ewing, R.C.; Ramana, C.V.; Becker UKesler, S.E. A proposed new type of arsenian pyrite: Composition, nanostructure and geological significance. *Geochim. Cosmochim. Acta* **2008**, *72*, 2919–2933. [CrossRef]
69. Large, R.R.; Danyushevsky, L.; Hollit, C.; Maslennikov, V.; Meffre, S.; Gilbert, S.; Bull, S.; Scott, R.; Emsbo, P.; Thomas, H.; et al. Gold and trace element zonation in pyrite using a laser imaging technique: Implications for the timing of gold in orogenic and Carlinstyle sediment-hosted deposits. *Econ. Geol.* **2009**, *104*, 635–668. [CrossRef]

70. Reich, M.; Kesler, S.E.; Utsunomiya, S.; Palenik, C.S.; Chryssoulis, S.L.; Ewing, R.C. Solubility of gold in arsenian pyrite. *Geochim. Cosmochim. Acta* **2005**, *69*, 2781–2796. [CrossRef]
71. Barker, S.L.L.; Hickey, K.A.; Cline, J.S.; Dipple, G.M.; Kilburn, M.R.; Vaughan, J.R.; Longo, A.A. Uncloaking invisible gold: Use of NanoSIMS to evaluate gold, trace elements, and sulfur isotopes in pyrite from carlin-type gold deposits. *Econ. Geol.* **2009**, *104*, 897–904. [CrossRef]
72. Deditius, A.P.; Utsunomiya, S.; Ewing, R.C.; Chryssoulis, S.L.; Venter, D.; Kesler, S.E. Decoupled geochemical behavior of As and Cu in hydrothermal systems. *Geology* **2009**, *37*, 707–710. [CrossRef]
73. Cook, N.J.; Ciobanu, C.L.; Mao, J.W. Textural control on gold distribution in As-free pyrite from the Dongping, Huangtuliang and Hougou gold deposits, North China Craton (Hebei Province, China). *Chem. Geol.* **2009**, *264*, 101–121. [CrossRef]
74. Thomas, H.V.; Large, R.R.; Bull, S.W.; Maslennikov, V.; Berry, R.F.; Fraser, R.; Froud, S.; Moye, R. Pyrite and pyrrhotite textures and composition in sediments, laminated quartz veins, and reefs at Bendigo Gold Mine, Australia: Insights for ore genesis. *Econ. Geol.* **2011**, *106*, 1–31. [CrossRef]
75. Deditius, A.P.; Reich, M.; Kesler, S.E.; Utsunomiya, S.; Chryssoulis, S.L.; Walshe, J.; Ewing, R.C. The coupled geochemistry of Au and As in pyrite from hydrothermal ore deposits. *Geochim. Cosmochim. Acta* **2014**, *140*, 644–670. [CrossRef]
76. Muntean, J.L.; Cline, J.S.; Simon, A.C.; Longo, A.A. Magmatic-hydrothermal origin of Nevada's Carlin-type gold deposits. *Nat. Geosci.* **2011**, *4*, 122–127. [CrossRef]
77. Gregory, D.D.; Large, R.R.; Bath, A.B.; Steadman, J.A.; Wu, S.; Danyushevsky, L.; Bull, S.W.; Holden, P.; Ireland, T.R. Trace element content of pyrite from the Kapai Slate, St. Ives Gold District, Western Australia. *Econ. Geol.* **2016**, *111*, 1297–1320. [CrossRef]
78. Tardani, D.; Reich, M.; Deditius, A.P.; Chryssoulis, S.; Sánchez-Alfaro, P.; Wrage, J.; Roberts, M.P. Copper-arsenic decoupling in an active geothermal system: A link between pyrite and fluid composition. *Geochim. Cosmochim. Acta* **2017**, *204*, 179–204. [CrossRef]
79. Loftus-Hills, G.; Solomon, M. Cobalt, nickel and selenium in sulphides as indicators of ore genesis. *Miner. Depos.* **1967**, *2*, 228–242. [CrossRef]
80. Clark, C.; Grguric, B.; Mumm, A. Genetic implications of pyrite chemistry from the Palaeoproterozoic Olary Domain and overlying Neoproterozoic Adelaidean sequences, northeastern South Australia. *Ore Geol. Rev.* **2004**, *25*, 237–257. [CrossRef]
81. Migdisov, A.A.; Zezin, D.; Williams-Jones, A.E. An experimental study of cobalt (II) complexation in Cl-and H2 S-bearing hydrothermal solutions. *Geochim. Cosmochim. Acta* **2011**, *75*, 4065–4079. [CrossRef]
82. Grant, H.L.J.; Hannington, M.D.; Petersen, S.; Frische, M.; Fuchs, S.H. Constraints on the behavior of trace elements in the activelyforming TAG deposit, Mid-Atlantic Ridge, based on LA-ICP-MS analyses of pyrite. *Chem. Geol.* **2018**, *498*, 45–71. [CrossRef]
83. Keith, M.; Smith, D.J.; Jenkin, G.R.T.; Holwell, D.A.; Dye, M.D. A review of Te and Se systematics in hydrothermal pyrite from precious metal deposits: Insights into ore-forming processes. *Ore Geol. Rev.* **2018**, *96*, 269–282. [CrossRef]
84. Wang, K.X.; Zhai, D.G.; Liu, J.J.; Wu, H. LA-ICP-MS trace element analysis of pyrite from the Dafang gold deposit, South China: Implications for ore genesis. *Ore Geol. Rev.* **2021**, *139*, 104507. [CrossRef]
85. Simon, G.; Huang, H.; Penner-Hahn, J.E.; Kesler, S.E.; Kao, L.S. Oxidation state of gold and arsenic in gold-bearing arsenian pyrite. *Am. Mineral.* **1999**, *84*, 1071–1079. [CrossRef]
86. Campbell, F.A.; Ethier, V.G. Nickel and cobalt in pyrrhotite and pyrite from the Faro and Sullivan orebodies. *Can. Mineral.* **1984**, *22*, 503–506.
87. Bajwah, Z.U.; Seccombe, P.K.; Offler, R. Trace element distribution, Co:Ni ratios and genesis of the Big Cadia iron-copper deposit, New South Wales, Australia. *Miner. Depos.* **1987**, *22*, 292–300. [CrossRef]
88. Koglin, N.; Frimmel, H.E.; Minter, W.E.L.; Brätz, H. Trace-element characteristics of different pyrite types in Mesoarchaean to Palaeoproterozoic placer deposits. *Miner. Depos.* **2010**, *45*, 259–280. [CrossRef]
89. Reich, M.; Simon, A.C.; Deditius, A.; Barra, F.; Chryssoulis, S.; Lagas, G.; Tardani, D.; Knipping, J.; Bilenker, L.; Sánchez-Alfaro, P.; et al. Trace element signature of pyrite from the Los Colorados Iron Oxide-Apatite (IOA) deposit, Chile: A missing link between Andean IOA and Iron Oxide Copper-Gold systems? *Econ. Geol.* **2016**, *111*, 743–761. [CrossRef]
90. Meyer, F.M.; Robb, L.J.; Oberthür, T.; Saager, R.; Stupp, H.D. Cobalt, nickel, and gold in pyrite from primary gold deposits and Witwatersrand reefs. *S. Afr. J. Geol.* **1990**, *93*, 70–82.
91. Ohmoto, H. Systematics of sulfur and carbon isotopes in hydrothermal ore deposits. *Econ. Geol.* **1972**, *67*, 551–578. [CrossRef]
92. Peterson, E.C.; Mavrogenes, J.A. Linking high-grade gold mineralization to earthquake-induced fault-valve processes in the Porgera gold deposit, Papua New Guinea. *Geology* **2014**, *42*, 383–386. [CrossRef]
93. Rye, R.O.; Ohmoto, H. Sulfur and carbon isotopes and ore genesis: A review. *Econ. Geol.* **1974**, *69*, 826–842. [CrossRef]
94. Ohmoto, H. Stable isotope geochemistry of ore deposits. *Rev. Mineral. Geochem.* **1986**, *16*, 491–559.
95. Ohmoto, H.; Goldhaber, M.B. Sulfur and carbon isotopes. In *Geochemistry of Hydrothermal ore Deposits*, 2nd ed.; Barnes, H.L., Ed.; John Wiley & Sons Inc.: New York, NY, USA, 1997; pp. 517–612.
96. Seal, R.R. Sulfur isotope geochemistry of sulfide minerals. *Rev. Mineral. Geochem.* **2006**, *61*, 633–677. [CrossRef]
97. Wang, Y.; Zhu, X.K. Fe isotope systematics and its implications in ore deposit geology. *Acta Petrol. Sin.* **2012**, *28*, 3638–3654. (In Chinese with English Abstract)
98. Zhu, X.K. Fe isotope geochemistry of magmatic system. *J. Earth Sci. Environ.* **2016**, *38*, 1–10. (In Chinese with English Abstract)
99. He, Y.S.; Hu, D.P.; Zhu, C.W. Progress of iron isotope geochemistry in geoscience. *Earth Sci. Front.* **2015**, *22*, 54–71. (In Chinese with English Abstract)

100. Hill, P.S.; Schauble, E.A.; Young, E.D. Effects of changing solution chemistry on Fe^{3+}/Fe^{2+} isotope fractionation in aqueous Fe-Cl solutions. *Geochim. Cosmochim. Acta* **2001**, *74*, 6669–6689. [CrossRef]
101. Polyakov, V.B.; Soultanov, D.M. New data on equilibrium iron isotope fractionation among sulfides: Constraints on mechanisms of sulfide formation in hydrothermal and igneous systems. *Geochim. Cosmochim. Acta* **2011**, *75*, 1957–1974. [CrossRef]
102. Saunier, G.; Pokrovski, G.S.; Poitrasson, F. First experimental determination of iron isotope fractionation between hematite and aqueous solution at hydrothermal conditions. *Geochim. Cosmochim. Acta* **2011**, *75*, 6629–6654. [CrossRef]
103. Syverson, D.D.; Borrok, D.M.; Seyfried, W.E. Experimental determination of equilibrium Fe isotopic fractionation between pyrite and dissolved Fe under hydrothermal conditions. *Geochim. Cosmochim. Acta* **2013**, *122*, 170–183. [CrossRef]
104. Syverson, D.D.; Luhmann, A.J.; Tan, C.Y.; Borrok, D.M.; Ding, K.; Seyfried, W.E. Fe isotope fractionation between chalcopyrite and dissolved Fe during hydrothermal recrystallization: An experimental study at 350 °C and 500 bars. *Geochim. Cosmochim. Acta* **2017**, *200*, 87–109. [CrossRef]
105. Wu, H.J.; He, Y.S.; Bao, L.E.; Zhu, C.W.; Li, S.G. Mineral composition control on inter-mineral iron isotopic fractionation in granitoids. *Geochim. Cosmochim. Acta* **2017**, *198*, 208–217. [CrossRef]
106. Beard, B.L.; Johnson, C.M.; Skulan, J.L.; Nealson, K.H.; Sun, H. Application of Fe isotopes to tracing the geochemical and biological cycling of Fe. *Chem. Geol.* **2003**, *195*, 87–117. [CrossRef]
107. Rouxel, O.; Toner, B.; Germain, Y.; Glazer, B. Geochemical and iron isotopic insights into hydrothermal iron oxyhydroxide deposit formation at Loihi Seamount. *Geochim. Cosmochim. Acta* **2018**, *220*, 449–482. [CrossRef]
108. Zhu, X.K.; O'nions, R.K.; Guo, Y.L.; Reynolds, B.C. Secular variation of iron isotopes in North Atlantic deep water. *Science* **2000**, *287*, 2000–2002. [CrossRef] [PubMed]
109. Xiang, L.; Schoepfer, S.D.; Zhang, H.; Chen, Z.W.; Cao, C.Q.; Shen, S.Z. Deep-water dissolved iron cycling and reservoir size across the Ediacaran-Cambrian transition. *Chem. Geol.* **2020**, *541*, 119575. [CrossRef]
110. Du, D.H.; Wang, X.L.; Yang, T.; Chen, X.; Li, J.Y.; Li, W.Q. Origin of heavy Fe isotope compositions in high-silica igneous rocks: Arhyolite perspective. *Geochim. Cosmochim. Acta* **2017**, *218*, 58–72. [CrossRef]
111. Li, Q.W.; Zhao, J.H.; Wang, Q.; Zhang, Z.F.; An, Y.J.; He, Y.T. Iron isotope fractionation in hydrous basaltic magmas in deep crustal hot zones. *Geochim. Cosmochim. Acta* **2020**, *279*, 29–44. [CrossRef]
112. Li, W.Q.; Huberty, J.M.; Beard, B.L.; Kita, N.T.; Valley, J.W.; Johnson, C.M. Contrasting behavior of oxygen and iron isotopes in banded iron formations revealed by in situ isotopic analysis. *Earth Planet. Sci. Lett.* **2013**, *384*, 132–143. [CrossRef]
113. Yang, X.Q.; Mao, J.W.; Jiang, Z.S.; Santosh, M.; Zhang, Z.H.; Duan, S.G.; Wang, D.C. The Carboniferous Shikebutai iron deposit in Western Tianshan, Northwestern China: Petrology, Fe-O-C-Si isotopes, and implications for iron pathways. *Econ. Geol.* **2019**, *114*, 1207–1222. [CrossRef]
114. Zhu, X.K.; Sun, J.; Li, Z.H. Iron isotopic variations of the Cryogenian banded iron formations: A new model. *Precambrian Res.* **2019**, *331*, 105359. [CrossRef]
115. Wang, S.J.; Sun, W.D.; Huang, J.; Zhai, S.K.; Li, H.M. Coupled FeS isotope composition of sulfide chimneys dominated by temperature heterogeneity in seafloor hydrothermal systems. *Sci. Bull.* **2020**, *65*, 1767–1774. [CrossRef] [PubMed]
116. Liu, P.P.; Zhou, M.F.; Luais, B.; Cividini, D.; Bard, C.R. Disequilibrium iron isotopic fractionation during the high-temperature magmatic differentiation of the Baima Fe-Ti oxide-bearing mafic intrusion, SW China. *Earth Planet. Sci. Lett.* **2014**, *399*, 21–29. [CrossRef]
117. Ding, X.; Ripley, E.M.; Wang, W.Z.; Li, C.H.; Huang, F. Iron isotope fractionation during sulfide liquid segregation and crystallization at the Lengshuiqing Ni-Cu magmatic sulfide deposit, SW China. *Geochim. Cosmochim. Acta* **2019**, *261*, 327–341. [CrossRef]
118. Wang, Y. Fe and Cu Isotopic Geochemistry in Tongling Ore Concentration Area. Ph.D. Thesis, Chinese Academy of Geological Sciences, Beijing, China, 2011; 128p. (In Chinese with English Abstract).
119. He, Z.W.; Zhang, X.C.; Deng, X.D.; Hu, H.; Li, Y.; Yu, H.M.; Archer, C.; Li, J.W.; Huang, F. The behavior of Fe and S isotopes in porphyry copper systems: Constraints from the Tongshankou Cu-Mo deposit, eastern China. *Geochim. Cosmochim. Acta* **2020**, *270*, 61–83. [CrossRef]
120. Liu, C.H.; Liu, J.J.; Carranza, M.J.E.; Wang, J.P.; Zhai, D.G.; Zhang, F.F.; Wang, Y.H.; Liu, Z.J. Iron and magnesium isotope systematics from the Shuangwang gold deposit in the Qinling orogen, Central China. *Ore Geol. Rev.* **2021**, *134*, 104123. [CrossRef]
121. Zhang, Y.W.; Fan, H.R.; Hu, F.F.; Liu, X.; Xie, L.W.; Li, X.H. Extreme iron isotope variation of pyrite in the Muping gold deposit, Jiaodong: Implication for tracing metal origin. *Ore Geol. Rev.* **2021**, *139*, 104431. [CrossRef]
122. Cheng, Y.B.; Mao, J.W.; Zhu, X.K.; Wang, Y. Iron isotope fractionation during supergene weathering process and its application to constrain ore genesis in Gaosong deposit, Gejiu district, SW China. *Gondwana Res.* **2015**, *27*, 1283–1291. [CrossRef]
123. Du, D.W.; Ye, C.Z.; Zhang, J.L. *Survey Report of Dapoding Gold Deposit, Dawu County, Hubei Province*; Northeast Hubei Geological Team: Xiaogan, China, 2008.
124. Cao, Z.Q.; Xie, Y.B.; Zhou, X.H.; Li, H.L.; Cai, Y.T.; Li, C.K.; Song, H.; Wu, B.; Li, L.J. Origin and ore-controlling structure of the Daleishan gold deposit in Dawu area, Hubei Province. *Acta Petrol. Miner.* **2017**, *36*, 733–742. (In Chinese with English Abstract)

MDPI AG
Grosspeteranlage 5
4052 Basel
Switzerland
Tel.: +41 61 683 77 34

Minerals Editorial Office
E-mail: minerals@mdpi.com
www.mdpi.com/journal/minerals

www.ingramcontent.com/pod-product-compliance
Lightning Source LLC
Chambersburg PA
CBHW061620210926
51850CB00009B/2383
* 9 7 8 3 7 2 5 8 4 9 1 5 4 *